ENZYME INHIBITORS AND ACTIVATORS

Edited by **Murat Sentürk**

Enzyme Inhibitors and Activators
http://dx.doi.org/10.5772/63325
Edited by Murat Sentürk

Contributors

Bruno Ricardo De Castro Leite Júnior, Alline Artigiani Lima Tribst, Marcelo Cristianini, Metin Turan, Nurgül Kıtır, Emrah Nikerel, Adem Günes, Bahar Soğutmaz, Negar Ebrahim Pour Mokhtari, Leyla Tarhan, Ayse Gul Mutlu, Josiani De Cassia Pereira, Ellen Giese, Marcia Maria de Souza Moretti, Ana Carolina Dos Santos Gomes, Eleni Gomes, Daniela Alonso Bocchini Martins, Olavo Micali Perrone, Mark Chappell, Brian Rybarczyk, Nildris Cruz-Diaz, Abdur Rauf, Ganapati D Yadav, Deepali Magadum, Olga Lopina, Xianyong Ma, Dun Deng, Jonas Contiero, Katia Valevski Sales Fernandes, Jorge H. Fernandez, Gustavo L. Rocha, Antonia E.A. Oliveira

Published by InTech

Janeza Trdine 9, 51000 Rijeka, Croatia

Publishing Process Manager Edi Lipovic
Technical Editor SPi Global
Cover InTech Design team

First published March, 2017
Printed in Croatia
Legal deposit, Croatia: National and University Library in Zagreb

Additional hard copies can be obtained from orders@intechopen.com

Enzyme Inhibitors and Activators, Edited by Murat Sentürk
p. cm.
Print ISBN 978-953-51-3057-4
Online ISBN 978-953-51-3058-1

For EU product safety concerns:
IN TECH d.o.o., Prolaz Marije Krucifikse Kozulić 3, 51000 Rijeka, Croatia,
info@intechopen.com or visit our website at intechopen.com.

Meet the editor

Dr. Murat Şentürk obtained a baccalaureate degree in Chemistry in 2002, a master's degree in Biochemistry in 2006, and a doctorate degree in Biochemistry in 2009 from Atatürk University, Turkey. Dr. Şentürk currently works as an associate professor of Biochemistry in the Department of Chemistry, Faculty of Science and Art, Ağri İbrahim Çeçen University, Turkey. Dr. Şentürk published over 90 scientific papers, reviews, and book chapters and presented several conferences to scientists. His research interests span enzyme inhibitor or activator, protein expression, purification and characterization, drug design and synthesis, toxicology, and pharmacology. His research work has focused on antioxidant and metabolic enzyme systems. Dr. Şentürk serves as the editorial board member of several international journals.

Contents

Preface

Enzymes are biological catalysts that can catalyze specific reactions with 100% efficiency. The substances that affect the activity of enzymes are called moderators. If a moderator reduces the activity of the enzyme and stops it completely, the inhibitor is called the activator if it increases the activity of the enzyme. Types of inhibition are divided into two main groups: reversible inhibition and irreversible inhibition. Reversible inhibition is itself classified as competitive, uncompetitive, and noncompetitive.

Many molecules are enzyme inhibitors or activators, so their discovery and improvement are an active area of research in biochemistry, pharmacology, toxicology, molecular biology, etc. Medicinal enzyme inhibitors or activators are often judged by their specificity and their potency. In the last few decades, years of discovery of specific enzyme inhibitors or activators have received great attention due to their potential to be used in pharmacological applications.

This book titled *Enzyme Inhibitors and Activators* contains a selection of chapters focused on the research area of enzyme inhibitors or activators, enzyme-catalyzed biotransformation, usage of microbial enzymes, enzymes associated with programmed cell death, natural products as potential enzyme inhibitors, protease inhibitors from plants in insect pest management, peptidases, and renin-angiotensin system. The book provides an overview on basic issues and some of the recent developments in medicinal science and technology. Especially, emphasis is devoted to both experimental and theoretical aspect of modern medicine. The primary target audience for the book includes students, researchers, chemists, molecular biologists, medical doctors, pharmacologists, and professionals who are interested in associated areas.

The textbook is written by international scientists with expertise in biochemistry, enzymology, molecular biology, and genetics, many of which are active in biochemical and pharmacological research. I would like to acknowledge the authors for their contribution to the book. We hope that the textbook will enhance the knowledge of scientists in the complexities of some medical approaches; it will stimulate both professionals and students to dedicate part of their future research in understanding relevant mechanisms and applications.

I would like to thank my wife Esra and my son Yusuf Fatih for their support in the design of the book.

Dr. Murat Şentürk
Associate Professor of Biochemistry,
Ağri İbrahim Çeçen University,
Turkey

Section 1

Enzyme Inhibition

Chapter 1

Peptidases and the Renin-Angiotensin System: The Alternative Angiotensin-(1-7) Cascade

Nildris Cruz-Diaz , Bryan A. Wilson and
Mark C. Chappell

Additional information is available at the end of the chapter

http://dx.doi.org/10.5772/65949

Abstract

The renin-angiotensin system (RAS) constitutes a key hormonal system in the physiological regulation of blood pressure via peripheral and central mechanisms. Dysregulation of the RAS is considered a major factor in the development of cardiovascular pathologies, and pharmacologic blockades of this system by the inhibition of angiotensin-converting enzyme (ACE) or antagonism of the angiotensin type 1 receptor (AT_1R) are effective therapeutic regimens. The RAS is now defined as a system composed of different angiotensin peptides with diverse biological actions mediated by distinct receptor subtypes. The classic RAS comprises the ACE-Ang II-AT_1R axis that promotes vasoconstriction, water intake, sodium retention and increased oxidative stress, fibrosis, cellular growth, and inflammation. The nonclassical or alternative RAS is composed primarily of the ACE2-Ang-(1-7)-AT_7R pathway that opposes the Ang II-AT_1R axis. In lieu of the complex aspects of this system, the current review assesses the enzymatic cascade of the alternative Ang-(1-7) axis of the RAS.

Keywords: angiotensin-(1-7), neprilysin, dipeptidyl peptidase 3, ACE, ACE2, renin

1. Introduction

The renin-angiotensin system (RAS) has been long defined as a circulating endocrine system composed of an enzymatic cascade that includes renin to initiate the RAS through the conversion of the large precursor protein angiotensinogen (>350 amino acids) to the inactive decapeptide Ang I and the subsequent generation of the bioactive octapeptide Ang II by angiotensin-converting enzyme (ACE) (**Figure 1**). Ang II binds to the angiotensin type 1 receptor (AT_1R) to mediate both peripheral and central mechanisms in the regulation of blood pressure. Activation of the ACE-Ang II-AT_1R pathway is also associated with various pathological responses including fibrosis, inflammation, metabolic dysregulation, heart failure, cancer, aging, and diabetic injury [1–5]. Indeed, the targeting of this pathway by biochemical approaches that block the formation of Ang II through ACE inhibitors or binding of Ang II to the AT_1R by receptor antagonists is an effective therapy for the treatment of cardiovascular disease. It is now evident that these therapies have revealed alternative pathways within the RAS that may contribute to the beneficial actions of the RAS blockade. For example, the targeting of ACE reduces Ang II expression, but markedly enhances the circulating levels of Ang-(1-7), a ligand for the AT_7/MasR that generally opposes the actions of the Ang II-AT_1R pathway [6, 7]. The chronic treatment with AT_1R antagonists may also increase the formation of Ang-(1-7) through ACE2, as well as shunt Ang II to the AT_2R pathway that shares similar properties to the Ang-(1-7) system [8, 9]. Since the RAS is now viewed as a more complex array of components that can be functionally partitioned into distinct receptors and peptide ligands, we present an overview of the peptidases the may constitute the alternative Ang-(1-7) axis of the RAS. We include a discussion of renin, ACE, and chymase as they pertain to the formation of the substrates Ang I and Ang II for the direct processing to Ang-(1-7), as the well as the role of ACE to metabolize Ang-(1-7).

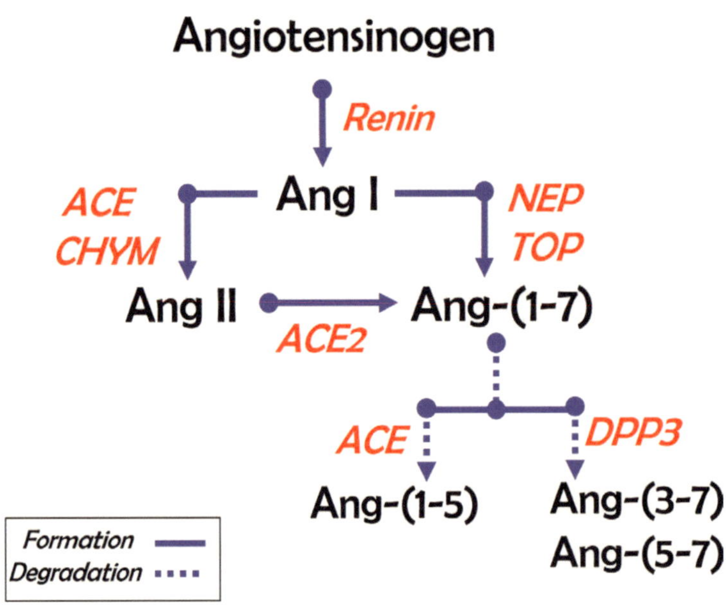

Figure 1. Processing cascade for angiotensin-(1-7). Renin cleaves angiotensinogen to Ang I which is further processed to the biologically active peptides Ang II by ACE or chymase (CHYM). Ang II undergoes further processing by the carboxypeptidase ACE2 to form Ang-(1-7). Ang-(1-7) is formed directly through non-Ang II pathways by the direct processing of Ang I by the endopeptidases neprilysin (NEP) and thimet oligopeptidase (TOP). Ang-(1-7) is subsequently metabolized by ACE to Ang-(1-5) and dipeptidyl peptidase 3 (DPP3) to Ang-(3-7) and Ang-(5-7). Adapted from Chappell [54].

2. Renin

Renin [EC 3.4.23.15] belongs to the family of aspartyl-type acid proteases (molecular size: 30–40 kDa); however, the protease exhibits a more neutral pH optima of 6.5–8.0. The only known substrate for renin is angiotensinogen, and renin cleaves off the N-terminal portion of angiotensinogen to form the inactive peptide Ang I. The formation of Ang I is considered the enzymatic event that initiates the RAS cascade (**Figure 1**). The enzyme is synthesized predominantly in the juxtaglomerular (JG) cells of the kidney and is stored in both inactive (prorenin) and active forms for subsequent release into the afferent arteriole to reach the circulation. The collecting duct (CD) cells are another source of renin within the kidney that primarily secretes the active form of renin from the apical aspect of the principal cells into the tubular fluid [10, 11]. The distal secretion of renin into the tubular fluid would presumably contribute to the processing of angiotensinogen that may arise from either proximal tubule synthesis or apical release or the glomerular filtration of the protein. Interestingly, the regulation of JG renin appears to be distinct from CD renin as Ang II reduces JG renin release and synthesis constituting the classic short-loop negative feedback system for RAS activation, whereas the peptide stimulates CD renin release suggesting a positive feedback system for renin in the distal nephron [12, 13].

There is evidence for alternative gene products of renin expressed in the kidney, brain, heart, and adrenal gland [14–17]. These products arise from an alternative start site for renin transcription that is situated downstream from the typical start site and yields a shorter, but active form of the enzyme. The truncated form of renin lacks the secretory signal of the protein and the enzyme is not secreted, but resides within the cell. Peters and colleagues find that truncated renin specifically localized to the mitochondria and that the active form of renin, but not prorenin, was internalized by mitochondria [14, 18, 19]. The overexpression of the active renin isoform protected the cells under high glucose conditions and this effect was not reversed by the renin inhibitor aliskerin or an AT_1R antagonist suggesting that the beneficial effects may not reflect activation of the Ang II-AT_1R axis [20]. In support of these findings, we recently reported the presence of active renin in isolated mitochondria from the sheep renal cortex [21]. In this case, we also quantified the mitochondrial levels of both Ang II and Ang-(1-7) (50–60 fmol/mg protein), which may indicate that the intracellular form of renin contributes to the cellular expression of angiotensin peptides within the kidney [21]. In the renal NRK-52 epithelial cell line, immunocytochemical staining for renin was evident in the nucleus of these cells [22]. Renin activity as measured by aliskerin-sensitive inhibition of Ang I formation from angiotensinogen was also evident in isolated nuclei in the renal epithelial NRK-52 cells, as well as quantifiable levels of Ang I, Ang II, and Ang-(1-7) [22]. In confirmation of intracellular renin within the kidney, Ishigami and colleagues report a truncated renin transcript expressed in the proximal tubules of the mouse kidney [16]. Overexpression of the renin isoform within the proximal tubules was associated with a sustained elevation in blood pressure and no change in the circulating levels of renin [16]. Although the intracellular distribution of the renin isoform was not ascertained, the fact that overexpression of tubular renin augmented blood pressure is consistent with previous studies demonstrating that the increased expression of proximal tubule angiotensinogen, the AT_1R, or intracellular Ang II also resulted in a sustained increase in blood pressure [23–28].

3. Angiotensin-converting enzyme

The predominant pathway of the classical RAS for the conversion of Ang I to the bioactive peptide Ang II is catalyzed by the metallopeptidase ACE [EC 3.4.15.1], a dipeptidyl carboxypeptidase that cleaves two residues from the carboxy terminus of Ang I (**Figure 1**). The peptidase is a membrane-bound, glycosylated protein (120–180 kDa) ubiquitously expressed in multiple tissues [29]. Soluble forms of the enzyme are present in the circulation, cerebrospinal fluid (CSF), lymph fluid, and urine that fully retain peptidase activity [29]. The soluble form of ACE arises from the hydrolysis of the membrane-anchoring or stalk region of the protein that may reflect the processing by A Disintegrin and Metalloproteinase (ADAM) family of metalloenzymes, although the precise role of ACE shedding in cardiovascular disease is presently unclear. Somatic ACE is characterized by two active sites termed N and C terminal domains that likely arose from the gene duplication of germinal or testicular ACE that contains only the single C terminal-active site. Importantly, in addition to forming Ang II, ACE degrades a number of other peptides that exhibit cardiovascular actions including bradykinin, substance P, acetyl-SDKP, and Ang-(1-7) [29]. Indeed, the cardioprotective effects of ACE inhibitors may reflect the protection of these peptides from metabolism in addition to the blockade of Ang II formation. We showed that ACE hydrolyzes Ang-(1-7) to the pentapeptide Ang-(1-5) and that ACE inhibition markedly reduced the clearance of the peptide [30, 31]. The reduced metabolism of Ang-(1-7) likely contributes to the elevation in circulating levels of Ang-(1-7) following the chronic treatment with ACE inhibitors in experimental animals and in humans [32]. Moreover, treatment with an ACE inhibitor was required to clearly demonstrate the accumulation of Ang-(1-7) derived from either Ang II or Ang I in isolated proximal tubules [33]. These data suggest a pivotal role for ACE to regulate the balance of Ang II and Ang-(1-7), two peptides within the RAS that exhibit strikingly different actions from one another [7].

4. Chymase

Chymases [EC 3. 4.21.39] comprise a family of serine peptidases that may generate Ang II by hydrolysis of the Phe^8-His^9 bond of Ang I and other peptide precursors [Ang-(1-12), Ang-(1-25)] (α-chymases) or metabolize Ang II at Tyr^4-Ile^5 to form Ang-(1-4) and Ang-(5-8) (β-chymases) [34–39]. Humans express α-chymase while rodents express primarily β-chymases, as well as other isoforms (mouse MCP-4 and rat MCP-5) that more closely resemble α-chymase in regard to the processing of Ang I to Ang II [35]. The human and mouse enzymes may also play a role in the conversion of the endothelin precursor to the active peptide, as well as the activation of various inflammatory cytokines [40]. Chymases (35 kDa) are synthesized and stored in an inactive proform within mast cells and neutrophils that are released with other proteases (cathepsin G, tryptases, and renin) upon degranulation following injury or inflam-

http://dx.doi.org/10.5772/65949

mation [41]. Although chymases are soluble enzymes, they associate with the cell membrane and may locate to the extracellular surface of tissues following release. The serine protease inhibitor chymostatin is typically used to demonstrate specificity; however, chymostatin inhibits other Ang II-generating enzymes (cathepsin G and elastase-2). Thus, chymostatin sensitivity for Ang II generation does not necessarily demonstrate the involvement of chymase and more selective approaches should be considered [42–45]. The extent that chymase or other peptidases participate in the formation of circulating or tissue Ang II through an ACE-independent pathway remains equivocal [46].

5. Neprilysin

Neprilysin [EC 3.4.24.11; ~95 kDa] is a metalloendopeptidase that is predominantly anchored to the plasma membrane. The peptidase was initially characterized in brain to hydrolyze the opiate peptide enkephalin, hence the original terminology of the enzyme as "enkephalinase." Neprilysin was subsequently found in a number of peripheral tissues with particularly high expression within the brush border on the apical region of the renal proximal tubules. Interest in the renal expression of neprilysin reflected the potential role of the peptidase to metabolize natriuretic peptides, and selective inhibitors alone or combined with an ACE inhibitor ("vasopeptidase inhibitor") were developed as cardiovascular therapies to prolong the renal actions of these peptides [47–52]. Although the vasopeptidase inhibitor omapatrilat was clinically withdrawn regarding a greater incidence of angioedema that may reflect increased levels of bradykinin, a new generation of agents that combine a neprilysin inhibitor and an AT_1R antagonist may be a promising therapeutic approach for the treatment of heart failure [53]. One potential caveat to this approach would be that the neprilysin inhibitor component may reduce circulating levels of Ang-(1-7) and potentially attenuate the cardioprotective effects of the combined antagonist/inhibitor.

Vascular neprilysin is responsible for the extracellular conversion of Ang I to Ang-(1-7) in the circulation, particularly under conditions of chronic ACE inhibition (**Figure 1**) [54]. Neprilysin hydrolyzes the Pro^7-Phe^8 bond of Ang I to form Ang-(1-7), as well as the Tyr^4-Ile^5 bond to generate Ang-(1-4) that reflects the enzyme's preference for aromatic and hydrophobic residues. We further showed that renal neprilysin on the proximal tubules and the renal cortex readily converted both Ang I and Ang-(1-12) to Ang-(1-7), as well as that a neprilysin inhibitor attenuated the blood pressure-lowering effects of the ACE inhibitor lisinopril in the spontaneously hypertensive rat [33, 55, 56]. Apart from the conversion of Ang I to Ang-(1-7), neprilysin is capable of metabolizing a number of other peptides including adrenomedullin, Ang II [to Ang-(1-4)], endothelin, bradykinin, and the natriuretic peptides ANP and BNP.

6. Thimet oligopeptidase

Thimet oligopeptidase [EC 3.4.24.15, 80 kDa] is primarily a soluble metalloendopeptidase that resides within the cell. Similar to neprilysin, thimet oligopeptidase prefers aromatic and hydrophobic residues and cleaves multiple peptide substrates. In regard to the RAS, thimet oligopeptidase hydrolyzes Ang I exclusivity at the Pro-Phe bond to form Ang-(1-7) (**Figure 1**). We recently showed that both neprilysin and thimet oligopeptidase contributed to the processing of Ang I to Ang-(1-7) within isolated mitochondria [21]. In the human proximal tubule HK-2 cell line, cytosolic thimet oligopeptidase was the sole activity responsible for the generation of Ang-(1-7) from exogenous Ang I [57]. Moreover, thimet oligopeptidase activity in isolated nuclei processed Ang I exclusively to Ang-(1-7) and may be a potential candidate that contributes to the nuclear levels of Ang-(1-7) within the cell [22]. The RAS was originally characterized as a classic endocrine or circulating system, but there is very compelling evidence for the intracellular expression of the RAS peptides, receptors, and synthetic components [7, 54, 58–60]. It is presently unclear the cellular mechanisms for the intracellular expression of Ang II or Ang-(1-7), although the intracellular peptidase thimet oligopeptidase may play a role in the cellular Ang-(1-7) axis.

7. Angiotensin-converting enzyme 2

ACE2 is a membrane-bound monocarboxypeptidase [EC 3.4.17.23; 120 kDa] that converts Ang II directly to Ang-(1-7) (**Figure 1**). ACE2 was initially characterized as a homolog to ACE and exhibits approximately ~40% homology with ACE [60–62]. ACE2 was initially reported to cleave Ang I not to Ang II, but to the nonapeptide Ang-(1-9) [63]. Subsequent studies found that Ang II exhibits far better kinetic values for ACE2 that would argue that the endogenous substrate is Ang II, not Ang I [64]. Among a number of peptide substrates (>100) that were screened for ACE2, Vickers et al. reported that only apelin 13 exhibited comparable kinetic values to that of Ang II [65]. In the murine heart, Ang II was primarily converted to Ang-(1-7) by ACE2 and that in the presence of the ACE2 inhibitor MLN-4760 or in ACE2 null mice there was essentially no metabolism of Ang II in the heart [66]. By contrast, under identical kinetic conditions, Ang I was primarily converted to Ang-(1-9) by carboxypeptidase A and not ACE2 in both the wild-type and ACE2 knockout mice.

In comparison to ACE, the circulating levels of ACE2 are typically quite low. Whether this reflects a reduced degree of shedding, lower vascular expression or both is not presently clear. Rice et al. reported that the molar concentration of ACE in human serum averaged 7 nM in over 500 subjects while ACE2 content was *200-fold lower* (33 pM) and was detectable in <10% of the patient population [67]. Circulating NEP content (290 pM) was also lower than ACE and evident in <30% of these patients [67]. Serum and urinary ACE2 activities are elevated in diabetes, heart failure, and hypertension [68–70]. Circulating ACE2 activity increased approximately threefold in the diabetic hypertensive mRen2.Lewis rat; however, serum ACE activity also increased in the diabetic rats. As measured under identical kinetic conditions, serum ACE was far higher than ACE2 suggesting that the capacity to generate Ang II (or metabolize Ang-(1-7) and bradykinin) remains greater than the capability to form Ang-(1-7) from Ang II [68].

Similar to ACE, ACE2 has a potentially significant role in the RAS pathway as a single catalytic step metabolizes Ang II to attenuate the Ang II-AT1 receptor pathway, and generate Ang-(1-7) that would stimulate the Ang-(1-7)-AT7/Mas receptor axis (**Figure 1**) [71, 72]. Indeed, ACE2-deficient animals exhibit exaggerated responses to Ang II or under chronic conditions of an activated RAS [73–76]. In turn, overexpression of ACE2 or administration of the soluble form of the peptidase, which retains full enzymatic activity, attenuates the Ang II-dependent increase in blood pressure and indices of target organ injury [77–85]. In regard to the benefits of an activated ACE2 pathway, several compounds have been identified that may act as allosteric activators of ACE2 including xanthenone (XNT) and diminazene aceturate (DIZE) to promote a higher ratio of Ang-(1-7) to Ang II [86]. Chronic treatment with DIZE ameliorated the extent of pulmonary hypertensin and fibrosis, renal tissue injury, and myocardial infarction consistent with enhanced levels of Ang-(1-7) and a reduction in Ang II [84, 87–90]. Interestingly, DIZE treatment was also associated with increased mRNA levels of ACE2 suggesting that DIZE may exhibit actions apart from the direct activation of the peptidase [87, 91]. However, it should be noted that the effects of DIZE on ACE2 activity or expression have not been confirmed by others. Haber et al. [92] found no effect of DIZE on soluble ACE2 activity or an influence on Ang II-dependent hypertension using similar doses of DIZE as previously reported. Velkosa et al. [93] also found no direct effect of various concentrations of DIZE on renal ACE2 activity and slightly increased renal ACE2 activity in the 5/6 nephrectomized rat following a 2-week administration. Of particular interest, the Velkosa study reported that DIZE normalized the marked increase in renal ACE activity suggesting that ACE may be a more relevant target than ACE2 [93]. DIZE also failed to increase ACE2 activity or enhance the local vascular actions of Ang-(1-7) in the isolated pig coronary arteries [94].

Conceptually, the use of ACE2 as a therapeutic agent to chronically alter the balance of Ang II and Ang-(1-7) is challenging. ACE activity in the circulation and the vasculature surface is significant with a very high capacity to generate Ang II. This reflects not simply the abundance of ACE but the significant capability of angiotensinogen and renin to generate the ACE substrate Ang I. Moreover, reduced Ang II levels by exogenous ACE2 should stimulate the generation of Ang II that reflects the activation of negative feedback mechanisms on renin. Therefore, it is difficult to conceive sufficiently high levels of ACE2 that can be achieved to chronically reduce Ang II *and* increase Ang-(1-7) except with the possible addition of an ACE inhibitor. In this case, supplementation of ACE2 may degrade residual levels of Ang II and the circulating levels of Ang-(1-7) may be augmented, particularly as the Ang-(1-7)-degrading pathway in the circulation is attenuated by the ACE inhibitor.

8. Dipeptidyl peptidase 3

We demonstrated a role for ACE in the metabolism of Ang-(1-7), but there are other potential pathways that may regulate endogenous levels of the peptide [95]. Marshall and colleagues reported that ACE and a second peptidase activity in the sheep cerebrospinal fluid degraded Ang-(1-7) [95–97]. Interestingly, the non-ACE-degrading activity accounted for a greater contribution of metabolism than ACE [96]. Moreover, this activity was inversely correlated to CSF levels of Ang-(1-7) in control and betamethasone-exposed sheep, a model of fetal programming that exhibits elevated blood pressure and an attenuated baroreflex [7]. Subsequent

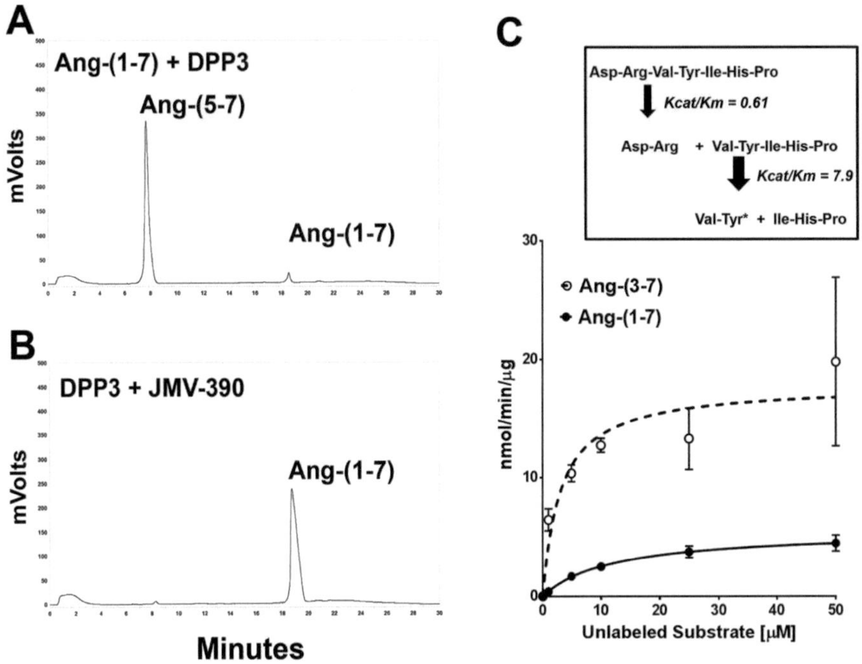

Figure 2. Kinetics of human DPP 3 hydrolysis of Ang-(1-7) and Ang-(3-7). (A) Chromatographs reveal that human DPP 3 cleaves Ang-(1-7) to the tripeptide Ang-(5-7). (B) Chromatograph reveals that the JMV-390 inhibitor attenuates Ang-(1-7) metabolism by human DPP 3. (C) Kinetic analysis of the hydrolysis of Ang-(1-7) and Ang-(3-7) by human DPP 3. Data adapted from Cruz-Diaz et al. [99].

studies found the Ang-(1-7)-degrading activity in sheep brain and kidney cortex, as well as in the human proximal tubule HK-2 cell line [57, 98]. The enzyme activity exhibited unusual characteristics as Ang I and other peptides equal to or greater than 10 residues were not substrates for the peptidase [57, 98]. In addition, the peptidase was sensitive to both chelating agents such as o-phenanthroline and EDTA, and the sulfhydryl inhibitors APMA and PCMB [57, 98]. Additional studies identified the inhibitor JMV-390, originally developed to block the metalloendopeptidases neprilysin, thimet oligopeptidase, and neurolysin, that potently inhibited the Ang-(1-7)-degrading activity in the brain and kidney [IC_{50} <1 nM] [57, 98]. Conversely, specific inhibitors against these endopeptidases did not attenuate the Ang-(1-7)-degrading activity [98]. Interestingly, the peptidase activity accounted for the sole degradative pathway in the cytosolic fraction and in the media of the HK-2 cells [57]. Utilizing the HK-2 cells as the source of the Ang-(1-7)-degrading activity, we recently purified the peptidase from the cell cytosol by ion exchange and hydrophobic interaction chromatography and identified the enzyme as dipeptidyl peptidase 3 [EC 3.4. 14.4, DPP 3] [99].

DPP 3 belongs to a family of metalloaminopeptidases that sequentially cleave two residues from the N-terminus of peptides no more than eight residues in length and this likely explains our previous results that Ang I, apelin-13, and neurotensin were not substrates for the Ang-(1-7)-degrading activity in the CSF and brain [98]. We obtained a human recombinant form of

http://dx.doi.org/10.5772/65949

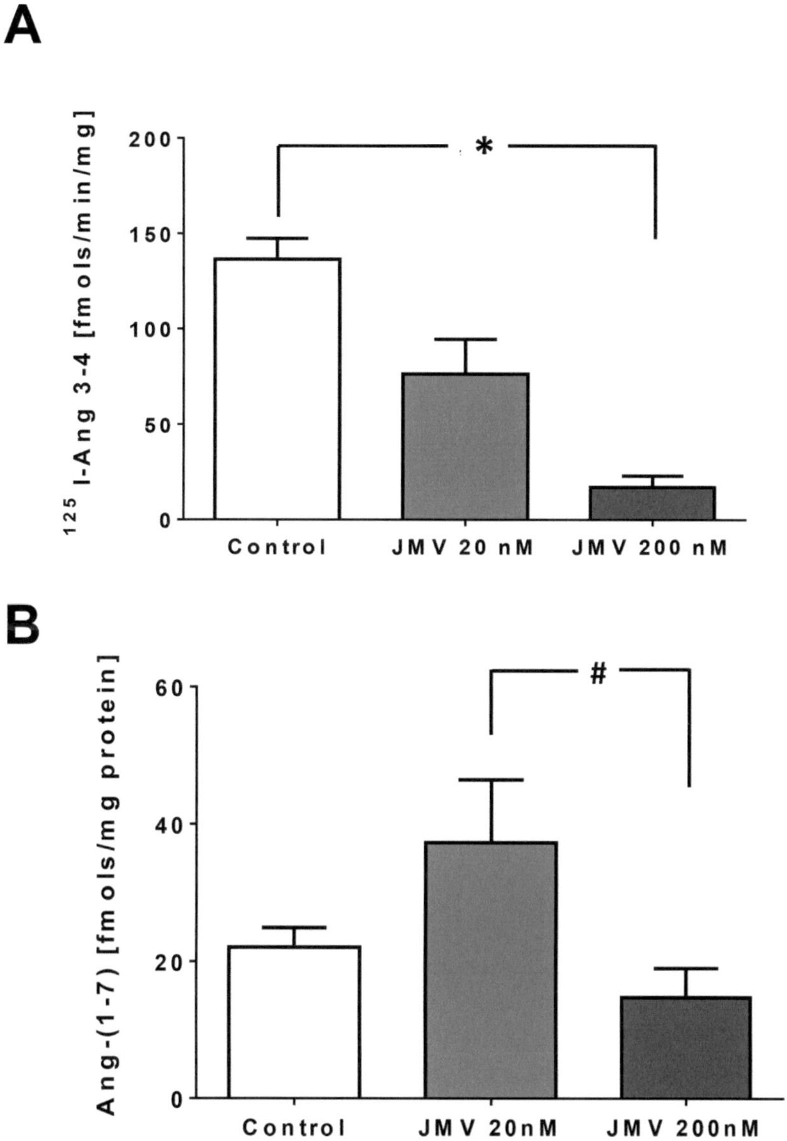

Figure 3. The JMV-390 inhibitor reduces DPP 3 activity and alters Ang-(1-7) cellular content of HK-2 cells. (A) Increasing doses of the JMV-390 inhibitor reduced DPP 3 activity (^{125}I-Ang-(3-4) generation—fmol/min/mg protein) in HK-2 cells. (B) Low-dose JMV-390 (20 nM) tended to increase the cellular content of Ang-(1-7) (fmols/mg protein), but high dose (200 nM) significantly reduced Ang-(1-7). $^{*}P < 0.05$ versus control; $^{\#}P < 0.05$ versus 20 nM JMV. Data adapted from Cruz-Diaz et al. [99].

DPP 3 to show that the enzyme metabolized Ang-(1-7) in two cleavage steps [99]. DPP 3 initially hydrolyzes Ang-(1-7) at the Arg^2-Val^3 bond to form Ang-(3-7) and the dipeptide Arg^1-Asp^2. Ang-(3-7) is then very rapidly cleaved at Tyr^4-Ile^5 to form Ang-(5-7) and Val^3-Tyr^4. The kinetic analysis of DPP 3 hydrolysis revealed a higher efficiency constant (kcat/km) for Ang-(3-7) than

Ang-(1-7) (**Figure 2**) [99]. The preferred hydrolysis of Ang-(3-7) by DPP 3 would account for our inability to demonstrate the accumulation of Ang-(3-7) following the initial metabolism of Ang-(1-7). In regard to an *in vivo* role for DPP 3 to modulate Ang-(1-7), human HK-2 cells were treated with varying doses of the JMV-390 inhibitor and we assessed both the endogenous content of Ang-(1-7) and the intracellular DPP-3/Ang-(1-7)-degrading activity in the cells. As shown in **Figure 3**, 20- and 200-nM JMV-390 reduced DPP3 activity by >30 and >80%, respectively, as compared to control suggesting that the inhibitor effectively penetrates the cells [99]. The lower dose of JMV increased the cellular content Ang-(1-7) approximately twofold, although this did not reach statistical significance. The higher dose of JMV, however, significantly reduced the intracellular levels of the peptide [99]. We interpret the latter results that the high dose may spill over to block other peptidases including thimet oligopeptidase that may be involved in the generation of Ang-(1-7) in the renal cells [57]. Thus, the blockade of Ang-(1-7)-forming enzymes by the high-dose JMV may override any protective effects of DPP 3 inhibition. We are currently assessing alternative approaches to block DPP 3 within the HK-2 cells and assess both secreted and intracellular levels of Ang-(1-7), as well as Ang-(1-7)-dependent actions on these cells.

9. AGE-induced peptidase

Our recent studies assessed the potential role of Ang-(1-7) in the progression of epithelial-to-mesenchymal transition (EMT) in the renal epithelial NRK-52 cell line by either advanced glycation end products (AGEs) or the cytokine TGF-β [100]. Ang-(1-7) essentially abolished the cellular indices for EMT in the NRK-52 cells likely through the inhibition of the noncanonical ERK 1/2-signaling pathway stimulated by AGE (100). Interestingly, AGE exposure reduced the intracellular levels of Ang-(1-7) but did not alter the cellular content of Ang II [100]. Processing of Ang I to Ang-(1-7) by thimet oligopeptidase tended to be reduced by AGE; however, the intracellular metabolism of Ang-(1-7) to Ang-(1-4) was significantly increased by AGE exposure [100]. These data suggest that AGE-induced EMT may reflect lower Ang-(1-7) tone in the renal epithelial cells that may be permissive for the progression of EMT and the increase in tissue fibrosis. Our preliminary data suggest that DPP 3 is not responsible for the AGE-induced metabolism of Ang-(1-7) in these cells and that another peptidase activity may participate in the cellular metabolism of the peptide. As shown in **Figure 4A**, the NRK-52 cytosolic fraction readily hydrolyzed the quenched Ang-(1-7) fluorescent substrate Abz^1-Ang-(1-7)-$Tyr^7(NO_2)$ to Abz-Ang-(1-4). By contrast, DPP 3 in the HK-2 renal cells did not cleave the Ang-(1-7) fluorescent peptide (**Figure 4B**). The failure of DPP 3 to cleave the fluorescent Ang-(1-7) substrate is consistent with the inability of DPP 3 to hydrolyze blocked N-terminal residues of peptides such as the Abz moiety on the Ang-(1-7) substrate. Moreover, hydrolysis of the N-terminally blocked Ang-(1-7) fluorescent substrate by the cytosolic fraction of the NRK-52 cells suggests the peptidase is likely an endopeptidase. Apart from the identification of the Ang-(1-7)-degrading activity in the NRK-52E cells, it remains to be determined whether intracellular levels of Ang-(1-7) influence the EMT process.

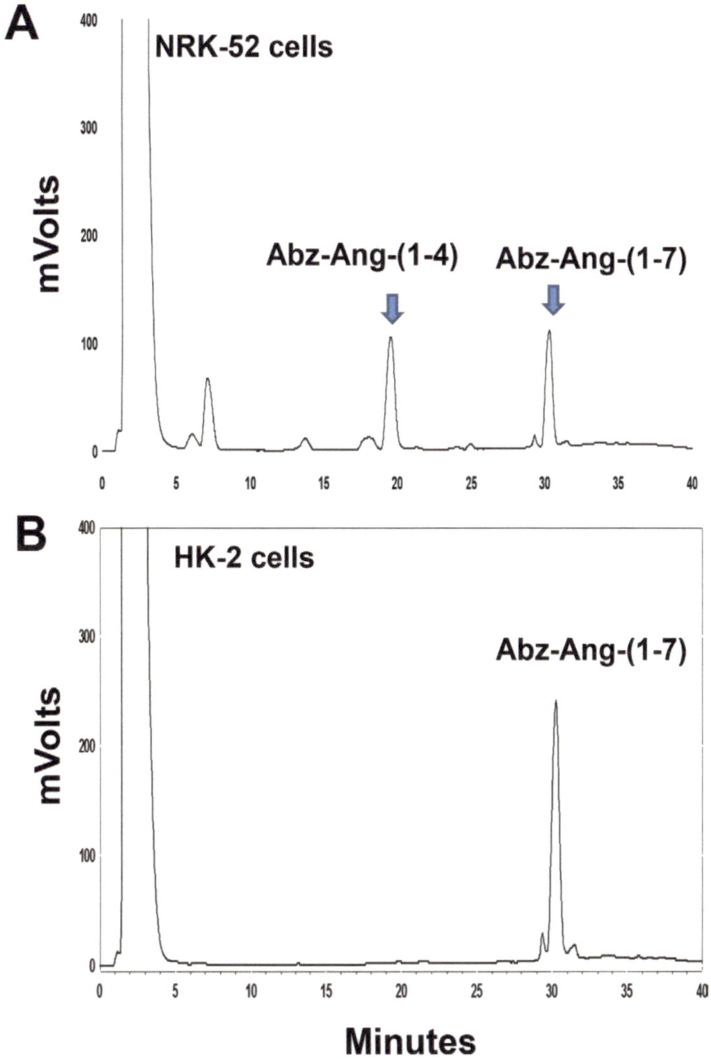

Figure 4. Metabolism of Abz[1]-Ang-(1-7)-Tyr$_7$ (NO_2). (A) HPLC chromatograph reveals that the cytosolic fraction of the NRK-52E cells cleaves the Ang-(1-7)-fluorescent substrate Abz-Ang-(1-7) to Abz-Ang-(1-4). (B) By contrast, HK-2 cytosolic fraction does not hydrolyze the Ang-(1-7)-fluorescent substrate suggesting an endopeptidase activity distinct from DPP 3. The large void peak is absorbance from the DMSO solvent to dissolve the fluorescent substrate. Substrate concentration and incubation conditions are identical for both cell types. The HK-2 cell data are adapted from Wilson et al. [57].

10. Characterization of peptidase pathways

In the review of the peptidases of the Ang-(1-7) axis, a brief discussion of the various biochemical approaches to characterize these enzymatic components is warranted. Ideally, utilization of the endogenous peptide substrates including Ang I, Ang II, and Ang-(1-7) should be assessed to identify the peptidase activities involved in the RAS [101]. One advantage is that the contribution of various peptidases for a given peptide is directly comparable to determine the predominant pathway in a particular tissue or treatment condition. Peptidase activities derived by different synthetic substrates are not comparable unless standardized to the enzyme concentration. Moreover, the use of endogenous peptide substrates may reveal novel peptidase activities involved in angiotensin processing [102, 103]. Peptidase assays developed in our laboratory typically utilize ^{125}I-radiolabled peptides coupled to high-performance liquid chromatography (HPLC)-based separation and automated in-line γ-detection. Advantages of this are that only microliter amounts of serum or microgram quantities of tissue are normally required that reflects detection sensitivity in the fmol range and the lack of detector interference or quenching [33].

More recent studies have incorporated mass spectroscopy (MS) detection of angiotensin metabolism in tissues, cells, and plasma, as well as the derivation of processing networks [104]. Velez and colleagues applied HPLC-MS analysis of Ang I processing in rat glomeruli to reveal the predominant processing of Ang I to Ang-(1-7) catalyzed by neprilysin [104]. Interestingly, the authors could not demonstrate an Ang I to Ang II pathway even following the blockade of the Ang-(1-7) pathway with a neprilysin inhibitor. Hildesbrand et al. [103] utilized a HPLC-tandem quadrupole system (HPLC-MS/MS) to reveal multiple metabolism pathways from Ang I to its N-terminal metabolites Ang-(5-10) and Ang-(4-10), as well as Ang II and Ang-(1-7) in immobilized proteins from human plasma. Suski et al. [105] reported that Ang I was primarily converted to Ang-(1-7) in vascular smooth muscle cells (VSMCs) as characterized by HPLC-MS/MS and confirms our earlier study that thimet oligopeptidase directly processed Ang I to Ang-(1-7) in rat VSMC [106]. Grobe and colleagues have applied "in situ" MALDI to characterize both renal and cardiac metabolism of exogenous Ang II [107, 108]. Ang-(1-7) was the primary product from Ang II in the renal cortex while Ang III was the major metabolite in the medulla [108]. In the heart, Ang III and Ang-(1-7) were products of Ang II metabolism catalyzed by APA and ACE2, respectively [107]. These data confirm earlier HPLC-based studies on the contribution of ACE2 to Ang-(1-7) formation in the mouse and human heart [66, 109]. Although this approach cannot distinguish intracellular versus membrane or extracellular processing and requires relatively high-substrate concentrations, it is likely that these systems will develop the required sensitivity and resolution to detect peptides *in situ*, as well as characterize the extent of enzymatic processing.

Acknowledgements

Portions of this chapter are reproduced from the author's recent publication [54]. These studies were supported in part by grants from the National Institute of Health grants (HL-56973, HL-51952, HD084227, HD-047584, and HD-017644) and the American Heart Association (AHA-151521 and AHA-355741). An unrestricted grant from the Farley-Hudson Foundation (Jacksonville, NC), Groskert Heart Fund, and the Wake Forest Venture Fund is also acknowledged.

http://dx.doi.org/10.5772/65949

Author details

Nildris Cruz-Diaz[1], Bryan A. Wilson[2] and Mark C. Chappell[1*]

*Address all correspondence to: mchappel@wakehealth.edu

1 The Hypertension & Vascular Research Center, Wake Forest University School of Medicine, Winston-Salem, NC, USA

2 McAllister Heart Institute, University of North Carolina at Chapel Hill, Chapel Hill, NC, USA

References

[1] Aroor, A. R.; Demarco, V. G.; Jia, G.; Sun, Z.; Nistala, R.; Meininger, G. A.; Sowers, J. R. The role of tissue renin-angiotensin-aldosterone system in the development of endothelial dysfunction and arterial stiffness. Front Endocrinol (Lausanne) 2013;4:161.

[2] Ferrario, C. M.; Strawn, W. B. Role of the renin-angiotensin-aldosterone system and proinflammatory mediators in cardiovascular disease. Am J Cardiol 2006;98:121–128.

[3] Te, R. L.; van Esch, J. H.; Roks, A. J.; van den Meiracker, A. H.; Danser, A. H. Hypertension: renin-angiotensin-aldosterone system alterations. Circ Res 2015;116(6):960–975.

[4] Kobori, H.; Nangaku, M.; Navar, L. G.; Nishiyama, A. The intrarenal renin-angiotensin system: from physiology to the pathobiology of hypertension and kidney disease. Pharmacol Rev 2007;59(3):251–287.

[5] Navar, L. G.; Prieto, M. C.; Satou, R.; Kobori, H. Intrarenal angiotensin II and its contribution to the genesis of chronic hypertension. Curr Opin Pharmacol 2011;11(2): 180–186.

[6] Santos, R. A. Angiotensin-(1-7). Hypertension 2014;63(6):1138–1147.

[7] Chappell, M. C.; Marshall, A. C.; Alzayadneh, E. M.; Shaltout, H. A.; Diz, D. I. Update on the Angiotensin converting enzyme 2-Angiotensin (1-7)-MAS receptor axis: fetal programming, sex differences, and intracellular pathways. Front Endocrinol (Lausanne) 2014;4:201.

[8] Carey, R. M.; Padia, S. H. Role of angiotensin AT(2) receptors in natriuresis: Intrarenal mechanisms and therapeutic potential. Clin Exp Pharmacol Physiol 2013;40(8):527–534.

[9] Sumners, C.; de Kloet, A. D.; Krause, E. G.; Unger, T.; Steckelings, U. M. Angiotensin type 2 receptors: blood pressure regulation and end organ damage. Curr Opin Pharmacol 2015;21:115–121.

[10] Gonzalez, A. A.; Lara, L. S.; Luffman, C.; Seth, D. M.; Prieto, M. C. Soluble form of the (pro)renin receptor is augmented in the collecting duct and urine of chronic angiotensin II-dependent hypertensive rats. Hypertension 2011;57(4):859–864.

[11] Navar, L. G.; Kobori, H.; Prieto, M. C.; Gonzalez-Villalobos, R. A. Intratubular renin-angiotensin system in hypertension. Hypertension 2011;57(3):355–362.

[12] Gonzalez, A. A.; Liu, L.; Lara, L. S.; Seth, D. M.; Navar, L. G.; Prieto, M. C. Angiotensin II stimulates renin in inner medullary collecting duct cells via protein kinase C and independent of epithelial sodium channel and mineralocorticoid receptor activity. Hypertension 2011;57(3):594–599.

[13] Prieto-Carrasquero, M. C.; Harrison-Bernard, L. M.; Kobori, H.; Ozawa, Y.; Hering-Smith, K. S.; Hamm, L. L.; Navar, L. G. Enhancement of collecting duct renin in angiotensin II-dependent hypertensive rats. Hypertension 2004;44(2):223–229.

[14] Peters, J.; Clausmeyer, S. Intracellular sorting of renin: cell type specific differences and their consequences. J Mol Cell Cardiol 2002;34(12):1561–1568.

[15] Lavoie, J. L.; Liu, X.; Bianco, R. A.; Beltz, T. G.; Johnson, A. K.; Sigmund, C. D. Evidence supporting a functional role for intracellular renin in the brain. Hypertension 2006;47(3):461–466.

[16] Ishigami, T.; Kino, T.; Chen, L.; Minegishi, S.; Araki, N.; Umemura, M.; Abe, K.; Sasaki, R.; Yamana, H.; Umemura, S. Identification of bona fide alternative renin transcripts expressed along cortical tubules and potential roles in promoting insulin resistance in vivo without significant plasma renin activity elevation. Hypertension 2014;64(1):125–133.

[17] Peters, J.; Wanka, H.; Peters, B.; Hoffmann, S. A renin transcript lacking exon 1 encodes for a non-secretory intracellular renin that increases aldosterone production in transgenic rats. J Cell Mol Med 2008;12(4):1229–1237.

[18] Clausmeyer, S.; Sturzebecher, R.; Peters, J. An alternative transcript of the rat renin gene can result in a truncated prorenin that is transported into adrenal mitochondria. Circ Res 1999;84(3):337–344.

[19] Wanka, H.; Kessler, N.; Ellmer, J.; Endlich, N.; Peters, B. S.; Clausmeyer, S.; Peters, J. Cytosolic renin is targeted to mitochondria and induces apoptosis in H9c2 rat cardiomyoblasts. J Cell Mol Med 2009;13(9A):2926–2937.

[20] Wanka, H.; Staar, D.; Lutze, P.; Peters, B.; Hildebrandt, J.; Beck, T.; Baumgen, I.; Albers, A.; Krieg, T.; Zimmermann, K.; Sczodrok, J.; Schafer, S.; Hoffmann, S.; Peters, J. Anti-necrotic and cardioprotective effects of a cytosolic renin isoform under ischemia-related conditions. J Mol Med (Berl) 2016;94(1):61–69.

[21] Wilson, B. A.; Nautiyal, M.; Gwathmey, T. M.; Rose, J. C.; Chappell, M. C. Evidence for a Mitochondrial Angiotensin-(1-7) System in the kidney. Am J Physiol Renal Physiol 2016; 308(6):F594-F601.

[22] Alzayadneh, E. M.; Chappell, M. C. Nuclear expression of renin-angiotensin system components in NRK-52E renal epithelial cells. J Renin Angiotensin Aldosterone Syst 2014;307(5):R487-R489.

[23] Lavoie, J. L.; Lake-Bruse, K. D.; Sigmund, C. D. Increased blood pressure in transgenic mice expressing both human renin and angiotensinogen in the renal proximal tubule. Am J Physiol Renal Fluid Electrolyte Physiol 2004;286(5):F965–F971.

[24] Crowley, S. D.; Gurley, S. B.; Herrera, M. J.; Ruiz, P.; Griffiths, R.; Kumar, A. P.; Kim, H. S.; Smithies, O.; Le, T. H.; Coffman, T. M. Angiotensin II causes hypertension and cardiac hypertrophy through its receptors in the kidney. Proc Natl Acad Sci U S A 2006;103(47): 17985–17990.

[25] Li, X. C.; Cook, J. L.; Rubera, I.; Tauc, M.; Zhang, F.; Zhuo, J. L. Intrarenal transfer of an intracellular fluorescent fusion of angiotensin II selectively in proximal tubules increases blood pressure in rats and mice. Am J Physiol Renal Physiol 2011;300(5):F1076–F1088.

[26] Li, X. C.; Hopfer, U.; Zhuo, J. L. Novel signaling mechanisms of intracellular angiotensin II-induced NHE3 expression and activation in mouse proximal tubule cells. Am J Physiol Renal Physiol 2012;303(12):F1617–F1628.

[27] Li, X. C.; Zhuo, J. L. Proximal tubule-dominant transfer of AT1a receptors induces blood pressure responses to intracellular angiotensin II in AT1a receptor-deficient mice. Am J Physiol Regul Integr Comp Physiol 2013;304(8): R588-R598.

[28] Davisson, R. L.; Ding, Y.; Stec, D. E.; Catterall, J. F.; Sigmund, C. D. Novel mechanism of hypertension revealed by cell-specific targeting of human angiotensinogen in transgenic mice. Physiol Genomics 1999;1:3–9.

[29] Bernstein, K. E.; Ong, F. S.; Blackwell, W. L.; Shah, K. H.; Giani, J. F.; Gonzalez-Villalobos, R. A.; Shen, X. Z.; Fuchs, S.; Touyz, R. M. A modern understanding of the traditional and nontraditional biological functions of angiotensin-converting enzyme. Pharmacol Rev 2013;65(1):1–46.

[30] Chappell, M. C.; Pirro, N. T.; Sykes, A.; Ferrario, C. M. Metabolism of angiotensin-(1-7) by angiotensin converting enzyme. Hypertension 1998;31(1 (Pt 2)):362–367.

[31] Yamada, K.; Iyer, S. N.; Chappell, M. C.; Ganten, D.; Ferrario, C. M. Converting enzyme determines the plasma clearance of angiotensin-(1-7). Hypertension 1998;98(32):496–502.

[32] Chappell, M. C. Nonclassical renin-angiotensin system and renal function. Compr Physiol 2012;2(4):2733–2752.

[33] Shaltout, H. A.; Westwood, B.; Averill, D. B.; Ferrario, C. M.; Figueroa, J.; Diz, D. I.; Rose, J. C.; Chappell, M. C. Angiotensin metabolism in renal proximal tubules, urine and serum of sheep: Evidence for ACE2-dependent processing of Angiotensin II. Am J Physiol Renal Physiol 2006;292:F82–F91.

[34] Park, S.; Bivona, B. J.; Kobori, H.; Seth, D. M.; Chappell, M. C.; Lazartigues, E.; Harrison-Bernard, L. M. Major role for ACE-independent intrarenal ANG II formation in type II diabetes. Am J Physiol Renal Physiol 2010;298(1):F37–F48.

[35] Caughey, G. H.; Raymond, W. W.; Wolters, P. J. Angiotensin II generation by mast cell alpha- and beta-chymases. Biochim Biophys Acta 2000;1480(1–2):245–257.

[36] Nagata, S.; Hatakeyama, K.; Asami, M.; Tokashiki, M.; Hibino, H.; Nishiuchi, Y.; Kuwasako, K.; Kato, J.; Asada, Y.; Kitamura, K. Big angiotensin-25: a novel glycosylated angiotensin-related peptide isolated from human urine. Biochem Biophys Res Commun 2013;441(4):757–762.

[37] Urata, H.; Kinoshita, A.; Misono, K. S.; Bumpus, F. M.; Husain, A. Identification of a highly specific chymase as the major angiotensin II-forming enzyme in the human heart. J Biol Chem 1990;265:22348–22357.

[38] Ahmad, S.; Simmons, T.; Varagic, J.; Moniwa, N.; Chappell, M. C.; Ferrario, C. M. Chymase-Dependent Generation of Angiotensin II from Angiotensin-(1-12) in Human Atrial Tissue. PLoS One 2011;6(12):e28501.

[39] Wintroub, B. U.; Kaempfer, C. E.; Schechter, N. M.; Proud, D. A human lung mast cell chymotrypsin-like enzyme: identification and partial characterization. J Clin Invest 1986;77:196–201.

[40] Semaan, W.; Desbiens, L.; Houde, M.; Labonte, J.; Gagnon, H.; Yamamoto, D.; Takai, S.; Laidlaw, T.; Bkaily, G.; Schwertani, A.; Pejler, G.; Levesque, C.; Desjardins, R.; Day, R.; D'Orleans-Juste, P. Chymase inhibitor-sensitive synthesis of endothelin-1 (1-31) by recombinant mouse mast cell protease 4 and human chymase. Biochem Pharmacol 2015;94(2):91–100.

[41] Caughey, G. H. Mast cell proteases as protective and inflammatory mediators. Adv Exp Med Biol 2011;716:212–234.

[42] Santos, C. F.; Paula, C. A.; Salgado, M. C.; Oliveira, E. B. Kinetic characterization and inhibition of the rat MAB elastase-2, an angiotensin I-converting serine protease. Can J Physiol Pharmacol 2002;80(1):42–47.

[43] Santos, C. F.; Oliveira, E. B.; Salgado, M. C.; Greene, A. S. Molecular cloning and sequencing of the cDNA for rat mesenteric arterial bed elastase-2, an angiotensin II-forming enzyme. J Cardiovasc Pharmacol 2002;39(5):628–635.

[44] Santos, C. F.; Caprio, M. A.; Oliveira, E. B.; Salgado, M. C.; Schippers, D. N.; Munzenmaier, D. H.; Greene, A. S. Functional role, cellular source, and tissue distribution of rat elastase-2, an angiotensin II-forming enzyme. Am J Physiol Heart Circ Physiol 2003;285(2):H775–H783.

[45] Santos, C. F.; Greene, A. S.; Salgado, M. C.; Oliveira, E. B. Conversion of renin substrate tetradecapeptide to angiotensin II by rat MAB elastase-2. Can J Physiol Pharmacol 2004;82(11):1000–1005.

[46] Campbell, D. J. Angiotensin II generation in vivo: does it involve enzymes other than renin and angiotensin-converting enzyme? J Renin Angiotensin Aldosterone Syst 2012;13(2):314–316.

[47] Corti, R.; Burnett, J. C. J.; Rouleau, J. L.; Ruschitzka, F.; Luscher, T. F. Vasopeptidase inhibitors: a new therapeutic concept in cardiovascular disease? Circulation 1909;104(15):1856–1862.

[48] Cataliotti, A.; Boerrigter, G.; Chen, H. H.; Jougasaki, M.; Costello, L. C.; Tsuruda, T.; Lee, S. C.; Malatino, L. S.; Burnett, J. C., Jr. Differential actions of vasopeptidase inhibition versus angiotensin-converting enzyme inhibition on diuretic therapy in experimental congestive heart failure. Circulation 2002;105(5): 639–644.

[49] Ferrario CM, Averill DB, Brosnihan KB, Chappell MC, Iskandar SS, Dean RH, Diz DI. Vasopeptidase inhibition and angiotensin-(1-7) in the spontaneously hypertensive rat. Kidney Intl 2002;62(4):1349-1357.

[50] Tikkanen, I.; Tikkanen, T.; Cao, Z.; Allen, T. J.; Davis, B. J.; Lassila, M.; Casley, D.; Johnston, C. I.; Burrell, L. M.; Cooper, M. E. Combined inhibition of neutral endopeptidase with angiotensin converting enzyme or endothelin converting enzyme in experimental diabetes. J Hypertens 2002;20:707–714.

[51] Cao, Z.; Burrell, L. M.; Tikkanen, I.; Bonnet, F.; Cooper, M. E.; Gilbert, R. E. Vasopeptidase inhibition attenuates the progression of renal injury in subtotal nephrectomized rats. Kid Int 2001; 2:715-721.

[52] Kubota, E.; Dean, R. G.; Hubner, R. A.; Casley, D. J.; Johnston, C. I.; Burrell, L. M. Differential tissue and enzyme inhibitory effects of the vasopeptidase inhibitor omapatrilat in the rat. Clin Sci 2003;105:339–345.

[53] Hubers, S. A.; Brown, N. J. Combined angiotensin receptor antagonism and neprilysin inhibition. Circulation 2016;133(11):1115–1124.

[54] Chappell, M. C. Biochemical evaluation of the renin-angiotensin system: the good, bad, and absolute? Am J Physiol Heart Circ Physiol 2016;310(2):H137–H152.

[55] Iyer SN, Chappell MC, Averill DA, Diz DI, Ferrario CM. Vasodepressor actions of Angiotensin-(1-7) unmasked during combined treatment with lisinopril and losartan. Hypertension 1998:31:699-705.

[56] Westwood, B. M.; Chappell, M. C. Divergent pathways for the angiotensin-(1-12) metabolism in the rat circulation and kidney. Peptides 2012;35(2):190–195.

[57] Wilson, B. A.; Cruz-Diaz, N.; Marshall, A. C.; Pirro, N. T.; Su, Y.; Gwathmey, T. M.; Rose, J. C.; Chappell, M. C. An angiotensin-(1-7) peptidase in the kidney cortex, proximal tubules, and human HK-2 epithelial cells that is distinct from insulin-degrading enzyme. Am J Physiol Renal Physiol 2015;308(6):F594–F601.

[58] Zhuo, J. L.; Li, X. C. Novel roles of intracrine angiotensin II and signalling mechanisms in kidney cells. J Renin Angiotensin Aldosterone Syst 2007;8(1):23–33.

[59] Bkaily, G.; Nader, M.; Avedanian, L.; Choufani, S.; Jacques, D.; D'Orleans-Juste, P.; Gobeil, F.; Chemtob, S.; Al-Khoury, J. G-protein-coupled receptors, channels, and Na+-H+ exchanger in nuclear membranes of heart, hepatic, vascular endothelial, and smooth muscle cells. Can J Physiol Pharmacol 2006;84(3-4):431–441.

[60] Re, R. N. Intracellular renin and the nature of intracrine enzymes. Hypertension 2003;42(2):117–122.

[61] Tipnis, S. R.; Hooper, N. M.; Hyde, R.; Karran, E.; Christie, G.; Turner, A. J. A human homolog of angiotensin-converting enzyme. Cloning and functional expression as a captopril-insensitive carboxypeptidase. J Biol Chem 2000;275:33238–33243.

[62] Turner, A. J.; Hooper, N. M. The angiotensin-converting enzyme gene family: genomics and pharmacology. TIPS 2002;23(4):177–183.

[63] Donoghue, M.; Hsieh, F.; Baronas, E.; Godbout, K.; Gosselin, M.; Stagliano, N.; Donovan, M.; Woolf, B.; Robinson, K.; Jeyaseelan, R.; Breitbart, R. E.; Acton, S. A novel angiotensin-converting enzyme-related carboxypeptidase (ACE2) converts angiotensin I to angiotensin 1-9. Circ Res 2000;87:E1–E9.

[64] Rice, G. I.; Thomas, D. A.; Grant, P. J.; Turner, A. J.; Hooper, N. M. Evaluation of angiotensin-converting enzyme (ACE), its homologue ACE2 and neprilysin in angiotensin peptide metabolism. Biochem J 2004;383(Pt 1):45–51.

[65] Vickers, C.; Hales, P.; Kaushik, V.; Dick, L.; Gavin, J.; Tang, J.; Godbout, K.; Parsons, T.; Baronas, E.; Hsieh, F.; Acton, S.; Patane, M.; Nichols, A.; Tummino, P. Hydrolysis of biological peptides by human angiotensin-converting enzyme-related carboxypeptidase. J Biol Chem 2002;277(17):14838–14843.

[66] Garabelli, P. J.; Modrall, J. G.; Penninger, J. M.; Ferrario, C. M.; Chappell, M. C. Distinct roles for angiotensin converting enzyme 2 and carboxypeptidase A in the processing of angiotensins in the murine heart. Exp Physiol 2008;96:613-621.

[67] Rice, G. I.; Jones, A. L.; Grant, P. J.; Carter, A. M.; Turner, A. J.; Hooper, N. M. Circulating activities of angiotensin-converting enzyme, its homolog, angiotensin-converting enzyme 2, and neprilysin in a family study. Hypertension 2006;48:914–920.

[68] Yamaleyeva, L. M.; Gilliam-Davis, S.; Almeida, I.; Brosnihan, K. B.; Lindsey, S. H.; Chappell, M. C. Differential regulation of circulating and renal ACE2 and ACE in hypertensive mRen2.Lewis rats with early-onset diabetes. Am J Physiol Renal Physiol 2012;302(11):F1374–F1384.

[69] Epelman, S.; Shrestha, K.; Troughton, R. W.; Francis, G. S.; Sen, S.; Klein, A. L.; Tang, W. H. Soluble angiotensin-converting enzyme 2 in human heart failure: relation with myocardial function and clinical outcomes. J Card Fail 2009;15(7):565–571.

[70] Tikellis, C.; Bialkowski, K.; Pete, J.; Sheehy, K.; Su, Q.; Johnston, C.; Cooper, M.; Thomas, M. ACE2 deficiency modifies renoprotection afforded by ACE inhibition in experimental diabetes. Diabetes 2008;57:1018–1025.

[71] Bradford, C. N.; Ely, D. R.; Raizada, M. K. Targeting the vasoprotective axis of the renin-angiotensin system: a novel strategic approach to pulmonary hypertensive therapy. Curr Hypertens Rep 2010;12(4):212–219.

[72] Chappell, M. C. Emerging evidence for a functional angiotensin-converting enzyme 2-angiotensin-(1-7) mas receptor axis; more than regulation of blood pressure? Hypertension 2007;50(4):596–599.

[73] Soler, M. J.; Wysocki, J.; Ye, M.; Lloveras, J.; Kanwar, Y.; Batlle, D. ACE2 inhibition worsens glomerular injury in association with increased ACE expression in streptozotocin-induced diabetic mice. Kid Int 2007;72:614–623.

[74] Oudit, G. Y.; Herzenberg, A. M.; Kassiri, Z.; Wong, D.; Reich, H.; Khokha, R.; Crackower, M. A.; Backx, P. H.; Penninger, J. M.; Scholey, J. W. Loss of angiotensin-converting enzyme-2 leads to the late development of angiotensin II-dependent glomerulosclerosis. Am J Pathol 2006;168(6):1808–1820.

[75] Crackower, M. A.; Sarao, R.; Oudit, G. Y.; Yagil, C.; Kozieradzki, I.; Scanga, S. E.; Oliveira-dos-Santo, A. J.; da Costa, J.; Zhang, L.; Pei, Y.; Scholey, J.; Bray, M. R.; Ferrario, C. M.; Backx, P. H.; Manoukian, A. S.; Chappell, M. C.; Yagil, Y.; Penninger, J. M. Angiotensin-converting enzyme 2 is an essential regulator of heart function. Nature 2002;417:822–828.

[76] Oudit, G. Y.; Kassiri, Z.; Patel, M. P.; Chappell, M.; Butany, J.; Backx, P. H.; Tsushima, R. G.; Scholey, J. W.; Khokha, R.; Penninger, J. M. Angiotensin II-mediated oxidative stress and inflammation mediate the age-dependent cardiomyopathy in ACE2 null mice. Cardiovasc Res 2007;75(1):29–39.

[77] Chhabra, K. H.; Xia, H.; Pedersen, K. B.; Speth, R. C.; Lazartigues, E. Pancreatic angiotensin-converting enzyme 2 improves glycemia in angiotensin II-infused mice. Am J Physiol Endocrinol Metab 2013;304(8):E874–E884.

[78] Oudit, G. Y.; Liu, G. C.; Zhong, J.; Basu, R.; Chow, F. L.; Zhou, J.; Loibner, H.; Janzek, E.; Schuster, M.; Penninger, J. M.; Herzenberg, A. M.; Kassiri, Z.; Scholey, J. W. Human recombinant ACE2 reduces the progression of diabetic nephropathy. Diabetes 2010;59(2):529–538.

[79] Zhong, J.; Basu, R.; Guo, D.; Chow, F. L.; Byrns, S.; Schuster, M.; Loibner, H.; Wang, X. H.; Penninger, J. M.; Kassiri, Z.; Oudit, G. Y. Angiotensin-converting enzyme 2 suppresses pathological hypertrophy, myocardial fibrosis, and cardiac dysfunction. Circulation 2010;122(7):717–28, 18.

[80] Wysocki, J.; Ye, M.; Rodriguez, E.; Gonzalez-Pacheco, F. R.; Barrios, C.; Evora, K.; Schuster, M.; Loibner, H.; Brosnihan, K. B.; Ferrario, C. M.; Penninger, J. M.; Batlle, D. Targeting the degradation of angiotensin II with recombinant angiotensin-converting

enzyme 2: prevention of angiotensin II-dependent hypertension. Hypertension 2010;55(1):90–98.

[81] Ye, M.; Wysocki, J.; Gonzalez-Pacheco, F. R.; Salem, M.; Evora, K.; Garcia-Halpin, L.; Poglitsch, M.; Schuster, M.; Batlle, D. Murine recombinant angiotensin-converting enzyme 2: effect on angiotensin II-dependent hypertension and distinctive angiotensin-converting enzyme 2 inhibitor characteristics on rodent and human angiotensin-converting enzyme 2. Hypertension 2012;60(3):730–740.

[82] Huentelman, M. J.; Grobe, J. L.; Vazquez, J.; Stewart, J. M.; Mecca, A. P.; Katovich, M. J.; Ferrario, C. M.; Raizada, M. K. Protection from angiotensin II-induced cardiac hypertrophy and fibrosis by systemic lentiviral delivery of ACE2 in rats. Exp Physiol 2005;90(5):783–790.

[83] Diez-Freire, C.; Vasquez, J.; Correa de Adjounian, M. F.; Ferrari, M. F. R.; Yuan, L.; Silver, X.; Torres, R.; Raizada, M. K. ACE2 gene transfer attenuates hypertension-linked pathophysiological changes in the SHR. Physiol Genomics 2006;27:12–19.

[84] Shenoy, V.; Ferreira, A. J.; Qi, Y.; Fraga-Silva, R. A.; Diez-Freire, C.; Dooies, A.; Jun, J. Y.; Sriramula, S.; Mariappan, N.; Pourang, D.; Venugopal, C. S.; Francis, J.; Reudelhuber, T.; Santos, R. A.; Patel, J. M.; Raizada, M. K.; Katovich, M. J. The angiotensin-converting enzyme 2/angiogenesis-(1-7)/Mas axis confers cardiopulmonary protection against lung fibrosis and pulmonary hypertension. Am J Respir Crit Care Med 2010;182(8): 1065–1072.

[85] Yamazato, M.; Yamazato, Y.; Sun, C.; Diez-Freire, C.; Raizada, M. K. Overexpression of angiotensin-converting enzyme 2 in the rostral venterolateral medulla causes long-term decrease in blood pressure in the spontaneously hypertensive rats. Hypertension 2007;49:926–931.

[86] Hernandez Prada, J. A.; Ferreira, A. J.; Katovich, M. J.; Shenoy, V.; Qi, Y.; Santos, R. A.; Castellano, R. K.; Lampkins, A. J.; Gubala, V.; Ostrov, D. A.; Raizada, M. K. Structure-based identification of small-molecule angiotensin-converting enzyme 2 activators as novel antihypertensive agents. Hypertension 2008;51(5):1312–1317.

[87] Qi, Y.; Zhang, J.; Cole-Jeffrey, C. T.; Shenoy, V.; Espejo, A.; Hanna, M.; Song, C.; Pepine, C. J.; Katovich, M. J.; Raizada, M. K. Diminazene aceturate enhances angiotensin-converting enzyme 2 activity and attenuates ischemia-induced cardiac pathophysiology. Hypertension 2013;62(4):746–752.

[88] De Maria, M. L.; Araujo, L. D.; Fraga-Silva, R. A.; Pereira, L. A.; Ribeiro, H. J.; Menezes, G. B.; Shenoy, V.; Raizada, M. K.; Ferreira, A. J. Anti-hypertensive effects of diminazene aceturate: an angiotensin- converting enzyme 2 activator in rats. Protein Pept Lett 2016;23(1):9–16.

[89] Macedo, L. M.; Souza, A. P.; De Maria, M. L.; Borges, C. L.; Soares, C. M.; Pedrino, G. R.; Colugnati, D. B.; Santos, R. A.; Mendes, E. P.; Ferreira, A. J.; Castro, C. H. Cardio-

protective effects of diminazene aceturate in pressure-overloaded rat hearts. Life Sci 2016;155:63–69.

[90] Shenoy, V.; Gjymishka, A.; Jarajapu, Y. P.; Qi, Y.; Afzal, A.; Rigatto, K.; Ferreira, A. J.; Fraga-Silva, R. A.; Kearns, P.; Douglas, J. Y.; Agarwal, D.; Mubarak, K. K.; Bradford, C.; Kennedy, W. R.; Jun, J. Y.; Rathinasabapathy, A.; Bruce, E.; Gupta, D.; Cardounel, A. J.; Mocco, J.; Patel, J. M.; Francis, J.; Grant, M. B.; Katovich, M. J.; Raizada, M. K. Diminazene attenuates pulmonary hypertension and improves angiogenic progenitor cell functions in experimental models. Am J Respir Crit Care Med 2013;187(6):648–657.

[91] Velkoska, E.; Dean, R. G.; Griggs, K.; Burchill, L.; Burrell, L. M. Angiotensin-(1-7) infusion is associated with increased blood pressure and adverse cardiac remodelling in rats with subtotal nephrectomy. Clin Sci (Lond) 2011;120(8):335–345.

[92] Haber, P. K.; Ye, M.; Wysocki, J.; Maier, C.; Haque, S. K.; Batlle, D. Angiotensin-converting enzyme 2-independent action of presumed angiotensin-converting enzyme 2 activators: studies in vivo, ex vivo, and in vitro. Hypertension 2014;63(4): 774–782.

[93] Velkoska, E.; Dean, R. G.; Burchill, L.; Levidiotis, V.; Burrell, L. M. Reduction in renal ACE2 expression in subtotal nephrectomy in rats is ameliorated with ACE inhibition. Clin Sci (Lond) 2010;118(4):269–279.

[94] Raffai, G.; Khang, G.; Vanhoutte, P. M. Angiotensin-(1-7) augments endothelium-dependent relaxations of porcine coronary arteries to bradykinin by inhibiting angiotensin-converting enzyme 1. J Cardiovasc Pharmacol 2014;63(5):453–460.

[95] Wilson, B. A.; Marshall, A. C.; Alzayadneh, E. M.; Chappell, M. C. The ins and outs of angiotensin processing within the kidney. Am J Physiol Regul Integr Comp Physiol 2014;307(5):R487–R489.

[96] Marshall, A. C.; Shaltout, H. A.; Pirro, N. T.; Rose, J. C.; Diz, D. I.; Chappell, M. C. Antenatal betamethasone exposure is associated with lower ANG-(1-7) and increased ACE in the CSF of adult sheep. Am J Physiol Regul Integr Comp Physiol 2013;305(7):R679–R688.

[97] Marshall, A. C.; Shaltout, H. A.; Pirro, N. T.; Rose, J. C.; Diz, D. I.; Chappell, M. C. Enhanced activity of an angiotensin-(1-7) neuropeptidase in glucocorticoid-induced fetal programming. Peptides 2014;52:74–81.

[98] Marshall, A. C.; Pirro, N. T.; Rose, J. C.; Diz, D. I.; Chappell, M. C. Evidence for an angiotensin-(1-7) neuropeptidase expressed in the brain medulla and CSF of sheep. J Neurochem 2014;130(2):313–323.

[99] Cruz-Diaz, N.; Wilson, B. A.; Pirro, N. T.; Brosnihan, K. B.; Marshall, A. C.; Chappell, M. C. Identification of dipeptidyl peptidase 3 as the angiotensin-(1-7) degrading peptidase in human HK-2 renal epithelial cells. Peptides 2016;83:29–37.

[100] Alzayadneh, E. M.; Chappell, M. C. Angiotensin-(1-7) abolishes AGE-induced cellular hypertrophy and myofibroblast transformation via inhibition of ERK1/2. Cell Signal 2014;26(12):3027–3035.

[101] Lew, R. A. HPLC in the analysis of peptide metabolism. Methods Mol Biol 2004;251:275–290.

[102] Velez, J. C.; Ierardi, J. L.; Bland, A. M.; Morinelli, T. A.; Arthur, J. M.; Raymond, J. R.; Janech, M. G. Enzymatic processing of angiotensin peptides by human glomerular endothelial cells. Am J Physiol Renal Physiol 2012;302(12):F1583–F1594.

[103] Hildebrand, D.; Merkel, P.; Eggers, L. F.; Schluter, H. Proteolytic processing of angiotensin-I in human blood plasma. PLoS One 2013;8(5):e64027.

[104] Schwacke, J. H.; Spainhour, J. C.; Ierardi, J. L.; Chaves, J. M.; Arthur, J. M.; Janech, M. G.; Velez, J. C. Network modeling reveals steps in angiotensin peptide processing. Hypertension 2013;61(3):690–700.

[105] Suski, M.; Gebska, A.; Olszanecki, R.; Stachowicz, A.; Uracz, D.; Madej, J.; Korbut, R. Influence of atorvastatin on angiotensin I metabolism in resting and TNF-alpha - activated rat vascular smooth muscle cells. J Renin Angiotensin Aldosterone Syst 2014;15(4):378–383.

[106] Chappell, M. C.; Tallant, E. A.; Brosnihan, K. B.; Ferrario, C. M. Conversion of angiotensin I to angiotensin-(1-7) by thimet oligopeptidase (EC 3.4.24.15) in vascular smooth muscle cells. J Vasc Med Biol 1994;5(4):129–137.

[107] Alghamri, M. S.; Morris, M.; Meszaros, J. G.; Elased, K. M.; Grobe, N. Novel role of aminopeptidase-A in angiotensin-(1-7) metabolism post myocardial infarction. Am J Physiol Heart Circ Physiol 2014;306(7):H1032–H1040.

[108] Grobe, N.; Elased, K. M.; Cool, D. R.; Morris, M. Mass spectrometry for the molecular imaging of angiotensin metabolism in kidney. Am J Physiol Endocrinol Metab 2012;302(8):E1016–E1024.

[109] Zisman, L. S.; Keller, R. S.; Weaver, B.; Lin, Q.; Speth, R.; Bristow, M. R.; Canver, C. C. Increased angiotensin-(1-7)-forming activity in failing human heart ventricles: evidence for upregulation of the angiotensin-converting enzyme Homologue ACE2. Circulation 2003;108(14):1707–1712.

Chapter 2

Programmed Cell Death-Related Proteases in Plants

Gustavo Lemos Rocha, Jorge Hernandez Fernandez,
Antônia Elenir Amâncio Oliveira and
Kátia Valevski Sales Fernandes

Additional information is available at the end of the chapter

http://dx.doi.org/10.5772/65938

Abstract

From an ancient Greek term related to the "leavening of bread" (*en*, in; *zyme*, leaven), an enzyme can be defined as a substance showing the properties of a catalyst that is produced as a result of cellular activity. Every proteinaceous enzyme that performs hydrolysis of peptide bonds is appropriately termed "protease" (peptidase). All of them share aspects of catalytic strategy, but with some variation. As a result, the proteases are grouped into six different catalytic families: serine, threonine, cysteine, aspartic, glutamic and metallopeptidases (http://merops.sanger.ac.uk/). The larger families (cysteine, serine, aspartic and metallopeptidases) have a wide range of distribution on living organism groups, and are also present in the "controversial" viruses. As a well-represented family, the cysteine proteases play important roles in events such as signalling pathways, programmed cell death (PCD), nutrient mobilization, protein maturing, hormone synthesis and degradation. In the past two decades, an increased interest was driven to the study of the programmed cell death (PCD), mainly after the discovery of caspase-related proteins and caspase-like activities in organisms not metazoan. Caspases are cysteine proteases that cleave their substrate after aspartate residues and are part of signalling cascades of the apoptotic PCD process (also in inflammatory process), unique of metazoan. The caspase-related proteins are named paracaspases and metacaspases. Paracaspases are found on metazoan and *Dictyostelium*, whereas the metacaspases are present on plants, fungi and groups of protozoan. On plants, PCD has features that are distinct from that of animals and is an important pathway on developmental events, defensive and stress response (biotic and abiotic). All these events have their own particularities, but the participation of proteases seems to be universal with those responsible for caspase-like activities and metacaspases having an increasing number of reports that put them as important for plant PCD. In this chapter, we tackle important aspects of the proteases, in special that involved in plant PDC, as well as their specific regulators. Aspects of function, catalytic mechanisms and interaction with ligands will be on focus.

Keywords: plants, programmed cell death, metacaspases

1. Introduction

Every proteinaceous enzyme that performs hydrolysis of peptide bonds is correctly designed as protease (peptidase) [1, 2]. This term was first used by Vines [3] based on direct and indirect evidences from studies with algae, some fungi and Phanerogams. Long before this work, the word "proteolytic" was applied by Roberts [4] to describe the digestive process on human stomach, and in the first years of the twentieth century, the digestive proteins trypsin and pepsin as well as other autocatalytic enzymes were known as proteases, being pepsin credited to do "proteolysis" since 1877 [5]. In 1928, Grassmann and Dyckerhoff [6] established important definitions concerning the nature of the catalytic activities of proteases. In resume, the peptide cleavage pattern by proteases can be internal, for endopeptidases; on the N-terminal portion, for aminopeptidases; or C-terminal, for carboxypeptidases [2].

Despite this difference, all proteases share the same catalytic strategy, as they polarize the carbonyl group on the peptide bond of the substrate by the stabilization of the oxygen atom on an oxyanionic hole, what makes the carbon atom more vulnerable to the attack of an activated nucleophile. The nature of the nucleophile presents some variation and is determinant to the employed mechanism for enzymatic catalysis. As a result of these variations, the proteases are grouped into six different catalytic families: serine, threonine, cysteine, aspartic, glutamic and metallopeptidases (http://merops.sanger.ac.uk/). The larger families (cysteine, serine, aspartic and metallopeptidases) have a wide range of distribution on living organisms, and are also present in the "controversial" group of viruses [7].

About their function, these enzymes are well known for the promotion of protein degradation on amino acid unities. Besides this function, they regulate the destination of other proteins based on their cleavage specificity and participate on important cellular pathways, being key regulators in different response processes to environmental factors and developmental signals [2].

One of these cellular pathways is the event of programmed cell death (PCD) that, since the 2000s, is being studied with increased interest on non-metazoan organisms, in particular on plant models. On metazoan, one of the main constituent of the pathway are cysteine proteases known as caspases. They are not found in plants, but an increasing number of reports shows strong evidences that caspase-related proteases that belong to the metacaspase group, as well as proteases with caspase-like activities are eminent on plant PCD [8].

2. Programmed cell death in plants

Programmed cell death (PCD) is a genetically controlled physiological innate mechanism, which involves the selective death of individual cells, tissues or entire organs. It is a process different from necrosis as it occurs passively in response to environmental perturbations [9]. Together with the chromatin remodelling machinery, the cell cycle regulation mechanisms, the nuclear envelope and the cytoskeleton, this process is one of the major eukaryotic innovative aspects, which allowed the development of more complex organisms [10].

Molecular evidences have pointed out that the PCD machinery has evolved since very early stages of the evolutive history, and that this evolution has been processed through expansion and innovation of protein recruitment domains, as well as through the derivation of effector domains and horizontal gene transference events [11].

The best-understood models in PCD are metazoan organisms and in concern to the cell morphology and involved organelles, two main categories are known: apoptosis and autophagy [12–14].

Apoptosis is defined by three main morphological characteristics: nuclear DNA fragmentation, apoptotic bodies' formation and degradation of the apoptotic bodies on the lysosome of a phagocytic cell [12–14].

Autophagy is the main system of degradation and recycling in eukaryotic cells, contributing to the clearing of cellular compounds and cytosolic portions. This process can occur in two forms: through the cytosolic sequestration by the vacuole or through the sequestration of large portions of cytosol by a structure called autophagosome [15–17].

On situations where a cellular set is under a more intense stress, so that the cells are not able to activate the apoptotic PCD pathway, cell death occurs through necrosis, characterized by a protoplasmic swelling due to the loss of the osmoregulation control capability, and consequent water and ions migration to the cell [18].

In plants, PCD is observed under diverse circumstances through the entire life cycle of many species, as well as in response to biotic and abiotic stimuli, what allows wide biochemical and developmental plasticity [19], as, for example:

- on the degeneration of cells from tissues with transitory functions, such as cotyledons, suspensor, certain leaves, petals [20] and secretory tissues [21];
- on the elimination of excessive produced cells, as in the case of some unisexual flowers that initially produce male and female organs, and must eliminate one of these groups to become functional [22].

3. Programmed cell death-related proteases

Concerning to PCD, the main group of proteases performing important roles is the subfamily C14 from the CD clan of cysteine proteases. Their representatives include the metacaspases, paracaspases and caspases.

The first discovered and the most known in terms of structure and function are the caspases. The early reports of caspases and related proteins are those with the genes *cd-3* and *cd-4*, as well as their encoded products from *Caenorhabditis elegans* and the enzyme caspase-1 from mammals [23–27].

To date, caspases have been proven to be "in the heart" of a pathway that mediates the highly ordered process of apoptotic cell suicide [28], and, indeed, they are the convergence point of biochemical pathways on cellular substrates which lead to the activation of a "protease cascade" (the caspase cascade).

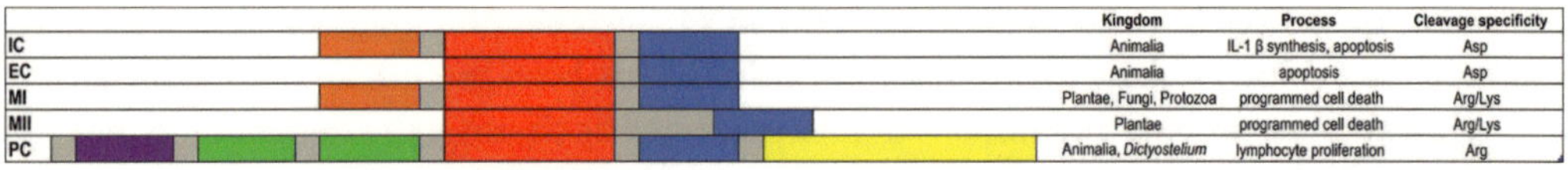

Figure 1. Representation of caspase, metacaspase and paracaspase major domains. IC: initiator caspase; EC: executor caspase; MI: type I metacaspase; MII: type II metacaspase; PC: paracaspase. Red: p20 domain; blue: p10 domain; orange: prodomain/recruitment domain; purple: death domain; green: immunoglobulin-like domain; yellow: paracaspase C-terminal region. Based on Vercammen et al. [8].

The importance of this cascade can be seen when disturbance of its regulation occurs on the cells, what causes immunodeficiency, carcinogenesis and other troubles related to aberrant PCD [29–31].

The cascade overview reveals two functional groups of caspases, concerning their position on the sequence of proteolytic events (**Figure 1**). The first caspases to be activated are the initiators which become active by the "induced proximity mechanism", often triggered by recruitment of adaptor protein complexes, such as recruitment domain of membrane receptors or by a huge protein complex called apoptosome. This induced proximity leads to the oligomerization of caspase molecules which form a heterodimer that becomes able to trigger the activation of the effector caspases by cleavage of a pro-domain, since these enzymes are synthetized as zymogens. After this cleavage, the effectors themselves also stay organized as heterodimer, and cleave the apoptotic substrates, leading to a typical cellular morphology [8].

Despite the positional and functional differences, all caspases belong to a group named ICE family (interleukin-1β-converting enzyme/caspase 1 family), since caspase 1 was the first characterized member.

About the phylogeny, these molecules are unique and well distributed among the Animalia Kingdom. Their presence ranges from vertebrate organisms such as *Homo sapiens* and *Danio rerio* to invertebrates such *C. elegans* and *Drosophila melanogaster* [32]. Also, their involvement with cell death processes is also well documented.

Alongside caspases, on clan CD, there are caspase relatives called paracaspases which are found in metazoans and in slime molds [4, 32] (**Figure 1**). In humans, paracaspase/MALT1 is associated to the lymphocyte activation by the NF-kB pathway [33, 34]. In *Dictyostelium*, the paracaspase gene disruption does not affect the developmental cell death process, what leads to the suggestion that the enzyme is not required for the phenomenon [35]. A function for human paracaspase MALT 1 on caspase-8 modulation is reinforced by Kawadler et al. [36], whose data show that MALT1 activates caspase-8 during TCR signalling but reduces the cleavage of caspase-3, avoiding apoptosis and inducing cell proliferation.

About its structural architecture, the human MALT 1 monomer has an apparent molecular mass of 41 kDa and the dimer, about 84 kDa and the overall predicted structure shows an N-terminal death domain, two immunoglobulin domains, a paracaspase domain and another immunoglobulin-like domain. The paracaspase domain is folded in a similar way to that of caspases and exhibits the ability to bind on substrate. Also, the enzyme, contrary to caspases, seems not to require cleavage of loop 4 to become active [37, 38].

The other member of CD clan of the cysteine proteases are the metacaspases (**Figure 1**). They are found in Fungi, Protozoa, Chromista and Plantae [8], and were first described in 2000 [32], on a study performing structural and sequence analysis, which revealed a great diversity of protease genes related to caspases in these Phyla. They were credited as strong candidates to perform central roles on PCD [11].

Based on their structural architecture, they were divided into two categories: type I and type II metacaspases. The type I subclass metacaspases have an N-terminal extension, a prodomain with a proline-rich repeat motif that is absent on the type II metacaspases, which, instead, present a 200 amino acid C-terminal extension. Also, many plant type I metacaspases have a zinc finger motif that is similar to that of the plant hypersensitive response/cell death protein lsd-1 [39].

Recently, a third group, named as type III metacaspases, has been proposed [40]. These proteins are found on many phytoplankton organisms and are not grouped with the other metacaspase groups, probably by a p10 domain rearrangement on which the motif SGCXDXQTSADV is located on the N-terminus rather than on the C-terminus, as usual for plant metacaspases. The studied organisms also revealed the presence of metacaspase-like proteases which possess only the p10 domain. These proteases are found on bacteria and may represent an evolutionary connection.

4. Metacaspases

4.1. Plant metacaspases

After the first finding of caspase-like activities on plants [41], as well as on other non-metazoan organisms, an extensive search for the enzymes responsible for these activities was performed [32]. For the model plant, *Arabidopsis thaliana*, for example, nine metacaspases encoded by its genome were found: 3, of type I and 6, of type II metacaspases [42]. The list is being extended, among other examples are: soybean (*Glycine max*) with 16 metacaspases (10, type I and 6, type II); wheat (*Zea mays*) with 12 metacaspases (8, type I and 4, type II) and *Oryza sativa* sp japonica with eight metacaspases (4, type I and 4, type II) [43].

The first observations using cell extracts suggested that metacaspases could be responsible for the caspase-like activities found in plants [44]. Nevertheless, recorded reports show that plant metacaspases are unable to cleave caspase synthetic substrates [45–47]. The caspases cleave their substrates after aspartate residues and metacaspases cleave after Asp or Lys residues at P1 position [43]. So, appears that the metacaspases are not directly responsible for the caspase-like activities found in plants.

Although metacaspases do not have caspase-like activities, many works suggest that they have a role in PCD [48]. The inhibition of a type II metacaspase (McIIPa) suppressed PCD in suspensor cells from an embryonic culture from *Picea abies*. During *P. abies* embryogenesis, it was shown that McIIPa is transported from cytosol to nuclei, where its presence was correlated to DNA fragmentation detection. Also, through experiments with a cell free system in the presence of a mutant form of this protein on its active site, the nuclear alterations were not detected [44]. These data reinforce that McIIPa is directly involved on a pathway which generates nuclear degradation, an event present on most eukaryotic PCD programmes. In this pathway, the protease can play a role on the cleavage of nuclear proteins, similar to what happens to the metazoan effector caspases [46].

The heterologous expression of the *A. thaliana* metacaspases MCP1b and MCP2b on a *Saccharomyces cerevisiae* strain, disrupted for its metacaspase YCA1 gene, was able to trigger an apoptosis-like phenotype under hydrogen peroxide treatment, and this effect was seen to be abolished on the presence of the pancaspase inhibitor based on the sequence VAD. These findings reinforce a metacaspase role on PCD, as well as the requirement of proteins with caspase-like activity [42].

The involvement of metacaspases on PCD is also suggested by works with plants under pathogen attack. Some examples are given by the detection of gene expression for MCA1 on *A. thaliana* leaves infected by *Pseudomonas syringae* [49], and by the increasing levels of a type II metacaspase on tomato leaves under *Botrytis cynerea* infection, with the detection of cell death phenotype [50].

It cannot be excluded, though, that metacaspases could not be directly involved on PCD regulation, but indirectly involved on signalling cascades that leads to PCD [48].

In face of this, the role of metacaspases are still under discussion, as well as their classification, since there are evidences which favour, and others contrary, to their groupment together with caspases [51, 52]. A cascade mechanism comparable to that of caspases was proposed for vegetable systems concerning cysteine proteases on senescent leaves and seeds on maturation. Bozhkov et al. [46] reinforced the idea that the execution of PCD in plants is controlled by two groups of enzymes with separated cellular localization. One of them is accumulated on lytic compartments and vacuoles, and the other has cytoplasmic-nuclear localization, as in the case of MCIIPa.

4.2. Yeast metacaspases

The first report of a metacaspase on yeast was made by Madeo et al. [53], where the overexpression of the protein codified by theYor197w gene stimulated PCD associated-caspase-like activity on *S. cerevisiae* cells, under H_2O_2 stress. The authors proposed the name Yeast Caspase-1 (YCA1) for the protein. Also, a *S. cerevisiae* mutant strain with an inactive gene encoding a deubiquitinating enzyme (UBP10) showed an apoptosis-related phenotype which appears to be related to the YCA1 presence, in a manner similar to that observed when the cells were submitted to external stimuli, such as treatment with H_2O_2. The overexpression of YCA1 in cells lacking UBP10 resulted on a decrease of about 53% in viability [54].

A study with frataxin-deficient yeast cells (Δ*yfh 1*), sensible to pro-oxidant chemicals, showed that YCA1 is induced under H_2O_2 stress. When performance of metacaspase-deficient yeasts (Δ*yca1*) was evaluated under oxidative stressing condition, a resistance to cell death, reflected by higher glutathione concentrations than the wild-type, was detected, probably caused by the absence of the metacaspase [55].

By using an original approach of combining the techniques of a digestome analysis (an *in vitro* assay aimed to search putative specific substrates of proteases), cleavage of recombinant GAPDH by metacaspase and evaluation of protein levels *in vivo*, wild-type and YCA1 overexpressing yeast cells upon H_2O_2-induced apoptosis were evaluated. Under these oxidative conditions, the enzyme glyceraldehyde-3-phosphate dehydrogenase (GAPDH), was shown, by mass spectrometric analysis, to be a specific target of metacaspase. This result suggests a link between programmed cell death and metabolism [56]. Besides the potential role in MCP, this metacaspase seems to have involvement on cell cycle regulation. Deficient *yca1* yeast revealed an increased level of proteins related to the vacuolar catabolism, including six peptidases, as well as chaperones involved on stress response and protein aggregates elimination. YCA1 was also copurified with chaperones that respond to protein aggregation and with a cell-cycle control protein; thus, a role on clearance of protein aggregates during mitosis and aging was proposed [57]. This potential on protein clearance is dependent on the presence of Hsp40 and Hsp104 chaperones, and is reduced by deletion of the proteasome regulator Rnp4, which reduces the proteasome levels. When overproduced, YCA1 extended the life span of the cells by 46–56% [58].

Metacaspase studies are also being performed with other fungi, as in the case of *Candida albicans* metacaspase 1 (CaMCA1). Mutant strains for this gene showed a lack of PCD under oxidative stress, as well as a decrease of energy metabolism intensity followed by accumulation of trehalose and increase on expression of genes related to trehalose biosynthesis, what is a well-known protective mechanism against oxidative stresses [59]. Other example is the study of two metacaspases from *Aspergillus fumigates,* whose deficiency was associated with loss of membrane phospholipid asymmetry and an increased growth under conditions of endoplasmic reticulum stress [60].

4.3. Metacaspases of protozoa

Among the first reports concerning metacaspases on protists was the work of Szallies et al. [61]. According to this, *Trypanosoma brucei* encodes five putative metacaspases. The expression of the *T. brucei* metacaspase MCA4 in *S. cerevisiae* caused a PCD-related phenotype, and the protein was found to be located on nuclei, alongside the yeast metacaspase MCA1. When overexpressed, MCA4 induced PCD.

Two metacaspases (LdMC1 and LdMC2) of *Leishmania donovani* were also characterized and displayed high homology among each other. The proteins were seen to be capable of cleaving trypsin substrates and were inhibited by classical trypsin inhibitors such as leupeptin and antipain. Despite not being able to cleave caspase substrates, when their genes were overexpressed in the parasite cells, a higher sensibility to PCD was observed [62]. The metacaspases MCA2, MCA3 and MCA 5 were detected *in vivo* in their full molecular form and seem not to suffer processing to become active in the parasite cells. They were found to be associated with recycling endosomes that contain recycling receptors and PCD-independent functions were proposed for this association with endosomes [63].

The capability to induce PCD was also investigated for the *Leishmania major* single metacaspase (LmjMCA), whose heterologous expression in a *S. cerevisiae* defective strain was successful on the induction of PCD by oxidative stress. The protease originated from the metacaspase gene revealed an arginine-specific activity and through approaches of site-directed mutagenesis, it was possible to determine that its activity is dependent of H147 and C202 [64]. This metacaspase was suggested to be essential for the parasite, since the tentative to obtain a null mutant was unsuccessful. The protein was detected in variable levels in different compartments at different stages of the cell cycle, in association with the dividing kinetoplast and the nuclear spindle, what may suggests a role on DNA segregation. Also, it was seen that the protein was associated with cytoskeleton filaments. A role on differentiation and proliferation has been additionally proposed [65].

The activity of a *Plasmodium falciparum* metacaspase 1 (PfMCA-1) increased under high concentrations of calcium and induced PCD under stress conditions [66]. The authors have also presented evidences of the ability of PfMCA1 of triggering a downstream enzyme that is sensible to the pancaspase inhibitor z-VAD-fmk.

5. The caspase fold of metacaspases

Caspases, metacaspases and paracaspases have a conserved pattern of tridimensional organization and are then considered as structural homologues. The degree of this conservation is variable, but the overall structure is related to a conformation named caspase fold that is characterized by a core formed by a contiguous six-stranded β-sheet (β1–β4, β7 and β8) and helices $\alpha1$–$\alpha5$ region, present in every caspase structure. Also, the presence of three well-ordered loops (L1, L2 and L4) is well characterized [67].

As it was early discussed, metacaspases are divided into two groups based on the presence of a pro domain or a linker region. Type I metacaspases possess an N-terminal prodomain with length of about 80–120 amino acids [32], with two CXXC-type LSD1-like zinc finger structures as well as a proline/glutamine rich region [45, 68]. The type II metacaspases do not have prodomains but, otherwise, contain a large loop (linker region) between the p10 and p20 domains which ranges about 90–150 amino acids [32, 69].

The p10 and p20 domains are present in all of these proteins. For p10, there is a SGCXDXQTSADV consensus sequence, as well as other conserved short sequences [40]. The p20 domain contains the conserved catalytic dyad histidine/cysteine as a remarkable feature, where the two amino acids are distant from each other in about 29–47 amino acids. There are also other conserved regions/amino acids that together give about 80% of consensus on the entire sequence [32]. A noticeable signature is the motif DSCHSG in the surroundings of the catalytic Cys, which is highly conserved among all type II plant metacaspases [70].

Differently of metacaspases, all caspases contain a conserved QACXG (where X can be R, Q, or G) pentapeptide active-site motif. The catalytic residues histidine 237 and cysteine 285, and those involved in forming the P1 carboxylate binding pocket on caspase 1 (Arg-179, Gln-283, Arg-341 and Ser-347), are also conserved in all other caspases, except for the conservative substitution of the threonine for the serine 347 in caspase 8. This explains the requirement for an aspartate in the substrate P1 position. The residues that form the P2–P4 binding pocket are not well conserved, suggesting that they may determine the substrate specificities of the different caspases [71]. The metacaspases do not have these features and present cleavage specificity for lysine/arginine on the P1 position on the substrate, so their binding residues seem to be of opposite chemical nature from those of caspases.

The type II metacaspases present autoprocessing sites, whose cleavage seems to be necessary for their full activation: the residues Lys 260 and Arg 214, on the wheat type II metacaspase

[70]; Lys 269, from MCIIPa [46]; and the Arg 214, from AtMC9 [45]. Caspases, otherwise, have different cleavage site, always after Asp residues. Caspase 1 is cleaved after Asp-103, Asp-119, Asp-297 and Asp-316 [71]; caspase 7, after Asp-23, Asp-198 [72]; and caspase 3, after Asp-9/ Asp-28 and Asp-175 [73].

Actually, there are only two metacaspases with elucidated structural organization. MCA2, from *T. brucei*, the first metacaspase with elucidated structure, is topologically more related to other metacaspases, such as YCA1, than to caspases, presenting the same number and organization of β sheets and α helix, although the enzyme showed 65% of secondary elements similarly to caspase 7. Also, the β-sheet region of MCA2 is sized two strands longer than caspases. As for YCA1, this characteristic prevents dimerization. Autocatalytic processing occurs only *in vitro* and is not required for enzyme activity, which was also shown to be dependent of calcium as in other metacaspases [74].

The crystal structure from the yeast metacaspase Yca1 shows the general patterns of the caspase fold, with the well-ordered loops, being L1 and L4 in opposing sides of the substrate interaction site, and the conserved caspase core. Concerning β conformations, yeast YCA1 presents two β-strands (β5 and β6) which are absent in caspases (caspase 3 and caspase 9), and these are located in a way that blocks dimerization. As a result, YCA1 cannot form dimer, as caspases do. The catalytic dyad consists on the residues Cys276 and His220 which are well conserved among other proteins with caspase fold. Also, the identity between the YCA1 and the caspases 3 and 9 is lower than 12%, and the sequence divergence greatly affected the root-mean-square deviation (RMSD) analysis. The superimposition of YCA1 and caspase 3 showed a higher structural variation than the superimposition of caspase 3 and 9. On *in vitro* assays, the presence of Ca^{2+}, and not of other divalent cations, was required to an enhanced capability of auto processing. Under this condition, two peptide fragments were formed revealing that the processing occurred after Arg72 and Lys86. When compared with MCA2, YCA1 shows 24.9% of amino acid sequence identity and a very similar structure through RMSD analysis [75].

6. Metacaspase activators and regulators

Despite the growing knowledge about the structure and function of CD clan members of cysteine proteases, the molecules involved on the control of their activities are still to be more unravelled. As discussed before, one of the main activation mechanisms of the caspases is its cleavage on specific sites, promoted by other caspases. This processing results on conformational changes and dimerization that enhance the substrate cleavage activity of the target caspase.

For this to happen, it is necessary the assemblage of huge protein complexes, which function as activation platforms. Examples of these are the Fas death-induced signalling complex (DISC), whose association is required for the caspase 8 (pro-caspase 8) self-activation and the apoptosome, which binds to caspase 9 (pro-caspase 9) prior to its activation [11, 28, 31]. It is not clear, if in the case of metacaspases, the auto processing and dimerization are always required for their activation as well as the formation of protein complexes [8].

Until now, biochemical studies have demonstrated that only type II metacaspases undergo autocatalytic activation, similar to the phenomenon observed for caspases. Contrary to that,

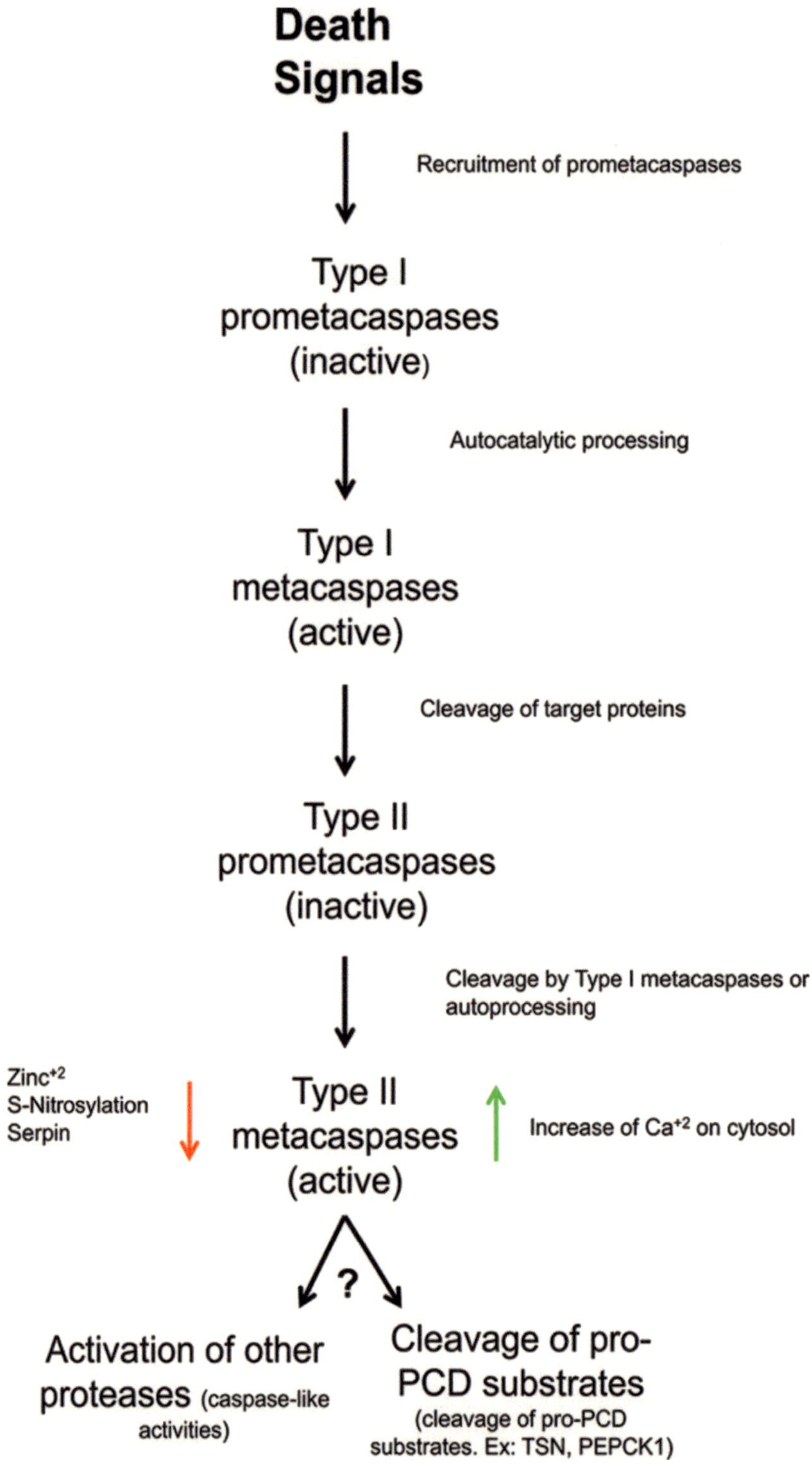

Figure 2. Proposed cascade model of plant proteolytic events triggered by a death inductor signal. The type II metacaspase inhibitors (red arrow) and stimulators (green arrow) are displayed, as well as possible substrates.

no proteolytic activation was observed for type I metacaspases. So a scenery, where type I prometacaspases become active by the action of death signals coupled to the activation of type II prometacaspases, which by their turn become active and able to cleave other proteases and trigger the degradation of cellular components, was proposed as a probable signal transduction pathway during PCD proteolysis in plants [69] (**Figure 2**).

Alongside with the proteolysis processing requirements, there are reports which show calcium dependence for type II metacaspases. By *in vitro* experiments, a type II recombinant metacaspase from *A. thaliana* exhibited Ca^{2+} dependence (10 mM) for its activation, on a pattern indicative of auto-processing [76]. A similar Ca^{2+} dependence was detected on *in vitro* assays with the *T. brucei* MCA2, whose activity peaks on the presence of 10 mM of Ca^{2+}, although the enzyme does not require autoprocessing [77]. This characteristic was shown to be valid for both *L. major* and *P. falciparum* metacaspases [66].

One candidate to be a metacaspase regulator is a serine protease inhibitor called Serpin. Serpin1 from *A. thaliana* was shown to have a potent cleavage activity towards the reactive centre loop of metacaspase 9, besides the ability of covalent binding to the target protein on *in vitro* assays. This study has also demonstrated that both proteins were localized on extracellular space, suggesting they could interact under *in vivo* conditions as well [78].

Other control mechanism proposed is the S-nitrosylation, that is used as a regulation strategy of certain proteins under basal NO levels. The S-nitrosylation of the catalytic cysteine (Cys-147) on *A. thaliana* metacaspase 9 (AtMC9) on its mature processed form does not affect its activity. This happens because other cysteine (Cys-29) residue can act as alternative nucleophile. Despite this, the enzyme can be kept inactive through S-nitrosylation, and otherwise, become active only under conditions of disturbance on cellular redox balance [79].

The effect of zinc on metacaspase activity was also investigated. The supplementation of plant embryos with extra zinc suppressed the terminal differentiation and death of the suspensors, delaying the embryo maturing, and also reducing the intensity of metacaspase activity between 96 and 168 h of development, which is the period when the suspensor death occurs [80]. These data, alongside the work of Bozhkov et al. [46], suggest that zinc may be part of a mechanism of posttranslational regulation of metacaspases, still to be further examined.

7. Metacaspase targets

Little is known about the metacaspase natural substrates, to the present date. This lack of information makes hard to construct a PCD pathway as was done for caspases. Despite this, the efforts aiming at the elucidation of this question are growing.

The first biological substrate discovered for a metacaspase is the Tudor Staphylococcal Nuclease (TSN), a protein involved on gene expression regulation, highly conserved phylogenetically. The cleavage of this protein prevents its function and is important for the execution of apoptosis; also, the protein is known to be a part of the human caspase 3 degradome. TSN was shown to be cleaved by the metacaspase McIIPa (type II metacaspase of *P. abies*) on an *in vitro* assay. *In vivo,* the McIIPa activity was shown to be simultaneous to the decrease of TSN activity, what was deduced from TUNEL positive embryonic cells [81].

Another substrate found for a metacaspase was the already mentioned GAPDH. This protein was detected as a digest product of an YCA1 metacaspase-enriched extract from *S. cerevisiae* cells subjected to H_2O_2 treatment. As recombinant YCA1 was shown to cleave GAPDH *in vitro,* the *in vivo* evaluation of the GAPDH performed during a comparison between the wild-type and a mutant yeast strain, disrupted for YCA1, both under PCD triggering condition, showed a reduction of the enzyme levels on the wild-type (resistant to PCD). The GAPDH is also a caspase substrate, but the cleavage again happens on different sites of those targeted by the metacaspases. Yet, this is another evidence for the existence of conserved molecular members of a PCD pathway in metazoan and in metacaspase-bearing organisms [56].

Recently, as a remarkable effort, a proteome-wide-level study of *A. thaliana* seedlings, focusing on the identification of physiologic substrates of metacaspase 9, has been performed employing a digestome analysis strategy. Important features of the target proteins were prospected, and it was possible to map the frequencies of the amino acids sitting at the neighbourhood of the Arg or Lys P1-specific cleavage sites. Along with other interesting features, the enzyme has shown a strong tendency to prefer acid residues, as Asp and Glu at the P1′ position. Among the identified substrates, was phosphoenolpyruvate carboxykinase 1 (PEPCK1), a gluconeogenesis enzyme. This protein was shown to be cleaved *in vivo* in such a manner that its activity was enhanced, and thus, the glucose *de novo* synthesis pathway may be stimulated during PCD [82].

8. Molecular modelling of the metacaspase 4 from *Glycine max* (type II metacaspase)

For the comprehension of the structural organization of a type II metacaspase, the delimitation of the p20 and p10 domains of the metacaspase 4 from *G. max* was performed by our group, as well as the analysis of its catalytic amino acids residues and the motifs conservation with other metacaspases and caspases, through protein alignment. Also, the tridimensional structure of the protein was predicted. Metacaspase and caspase sequences of organisms from different taxa (**Table 1**) were aligned using the software Clustal X [83] (http://www.clustal.org/). The sequences were obtained from the National Center of Biotechnology Information (http://www.ncbi.nlm.nih.gov/) data bank and given a treatment for removal of prodomains and loops, for adjustment to the alignment. In this process, the works of Vercammen et al. [8] and of Uren et al. [32] were used as a guide to the delimitation of the domains and catalytic residues.

The p20 and p10 domains of the *G. max* metacaspase were confronted to the Protein Data Bank (http://www.wwpdb.org/) for the search of templates for molecular prediction employing the software Swiss Model [84] (http://swissmodel.expasy.org/). The visualization, the analysis, the validation and the improving of the protein structures were performed with the assistance of the software NOC [85] (http://noch.sourceforge.net).

Protein	Specie	Identification code (gi)
Caspase8	*Homo sapiens*	12862693
Caspase8	*Mus musculus*	29436722
Caspase8	*Gallus gallus*	16555407
Caspase8	*Xenopus laevis*	148228484
Caspase8	*Danio rerio*	46397548
CED3/NEDD2-like	*Drosophila melanogaster*	220901640
CED3	*Caenorhabdits elegans*	11967321
Metacaspase4	*Glycine max*	356556698
Type II metacaspase	*Nicotiana tabacum*	195963550
Metacaspase8	*Arabidopsis thaliana*	32482822
Metacaspase9	*Arabidopsis thaliana*	332191328
Type II metacaspase	*Picea abies*	328887884
Type II metacaspase	*Medicago truncatula*	357451305
Type II metacaspase	*Triticum aestivum*	267850617
Latexprotein	*Hevea brasiliensis*	4235430
Type I metacaspase	*Saccharomyces cerevisiae*	151945290
Type I metacaspase	*Schizosaccharomyces pombe*	19076003

Table 1. List of proteins utilized on the analysis of sequence alignment.

9. Results

9.1. Protein alignment

The initial alignment of caspases and metacaspases sequences with the whole protein sequences (**Figure 3**) divided them into two groups: sequences with larger pro-domains and shorter loops; and sequences with smaller pro-domains and larger loops. Using the position of the catalytic dyad His-Cys as guide and after the removal of pro-domains and loops, the core of p20 and p10 domains became evident. The number of amino acid residues counted 235 residues, being 146 from p20 and 85 from p10 domain. It was also possible to underline the approximate segment borders for the domains of the *G. max* metacaspase, and it was seen that, besides the conserved catalytic dyad position, their adjacent residues (on primary sequence) shared the same chemical nature, encompassing all the protein sequences, despite the phylogenetic distance. The same was seen for the residues close to these regions, considering the p20 segment. In relation to p10, its catalytic residues presented conserved position, despite the chemical divergence (Asp for metacaspases and Arg for caspases). This segment also revealed conservation along its extension, such as the Ser residues close to the catalytic site, amino acids with shorter lateral groups and polar residues. Concerning the prediction

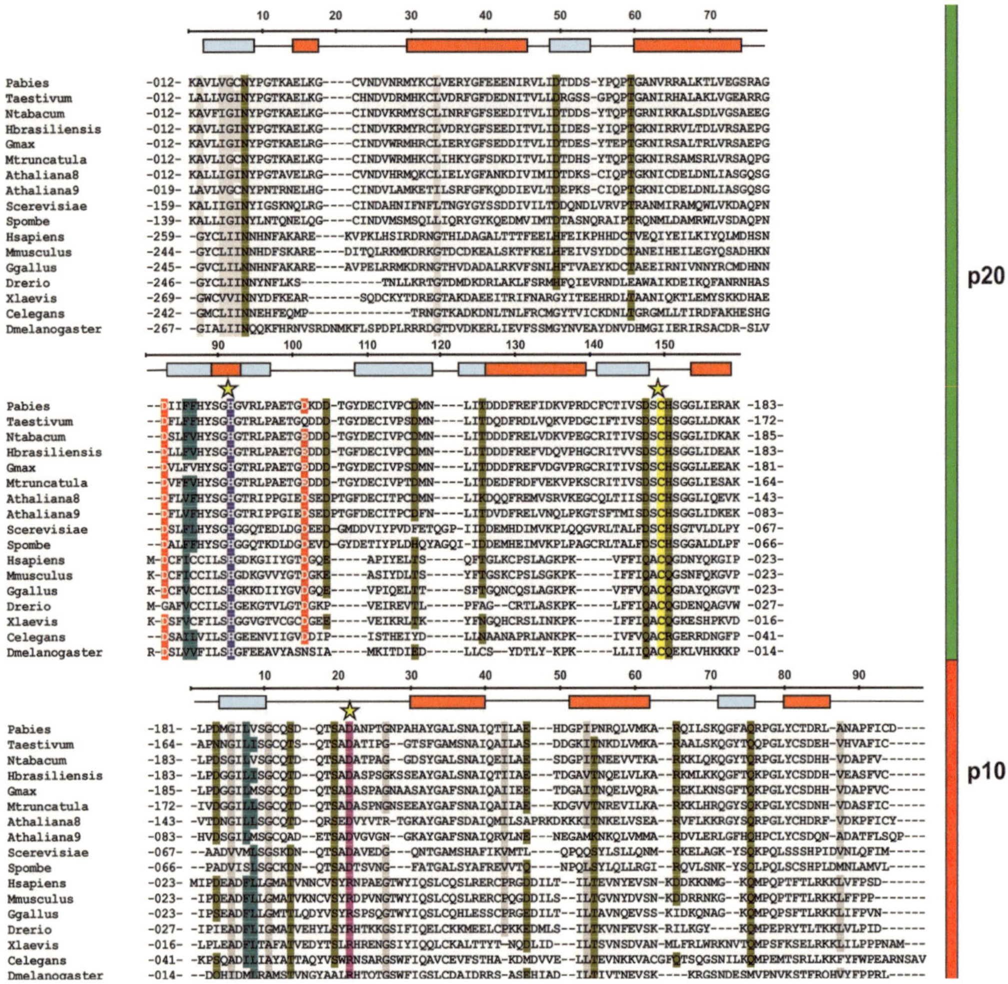

Figure 3. Sequence alignment of caspases and metacaspases, displaying the corresponding secondary structure. Blue: catalytic histidine; yellow: catalytic cysteine; brown: polar residues; grey: aliphatic residues; pink: contact aspartic acid and arginine residues; green: hydrophobic residues. Blue box: β conformation; red box: α-helix/3_{10}helix. Star: catalytic residues.

of secondary structure organization, it was possible to determine that about 33% of the sequence can form α-helix/3-10 helix and that about 23% can take part of β-sheet conformation (**Figure 3**).

9.2. Molecular structure prediction of *G. max* metacaspase 4

The search for a protein that could be applied as a template using the individual subunits from *G. max* metacaspase, resulted in only one choice with a significant sequence similarity rate for the p20 subunit: the chain A of a protein complex from *Geobacter sulfureduccens* (PDB-3BIJ). The identity and similarity rates between the template (from the 60th to the

148th residue) and p20 were of 35 and 55%, respectively. As the analysis of tertiary structure of the subunits and the total protein sequences were compatible, even with the high primary structure difference, 3BIJ was used as a template for the construction of a structural model for metacaspase 4. Only one α-helix from 3BIJ was removed for adjustment to the target protein sequence.

Concerning the established structural model (**Figure 4**), it was possible to note that the amino acids residues from the catalytic dyad (His/Cys) of *G. max* metacaspase were closer arranged, spatially. The contact Asp residue also kept this position, suggesting that they are, in fact, intimately associated to the enzymatic catalysis. The model also presented a tridimensional structure close to that of caspases and related proteins, with a core of β-conformations originated from both p20 and p10 being encircled by α-helixes, also originated from both domains. It totalized two β-conformations on its central position which are originated from p10, while p20 contributed with tree of these secondary structures. Other two β-sheets occupied the model extremity. With relation to the α-helixes and 3–10 helixes, there are eight of those that encircled the β-conformation core. Three of those are originated from p20; other two are from p10. The remaining helixes are disposed on the model extremities.

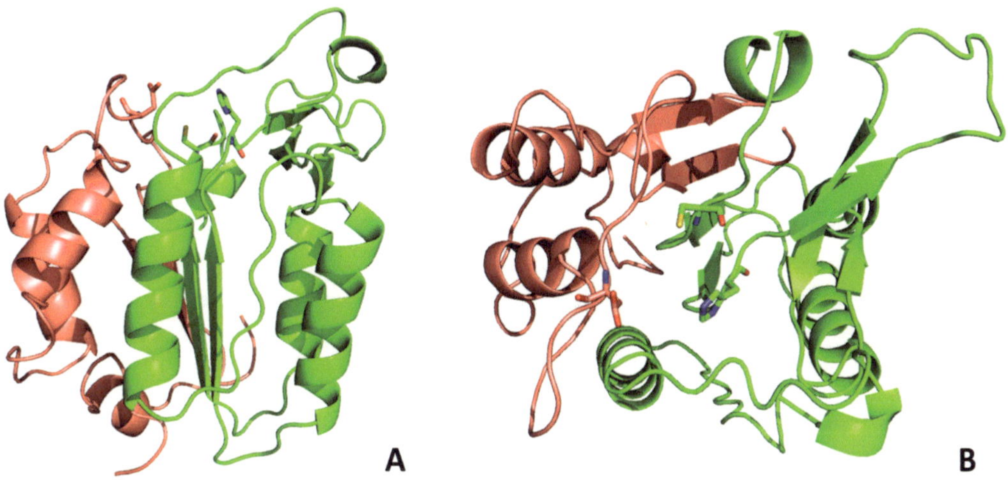

Figure 4. Structural model generated for metacaspase 4 of*Glycine max*. (A) Lateral and (B) top view from catalytic site. The p20 and p10 domains are colored in green and red respectively. The lateral chains from the catalytic amino acids are in evidence.

10. Discussion

The sequence comparison of metacaspases and caspases domains p20 and p10 clearly shows differences on the amino acid composition and disposition. Nevertheless, the segments, once aligned, displayed conserved positions of catalytic residues and of other amino acid residues with conserved physical-chemical properties, what is important to the arrangement on a similar secondary structure. Among the p20 secondary structure, six peptide sequences participate on α-helixes and 3–10 helixes, and seven composes β-conformations; for p10, three sequences form helix structures and two originate β-conformations. Together, these structures are organized in a similar way to that observed for the protein from the CD clan of C14 family of the cysteine proteases, which includes caspases and metacaspases [86]. The constructed *G. max* metacaspase model shows that these sequences are organized in form to present a core of β conformations encircled by five helixes, with the amino acid residues which compose the active site localized in one of the enzyme central axis poles, out of the β-sheets and α-helixes region.

The used template here was the chain A of the protein not functionally uncharacterized 3BIJ protein of *G. sulfureduccens*. This is the same protein used by Dudkiewicz and Piszczek [87], for the prediction of a model for *Triticum aestivum* type II metacaspase. Interestingly, 3BIJ was seen to be a better template for the considered metacaspase than the *Homo sapiens* caspase 7, whose similarity with soybean metacaspase was also high.

Curiously, a number of recent reports have demonstrated a difference of cleavage specificity among caspases and metacaspases. The recombinant metacaspase McIIPa, from *P. abies*, was shown to be efficient on the cleavage of the peptide sequence EGR and GRR, but not of VEID and YVAD, which are processed by caspases [46]. In 2008, He et al. [88] demonstrated that the recombinant metacaspase 8 from *Arabidopsis thaliana* was efficient on the cleavage of the sequence GRR, being unable to process DEVD, VEID, IETD and YVAD. This difference on enzymatic activity nature of metacaspases and caspases are generating an open discussion on literature. The denomination "caspase" itself gives clues to these discussions concerning to the particularities presented by caspases and metacaspases. Carmona-Gutierrez et al. [51] suggested that metacaspases and caspases present characteristics that fulfil the homology criteria, as they participate of a common program, share substrates and by the fact that metacaspase genes are present in all organisms, except superior animal taxa. In this scenario, the caspase genes could be derived from metacaspases.

In response, Enoksson and Salvesen [52] defended that yeasts and plants would employ PCD programmes other than apoptosis, what would be an innovation when compared to animals. Also they argue that, even if metacaspases and caspases share the tridimensional structure, the cleavage specificity displayed by them could show that they are derived from a common ancestor, which was neither caspase nor metacaspase.

This scenery is reinforced by data from the work of Koonin and Aravind [11], which showed that metacaspases have similarities with α-proteobacteria homologues, the group of endosymbiotic mitochondria ancestors, being the metacaspases from prokaryotic origin. Also, it was demonstrated that bacterial homologues of caspase-related proteins showed a greater diversity of phyletic distribution, domain architecture and sequence than their eukaryotic counterparts, suggesting that events of gene transference from prokaryote to eukaryotes could be an explanation for the distribution of caspase-related genes, what could have been assured by multiple bacterial gene infusions [87].

Author details

Gustavo Lemos Rocha, Jorge Hernandez Fernandez, Antônia Elenir Amâncio Oliveira and Kátia Valevski Sales Fernandes*

*Address all correspondence to: cowpkat@uenf.br

Centro de Biociências e Biotecnologia, Laboratório de Química e Função de Proteínas e Peptídeos - LQFPP, Universidade Estadual do Norte Fluminense Darcy Ribeiro, Campos dos Goytacazes, RJ, Brazil

References

[1] Horowitz, B. What are enzymes? The Scientific Monthly. 1918;**8**:253-259.

[2] Van der Hoorn, R.A.L. Plant proteases: from phenotypes to molecular mechanisms. Annual Review of Plant Biology. 2008;**59**:191-223.

[3] Vines, S.H. Proteolytic enzymes in plants. Annals of Botany. 1903;**17**(65):237-264.

[4] Roberts, W. On the Digestive Ferments and the Preparation and Use of Artificially Digested Food. London: Smith, Elder. 1881.

[5] Foster, M. A Textbook of Animal Physiology. London : Macmillan and Co. 1877.

[6] Grassmann, W.; Dyckerhoff, H. Über die Proteinase und die Polypeptidase der Hefe. 13. Abhandlungüber Pflanzenproteasen in der von R. Willstätter und Mitarbeiternbegonnenen Untersuchungsreihe. Biological chemistry Hoppe-Seyler. 1928;**179**(1-3):41-78.

[7] Rawlings, N.D. Peptidase specificity from the substrate cleavage collection in the merops database and a tool to measure cleavage site conservation. Biochimie. 2016;**122**:5-30.

[8] Vercammen, D.; Declercq, W.; Vandenabeele, P.; Van Breusegem, F. Are metacaspases caspases? The Journal of Cell Biology. 2007;**179**(3):375-380.

[9] Pennell, R.I; Lamb, C. Programmed cell death in plants. The Plant Cell. 1997;**9**(7):1157.

[10] Smith, J.M., Szathmary, E. The Major Transitions in Evolution. Oxford, England: Oxford University Press. 1997.

[11] Koonin, E.V; Aravind, L. Origin and evolution of eukaryotic apoptosis: the bacterial connection. Cell Death and Differentiation. 2002;**9**(4):394-404.

[12] Schweichel J.U.; Merkel, H.J. The morphology of various types of cell death in prenatal tissue. Teratology. 1973;**7**:253-266.

[13] Baehrecke, E.H. Autophagic programmed cell death in Drosophila. Cell Death & Differentiation. 2003;**10**(9):940-945.

[14] Lima, N.B.; Oliveira, A.E.A.; Fernandes, K.V.S. Programmed cell death in plants: the tip of the iceberg. Recent Research Developments in Phytochemistry. 2012;**10**(10):1-15.

[15] Otto, G.P.; Wu, M.Y.; Kazgan, N.; Anderson, O.R.; Kessin, R.H. Macroautophagy is required for multicellular development of the social amoeba *Dictyostelium discoideum*. Journal of Biological Chemistry. 2003;**278**(20):17636-17645.

[16] Yoshimori, T. Autophagy: a regulated bulk degradation process inside cells. Biochemical and Biophysical Research Communications. 2004;**313**(2):453-458.

[17] Reape, T.J.; Molony, E.M.; McCabe, P.F. Programmed cell death in plants: distinguishing between different modes. Journal of Experimental Botany. 2008;**59**(3):435-444.

[18] Lennon, S.V.; Martin, S.J.; Cotter, T.G. Dose-dependent induction of apoptosis in human tumour cell lines by widely diverging stimuli. Cell Proliferation. 1991;**24**(2):203-214.

[19] Dangl, J.L.; Dietrich, R.A.; Howard Thomas, H. Senescence and programmed cell death. In: Buchanan, B.B; Gruissem, W.; Jones, R.L. Biochemistry and Molecular Biology of Plants. Rockville, MD, USA: American Society of Plant Physiologists. Vol.1; 2000:1044-1100.

[20] Morgan, P. W.; Drew, M. C. Plant cell death and cell differentiation. In: Noodén, L.D. Plant Cell Death Processes. Cambridge, MA, USA: Academic Press. 2003; Vol. 1; 19-36.

[21] Bosabalidis, A.M.; Evert, R.F; Russin, W.A. Ontogeny of the vascular bundles and contiguous tissues in the maize leaf blade. American Journal of Botany. 1994;**81**(6):745-752.

[22] Caporali, E.; Spada, A.; Marziani, G.; Failla, O.; Scienza, A. The arrest of development of abortive reproductive organs in the unisexual flower of *Vitis vinifera* ssp. *silvestris*. Sexual Plant Reproduction. 2003;**15**(6):291-300.

[23] Yuan, J.; Horvitz, H.R. The *Caenorhabditis elegans* genes ced-3 and ced-4 act cell autonomously to cause programmed cell death. Developmental Biology. 1990;**138**(1):33-41.

[24] Yuan, J.; Robert Horvitz, H.R. The *Caenorhabditis elegans* cell death gene ced-4 encodes a novel protein and is expressed during the period of extensive programmed cell death. Development. 1992;**116**(2):309-320.

[25] Yuan, J.; Shaham, S.; Ledoux, S.; Ellis, H.M.; Horvitz, H.R. The *C. elegans* cell death gene ced-3 encodes a protein similar to mammalian interleukin-1β-converting enzyme. Cell. 1993;**75**(4):641-652.

[26] Cerretti, D.P; Kozlosky, C.J. Molecular cloning of the interleukin-1beta converting enzyme. Science. 1992;**256**(5053):97.

[27] Thornberry, N.A.; Bull, H.G.; Calaycay, J.R.; Chapman, K.T.; Howard, A.D.; Kostura, M.J., Elliston, K.O. A novel heterodimeric cysteine protease is required for interleukin-1β processing in monocytes. Nature. 1992;**356**(6372):768-774.

[28] Nicholson, D. W., Thornberry, N.A. Caspases: killer proteases. Trends in Biochemical Sciences. 1997;**22**(8):299-306.

[29] Parrish, A.B.; Freel, C.D; Kornbluth, S. Cellular mechanisms controlling caspase activation and function. Cold Spring Harbor Perspectives in Biology. 2013;**5**(6).

[30] Yuan, J.; Yankner, B.A. Apoptosis in the nervous system. Nature. 2000;**407**(6805):802-809.

[31] Li, J.; Yuan, J. Caspases in apoptosis and beyond. Oncogene. 2008;**27**(48):6194-6206.

[32] Uren, A.G.; O'Rourke, K.; Aravind, L.; Pisabarro, M.T.; Seshagiri, S.; Koonin, E.V.; Dixit, V.M. Identification of paracaspases and metacaspases: two ancient families of caspase-like proteins, one of which plays a key role in MALT lymphoma. Molecular Cell. 2000;**6**(4):961-967.

[33] Ruefli-Brasse, A.A.; French, D.M.; Dixit, V.M. Regulation of NF-κB-dependent lymphocyte activation and development by paracaspase. Science. 2003;**302**(5650):1581-1584.

[34] Che, T.; You, Y.; Wang, D.; Tanner, M.J.; Dixit, V.M.; Lin, X. MALT1/paracaspase is a signaling component downstream of CARMA1 and mediates T cell receptor-induced NF-κB activation. Journal of Biological Chemistry. 2004;**279**(16):15870-15876.

[35] Roisin-Bouffay, C.; Luciani, M.F.; Klein, G.; Levraud, J.P.; Adam, M.; Golstein, P. Developmental cell death in Dictyostelium does not require paracaspase. Journal of Biological Chemistry. 2004;**279**(12):11489-11494.

[36] Kawadler, H.; Gantz, M.A.; Riley, J.L.; Yang, X. The paracaspase MALT1 controls caspase-8 activation during lymphocyte proliferation. Molecular Cell. 2008;**31**(3):415-421.

[37] Snipas, S.J.; Wildfang, E.; Nazif, T.; Christensen, L.; Boatright, K.M.; Bogyo, M.; Salvesen, G.S. Characteristics of the caspase-like catalytic domain of human paracaspase. Biological Chemistry. 2004;**385**(11):1093-1098.

[38] Jong, W.Y.; Jeffrey, P.D.; Ha, J.Y.; Yang, X.; Shi, Y. Crystal structure of the mucosa-associated lymphoid tissue lymphoma translocation 1 (MALT1) paracaspase region. Proceedings of the National Academy of Sciences. 2011;**108**(52):21004-21009.

[39] Lamkanfi, M.; Declercq, W.;Kalai, M.; Saelens, X.; Vandenabeele, P. Alice in caspase land. A phylogenetic analysis of caspases from worm to man. Cell Death and Differentiation. 2002;**9**(4):358-361.

[40] Choi, C.J.; Berges, J.A. New types of metacaspases in phytoplankton reveal diverse origins of cell death proteases. Cell Death & Disease. 2013;**4**(2):e490.

[41] delPozo, O.; Lam, E. Caspases and programmed cell death in the hypersensitive response of plants to pathogens. Current Biology. 1998;**8**(20):1129-1132.

[42] Watanabe, N.; Eric Lam, E. Two Arabidopsis metacaspases AtMCP1b and AtMCP2b are arginine/lysine-specific cysteine proteases and activate apoptosis-like cell death in yeast. Journal of Biological Chemistry. 2005;**280**(15):14691-14699.

[43] Tsiatsiani, L.; Van Breusegem, F.; Gallois, P.; Zavialov, A.; Lam, E.; Bozhkovhkov, P.V. Metacaspases. Cell Death & Differentiation. 2011;**18**(8):1279-1288.

[44] Suarez, M.F.; Filonova, L.H.; Smertenko, A.; Savenkov, E.I.; Clapham, D.H.; von Arnold, S.; Bozhkovhkov, P.V. Metacaspase-dependent programmed cell death is essential for plant embryogenesis. Current Biology. 2004;**14**(9):339-340.

[45] Vercammen, D.; Van De Cotte, B.; De Jaeger, G.; Eeckhout, D.; Casteels, P.; Vandepoele, K.; Van Breusegem, F. Type II metacaspases Atmc4 and Atmc9 of Arabidopsis thaliana cleave substrates after arginine and lysine. Journal of Biological Chemistry. 2004;**279**(44): 45329-45336.

[46] Bozhkov, P.V.; Suarez, M.F.; Filonova, L.H.; Daniel, G.; Zamyatnin, A.A.; Rodriguez-Nieto, S.; Smertenko, A. Cysteine protease mcII-Pa executes programmed cell death during plant embryogenesis. Proceedings of the National Academy of Sciences of the United States of America. 2005;**102**(40):14463-14468.

[47] He, R.; Drury, G.E.; Rotari, V.I.; Gordon, A.; Willer, M.; Farzaneh, T.; Gallois, P. Metacaspase-8 modulates programmed cell death induced by ultraviolet light and H_2O_2 in Arabidopsis. Journal of Biological Chemistry. 2008;**283**(2):774-783.

[48] Bonneau, L.; Ge, Y.; Drury, G.E.; Gallois, P. What happened to plant caspases?. Journal of Experimental Botany. 2008;**59**(3):491-499.

[49] Hao, L.; Goodwin, P.H; Hsiang, T. Expression of a metacaspase gene of *Nicotiana benthamiana* after inoculation with *Colletotrichum destructivum* or *Pseudomonas syringae* pv. tomato, and the effect of silencing the gene on the host response. Plant Cell Reports. 2007;**26**(10):1879-1888.

[50] Hoeberichts, F.A.; Ten Have, A.; Woltering, E. J. A tomato metacaspase gene is upregulated during programmed cell death in *Botrytis cinerea*-infected leaves. Planta. 2003;**217**(3):517-522.

[51] Carmona-Gutierrez, D.; Fröhlich, K. U.; Kroemer, G.; Madeo, F. Metacaspases are caspases: doubt no more. Cell Death and Differentiation. 2010;**17**(3):377.

[52] Enoksson, M.; Salvesen, G.S. Metacaspases are not caspases—always doubt. Cell Death & Differentiation. 2010;**17**(8):1221.

[53] Madeo, F.; Herker, E.; Maldener, C.; Wissing, S.; Lächelt, S.; Herlan, M.; Fröhlich, K.U. A caspase-related protease regulates apoptosis in yeast. Molecular Cell. 2002;**9**(4):911-917.

[54] Bettiga, M.; Calzari, L.; Orlandi, I.; Alberghina, L.; Vai, M. Involvement of the yeast metacaspase Yca1 in ubp10Δ-programmed cell death. FEMS Yeast Research. 2004;**5**(2):141-147.

[55] Lefevre, S.; Sliwa, D.; Auchère, F.; Brossas, C.; Ruckenstuhl, C.; Boggetto, N.; Santos, R. The yeast metacaspase is implicated in oxidative stress response in frataxin-deficient cells. FEBS Letters. 2012;**586**(2):143-148.

[56] Silva, A.; Almeida, B.; Sampaio-Marques, B.; Reis, M.I.R.; Ohlmeier, S.; Rodrigues, F.; Ludovico, P. Glyceraldehyde-3-phosphate dehydrogenase (GAPDH) is a specific substrate of yeast metacaspase. Biochimica et Biophysica Acta. 2011;**1813**(12):2044-2049.

[57] Lee, R.E.; Brunette, S.; Puente, L.G.; Megeney, L.A. Metacaspase Yca1 is required for clearance of insoluble protein aggregates. Proceedings of the National Academy of Sciences. 2010;**107**(30):13348-13353.

[58] Hill, S.M.; Hao, X.; Liu, B.; Nyström, T. Life-span extension by a metacaspase in the yeast *Saccharomyces cerevisiae*. Science. 2014;**344**(6190):1389-1392.

[59] Cao, Y; Huang, S.; Dai, B.; Zhu, Z.; Lu, H.; Dong, L.; Jiang, Y. *Candida albicans* cells lacking CaMCA1-encoded metacaspase show resistance to oxidative stress-induced death and change in energy metabolism. Fungal Genetics and Biology. 2009;**46**(2):183-189.

[60] Richie, D.L.; Miley, M.D.; Bhabhra, R.; Robson, G.D.; Rhodes, J.C.; Askew, D.S. The *Aspergillus fumigatus* metacaspasesCasA and CasB facilitate growth under conditions of endoplasmic reticulum stress. Molecular Microbiology. 2007;**63**(2):591-604.

[61] Szallies, A.; Kubata, B.K.; Duszenko, M. A metacaspase of *Trypanosoma brucei* causes loss of respiration competence and clonal death in the yeast *Saccharomyces cerevisiae*. FEBS Letters. 2002;**517**(1-3):144-150.

[62] Lee, N.; Gannavaram, S.; Selvapandiyan, A.; Debrabant, A. Characterization of metacaspases with trypsin-like activity and their putative role in programmed cell death in the protozoan parasite Leishmania. Eukaryotic Cell. 2007;**6**(10):1745-1757.

[63] Helms, M. J.; Ambit, A.; Appleton, P. Tetley, L.; Coombs, G.H.; Mottram, J. C. Bloodstream form *Trypanosoma brucei* depend upon multiple metacaspases associated with RAB11-positive endosomes. Journal of Cell Science. 2006;**119**(6):1105-1117.

[64] González, I.J.; Desponds, C.; Schaff, C.; Mottram, J.C.; Fasel, N. Leishmania major metacaspase can replace yeast metacaspase in programmed cell death and has arginine-specific cysteine peptidase activity. International Journal for Parasitology. 2007;**37**(2):161-172.

[65] Ambit, A.; Fasel, N.; Coombs, G.H.; Mottram, J. C. An essential role for the Leishmania major metacaspase in cell cycle progression. Cell Death & Differentiation. 2008;**15**(1):113-122.

[66] Meslin, B.; Beavogui, A. H.; Fasel, N.; Picot, S. Plasmodium falciparum metacaspase PfMCA-1 triggers a z-VAD-fmk inhibitable protease to promote cell death. PLoS One. 2011;**6**(8):e23867.

[67] Shiozaki, E. N.; Chai, J.; Shi, Y. Oligomerization and activation of caspase-9, induced by Apaf-1 CARD. Proceedings of the National Academy of Sciences. 2002;**99**(7):4197-4202.

[68] Coll, N.S.; Vercammen, D.; Smidler, A.; Clover, C.; Van Breusegem, F.; Dangl, J.L.; Epple, P. Arabidopsis type I metacaspases control cell death. Science. 2010;**330**(6009):1393-1397.

[69] Piszczek, E.; Gutman, W. Caspase-like proteases and their role in programmed cell death in plants. Acta Physiologiae Plantarum. 2007;**29**(5):391-398.

[70] Piszczek, E.; Dudkiewicz, M.; Sobczak, M. Molecular cloning and phylogenetic analysis of cereal type II metacaspase cDNA from wheat. Biologia Plantarum. 2011;**55**(4):614-624.

[71] Cohen, G.M. Caspases: the executioners of apoptosis. Biochemical Journal. 1997;**326**(1):1-16.

[72] Chandler, J. M.; Cohen, G. M.; MacFarlane, M. Different subcellular distribution of caspase-3 and caspase-7 following Fas-induced apoptosis in mouse liver. Journal of Biological Chemistry. 1998;**273**(18):10815-10818.

[73] Nicholson, D.W.; Ali, A.; Thornberry, N.A.; Vaillancourt, J.P.; Ding, C.K.; Gallant, M.; Munday, N.A. Identification and inhibition of the ICE/CED-3 protease necessary for mammalian apoptosis. Nature. 1995;**376**(6535):37-43.

[74] McLuskey, K.; Rudolf, J.; Proto, W.R.; Isaacs, N.W.; Coombs, G.H.; Moss, C.X.; Mottram, J.C. Crystal structure of a *Trypanosoma brucei* metacaspase. Proceedings of the National Academy of Sciences. 2012;**109**(19):7469-7474.

[75] Wong, A.H.H.; Yan, C.; Shi, Y. Crystal structure of the yeast metacaspase Yca1. Journal of Biological Chemistry. 2012;**287**(35):29251-29259.

[76] Watanabe, N.; Lam, E. Arabidopsis metacaspase 2d is a positive mediator of cell death induced during biotic and abiotic stresses. The Plant Journal. 2011;**66**(6):969-982.

[77] Moss, C.X.; Westrop, G.D.; Juliano, L.; Coombs, G.H.; Mottram, J.C. Metacaspase 2 of *Trypanosoma brucei* is a calcium-dependent cysteine peptidase active without processing. FEBS Letters. 2007;**581**(29):5635-5639.

[78] Vercammen, D.; Belenghi, B.; Van De Cotte, B.; Beunens, T.; Gavigan, J. A.; De Rycke, R.; Van Breusegem, F. Serpin1 of *Arabidopsis thaliana* is a suicide inhibitor for metacaspase 9. Journal of Molecular Biology. 2006;**364**(4):625-636.

[79] Belenghi, B.; Romero-Puertas, M.C.; Vercammen, D.; Brackenier, A.; Inzé, D.; Delledonne, M.; Van Breusegem, F. Metacaspase activity of Arabidopsis thaliana is regulated by S-nitrosylation of a critical cysteine residue. Journal of Biological Chemistry. 2007;**282**(2):1352-1358.

[80] Helmersson, A.; von Arnold, S.; Bozhkov, P.V. The level of free intracellular zinc mediates programmed cell death/cell survival decisions in plant embryos. Plant Physiology. 2008;**147**(3):1158-1167.

[81] Sundström, J.F.; Vaculova, A.; Smertenko, A.P.; Savenkov, E.I.; Golovko, A.; Minina, E.; Saarikettu, J. Tudor staphylococcal nuclease is an evolutionarily conserved component of the programmed cell death degradome. Nature Cell Biology. 2009;**11**(11):1347-1354.

[82] Tsiatsiani, L.; Timmerman, E.; De Bock, P. J.; Vercammen, D.; Stael, S.; Van De Cotte, B.; Gevaert, K. The Arabidopsis metacaspase9 degradome. The Plant Cell. 2013;**25**(8):2831-2847.

[83] Larkin, M.A.; Blackshields, G.; Brown, N.P.; Chenna, R.; McGettigan, P.A.; McWilliam, H.; Thompson, J.D. Clustal W and Clustal X version 2.0. Bioinformatics. 2007;**23**(21):2947-2948.

[84] Kiefer, F.; Arnold, K.; Künzli, M.; Bordoli, L.; Schwede, T. The SWISS-MODEL repository and associated resources. Nucleic Acids Research. 2009;**37**(1):387-392.

[85] Chen, M.E.; Chang, H.X.; Nymeyer, H. NOC viewer version 3.01 [computer software]. Florida State University. Tallahassee, FL, USA. 2007.

[86] Trobacher, C.P.; Senatore, A.; Greenwood, J.S. Masterminds or minions? Cysteine proteinases in plant programmed cell death. Plant Cell Biology Botany. 2006;**84**(4):651-667.

[87] Dudkiewicz, M.Z.; Piszczek, E. Bacterial putative metacaspase structure from *Geobacter sulfureducens* as a template for homology modeling of type II *Triticum aestivum* metacaspase (TaeMCAII). Acta Biochimica Polonica. 2012;**59**(3):401.

[88] He, R.; Drury, G.E.; Rotari, V.I.; Gordon, A.; Willer, M.; Farzaneh, T.; Gallois, P. Metacaspase-8 modulates programmed cell death induced by ultraviolet light and H_2O_2 in Arabidopsis. Journal of Biological Chemistry. 2008;**283**(2):774-783.

Chapter 3

Effect of High-Pressure Technologies on Enzymes Applied in Food Processing

Bruno Ricardo de Castro Leite Júnior,
Alline Artigiani Lima Tribst and Marcelo Cristianini

Additional information is available at the end of the chapter

http://dx.doi.org/10.5772/66629

Abstract

High isostatic pressure (HIP) and high-pressure homogenization (HPH) are considered important physical technologies that able to induce changes on enzymes. HIP and HPH are emerging food processing technologies that involve the use of ultra high pressures (up to 1200 MPa for HIP and up to 400 MPa for HPH), where the first process is based on the principle that the maintenance of a product inside vessels at high pressures induces changes in the molecules conformation and, consequently, in the functionality of polysaccharides, proteins and enzymes. To the contrary, for HPH process, the high shear and sudden pressure drop are the responsible phenomena for the changes on the processed product. This chapter aims to evaluate comparatively the effects of HIP and HPH on the activity of enzymes currently applied in food industry and to identify the main structural changes induced by each process. The overall evaluation of the results shows that mild conditions of both processes were recently highlighted as able to improve the activity and the stability of several enzymes, whereas extreme process conditions (pressure, time and temperature) induce enzyme denaturation with consequent reduction of biological activity. Considering the complexity and diversity involved in the enzyme structure and its ability to react, it is not possible to determine specific conditions that each process is able to promote increase or reduction of enzyme activity, being necessary to evaluate HIP and HPH for each enzyme. Finally, in terms of molecular structure, the effect of HIP and HPH on enzymes can be explained by the alterations in the quaternary, tertiary and secondary structures of enzymes, which directly affects its active site configuration.

Keywords: emerging technologies, high pressure processing, high isostatic pressure, high pressure homogenization, food processing, food enzymes, enzymatic activity, molecular structure

1. Introduction

Enzymes are globular proteins that catalyze biochemical reactions. This occurs due to the spatial configuration of the enzymes and catalytic site, which is determined by the quantity and sequence of amino acids and the organization of these chains, with folds and twists induced by attractions and repulsions among the near amino acids, resulting in a structure with minimal energy content [1].

The enzyme reaction occurs due to the interaction of the catalytic site and substrate, forming a complex enzyme-substrate following by the product formation [2]. The maximum velocity of this reaction occurs at specific pH, temperature and salt concentration and, at non-optimum conditions, the enzyme reaction is highly affected [3]. However, in many times, the desirable industrial process conditions are different of the optimum enzyme activity, making difficult the industrial application. Additionally, high costs and low stability limits the extensive use of enzymes in many processes.

Several chemical and physical methods were studied aiming to improve the performance of enzymes, withdrawing limitations and consequently increasing the range of application. Among the physical methods proposed, high isostatic pressure (HIP) and high pressure homogenization (HPH) are considered an important way to induce interesting changes on enzymes [4, 5]. HIP and HPH are emerging food processing technologies that involve the use of pressures up to 1000 MPa for HIP and up to 400 MPa for HPH to cause desirable changes in food and other products. The HIP process is based on the principle that the maintenance of a product inside vessels at high pressures induces changes in the molecules. To the contrary, for HPH process, the high shear and sudden pressure drop are the responsible phenomena for changes on the processed product.

The overall evaluation of the results obtained for many authors shows that mild conditions of both processes were able to improve the activity and the stability of several enzymes, whereas extreme process conditions (pressure, time and temperature) induce enzyme denaturation with consequent reduction of biological activity [4–9]. Considering the complexity and diversity involved in the enzyme structure and its ability to react, it is not possible to determine specific conditions that each process is able to promote increase and reduction of enzyme activity, being necessary to evaluate HIP and HPH on each enzyme.

In terms of molecular structure, the effect of HIP and HPH on enzymes can be explained by the alterations in the quaternary, tertiary and secondary of enzymes, which directly affects the enzymes active site configuration, inducing exposure of hydrophobic amino acids, exposure of SH groups due to unfolding of the protein, a reduction in the total SH content due to new disulfide bonds formation and changes in the α-helix, β-sheet and β-turn ratio composition due to alterations of the secondary structure [4–6, 10, 11]. However, the occurrence of these phenomena—sequence of occurrence, intensity and required pressure—might be different for HIP and HPH.

The impact of each process on enzymes was evaluated by few published revisions [4, 6, 12], however, no one dedicated to compare the effect of HIP and HPH on the main enzymes used in the food processing, aiming to describe the differences among the process parameters and its consequences in the performance and structure of processed enzymes. Moreover, no revisions have already evaluated the impact of these processes on enzyme in different matrixes. Therefore, this chapter aimed to evaluate comparatively the effects of high isostatic pressure and high pressure homogenization in the activity of enzymes currently applied in food industry and to identify the main structural changes induced by each process pressure on the enzymes. Additionally, this work will be useful to challenge the scientific community to fulfill the information gaps in this area and for the industry, that will have access to a comparative evaluation of these two technologies, being an important way to decide which technology is better to be applied in order to have satisfactory results for different enzymes.

2. High isostatic pressure and high pressure homogenization technologies

The high isostatic pressure (HIP)—also known as high hydrostatic pressure or high pressure processing—and high pressure homogenization (HPH)—also called as ultra-high pressure homogenization or dynamic high pressure—are emerging process initially developed for food preservation by inactivation of microorganisms, with lower sensory and nutritional changes compared with the thermal process [12]. However, the studies of the consequences of these processes on food matrix highlighted that they were also able to induce changes on the food constituents, allowing the development of new applications and products/ingredients [13].

2.1. High isostatic pressure—principle and operation

The industrial application of HIP technology started in Japan in the early 1990s and has been gaining popularity and acceptance worldwide. Combining the interests of industry and consumers for these products, the development of new equipment with higher capacities helped to reduce costs and expand both the purchase intent of consumers (due to the perception of high quality products) as the search for new industries by new products, favoring the expansion of this technology [14].

The HIP is usually applied by subjecting the food, commonly vacuum sealed in flexible packaging, at a pressure up to 1000 MPa (10,000 bar), for a pre-defined time and at determined temperature to obtain the desired goal for each product. This process can be used to process liquid, semi-solid, or solid food. Equipment of higher volumes (687 L or 525 L) reaches pressures up to <310 MPa [15] or 600 MPa [16], respectively and temperatures up to <50°C, whereas equipment of lower volumes (<150 L) reaches temperatures up to 95°C and pressures up to 700 MPa [15]. The lab scale equipments reaching extreme pressures and temperature (900 MPa/110°C/chamber of 5L or 1400 MPa/110°C/chamber of 35 mL) [17]. During pressurization, the pressure is transferred instantaneously and evenly throughout the food (isostatic principle), regardless of the size and geometry of the product [18]. This is considered the main advantage of HIP processing, since there is no an equivalent to the so-called point of lower heating rate or "cold point" as in the case of heat conductive processes.

A typical system of HIP consists of a pressure chamber with closure and pressure generating system. Generally, it also have an apparatus coupled to the temperature control of the chamber. The batch process has three stages: the indirect pressurization using a liquid of low compressibility (e.g. water), the retention time at the desired pressure and depressurization. Semi-continuous processing can be obtained using multiple sequential chambers connected in series; while some cameras are under pressurization, others are being pressurized, unloaded, or loaded [14].

The pressurization is accompanied by a uniform temperature rise as consequence of the adiabatic heat of compression, being this specific to each compound [18]. For example, at 25°C, the water increases 3–5°C to every 100 MPa [18]. The adiabatic heating is completely reversed after the release of pressure. Although this temperature increase is relatively small, it can substantially contribute to the lethality of microorganisms in the overall process, resulting in significant implications when pressure is applied at elevated temperature [19]. On the other hand, this temperature increase can impact food structure, changing polysaccharides and proteins do to thermal effects. Therefore, when undesirable effects are observed due to adiabatic heating, the processes need to be carried out at lower temperature.

In molecular terms, the HIP breaks noncovalent bonds, such as ionic and hydrophobic bonds, but has little effect on covalent bonds. As a result, large biomolecules such as proteins and polysaccharides are affected by changes in its secondary, tertiary and quaternary structures (depending on the applied pressure), but small molecules are usually unaffected [6]. As the color components, flavor and vitamins are small molecules, the HIP process has little effect on these molecules in the food [6]. Furthermore, the process of pressurization followed the principle of "Le Chatelier," inducing a reduction in molecular volume and, consequently exponentially accelerating the occurrence of reactions favored by these conditions [20]. Thus, the rates of the chemical or physical reactions resulting in lower volume products are accelerated by the HIP, whereas the reactions that result in an increase in the total volume are retarded.

2.2. High pressure homogenization–Principle and operation

High pressure homogenization (HPH) is a nonthermal physical process applied for fluid foods [21]. This technology was introduced in the food field in the 1980s to improve the homogenization efficiency and emulsification of dairy products and emulsions, showing the same principle of operation of conventional homogenizers, however, using pressures around 10–15 times higher than usually applied, i.e., pressures up to 350 MPa [12].

In equipment, fluid is forced to pass through a homogenizing valve at high pressures [21]. The passage through the narrow gap (of micrometer order) and the abrupt decompression of the fluid generate an increase in speed (between 150 and 300 ms^{-1}) [22] and an increase in temperature (about 1.5–2.5°C every 10 MPa pressure increase) due to the intense friction in the homogenizing valve region [12]. In addition to shear effects, the fluid undergoes an intense drop of pressure, turbulence and cavitation, what leads to microbiological inactivation and modification of the constituents of the food [12, 13, 21, 22].

The main changes on food constituents are related to disruption of lipids globules, reduction of molecular weight of linear polysaccharides and modification of the quaternary and tertiary structure of proteins [4, 5, 13]. For some applications, these effects are positive and, therefore, the HPH emerged as a suitable operation to improve the versatility of biomolecules (such as polysaccharides and proteins) as food ingredients [13].

An important drawback of HPH technology is the difficult to be industrially implemented to do the small flow capacity of the available equipment at high pressures. Nowadays, the industrial equipment operates at pressures up to 150 MPa with maximum flow rates of 5.000 L/h [23], while equipment that reaches higher pressures (up to 400 MPa) work at maximum flow of 240 L/h [17]. However, due to the high industrial interest, new equipment with higher capacity has been developed, allowing industrial application of this technology.

3. Effect of high isostatic pressure and high pressure homogenization technologies on the activity and molecular structure of the food enzymes

The effects of HIP and HPH on enzymes were firstly studied considering the requirement of several enzymes inactivation (pectin methylesterase, polyphenoloxidase and peroxidases) to guarantee food product stability during storage [6], since these technologies were initially proposed to replace thermal processing in industrialized food [13, 18]. The results obtained in part of these researches showed that enzymes subjected to HIP or HPH processing had a behavior different from the commonly observed for thermal treatment, with activation at low pressures [4–13, 24]. Based on these results, both processes started to be considered as an interesting tool to change the performance of enzymes with commercial interest and several authors dedicated to study these effects on different enzymes, approaching functional and structural enzyme changes induced by HIP and HPH processing [4–13].

3.1. Enzymes activity

The results obtained by the authors who started to study the effect of HPH and HIP on enzymes of commercial interests corroborate the initial founds obtained with endogenous and deleterious enzymes of fruits, i.e., HIP and HPH could modify the activity of enzymes, inducing an activation at lower pressures and inactivation at higher [4–11, 24–26]. This behavior has been observed for the majority of studied enzymes, however, some of them showed high baroresistance, being not or almost not modified by HIP or HPH [27]. Moreover, although inactivation/activation be commonly observed by HPH and HIP, the level of pressure required for each process promotes that these changes are different, being the level of pressure in HPH necessary to induce inactivation lower than the normally required in HIP [4, 6]. Therefore, it is possible to suppose that the sample energy gain induced by the shear stress, cavitation and turbulence caused by the abrupt pressure dropped in HPH is higher than that induced by the maintenance of the same sample at the same pressure.

In addition to the process pressure, the effect on enzymes is also affected by the inlet temperature of the sample in the homogenizer and number of cycles applied in HPH [28] and by time, temperature and number of compression cycles in HIP [6]. For HPH, it is normally observed that increase of inlet temperature reduces the level of pressure required to start enzyme inactivation, probably due to the association of homogenization with thermal effects in the homogenization valve, while the effects of multiple cycles of homogenization are minimum for the majority of enzymes evaluated, i.e., the higher activity alteration occurs at the first cycle of pressurization [28]. For HIP processes, the majority of the enzymes show higher enzyme inactivation as higher were the temperature and time applied, being the effect of temperature more pronounced than the time effects [6]. Furthermore, in respect to the application of pulsed pressure cycles, it was observed that activation is probably stimulated when sort cycles (pulses) are applied at activation pressures, whereas inactivation rates increase when cycles are applied at inactivation pressures [29].

Other factors that also affected the effect of HIP and HPH on enzymes are the media used for enzymes solubilization (pH, salt concentration), presence or absence of substrate and enzyme concentration [4, 6, 8, 9]. The pH and salt concentration changes the native conformation of the enzyme, i.e., the enzyme can be processed at different initial configuration (high or less exposure of structure and active sites), impacting the process final effect [8]. Commonly, the use of acid pH has a synergic effect with pressure processing, reducing the activity of processed enzymes [6, 8]. The presence the subtract allows the occurrence of enzyme reaction under pressure (for HIP) or the process of enzymes partially configured as enzyme-substrate complex for both technologies, which alters the effect of process on enzyme and on the reaction products formation [30]. In this case, the observed effects are diverse, being not possible to establish a tendency of activation/inactivation induced by the presence/absence of substrate. Finally, the enzyme concentration can alter the effect of HPH processing, since higher enzyme concentration allows the occurrence of shear stress not only between molecules and equipment wall but also between molecules [31]. Therefore, it is possible to generally say that the HPH effects (enzyme activation or inactivation) are directly proportional to the enzyme concentration.

Another important point is that the impact of HIP and HPH on products can be measured directly on enzyme activity or in the changes caused by enzyme in the products (e.g. fruit products browning or softening). When enzyme activity was measured in laboratory using enzyme activity assays, several researches showed that higher increase of enzyme activity were observed at pH and temperature different from the optimum established for native (non-processed) samples [4, 8, 9, 32], indicating that process can alter the optimum conditions of enzymes, which can be very interesting from the industrial point of view. Additionally, when the enzyme is endogenous in the processed product, high differences in activation/inactivation level caused by process are observed, highlighting: (i) the importance of matrix and consequently the reaction media characteristics and (ii) the difficulty to compare results obtained for the same enzyme in different products or using different methods of enzyme activity quantification. These factors directly impact the establishment of general rules about the effect of HIP or HPH on enzymes, being necessary the enzyme evaluation in each food matrix and specific activity measurement conditions to determine the real impact of both processes on the enzymes. **Table 1** shows the results obtained for enzyme activity of the main food processed by HIP and HPH.

Matrix	Enzyme	Effect of HIP on enzymatic activity	Effect of HPH on enzymatic activity
		Fruit and vegetables	
Tomato juice	PME	Reduction of 70% at 800 MPa/15 min/65°C [33]	
	PG	Reduction of 100% at 500 MPa/10 min/55°C [33]	
	LOX	Reduction of 100% at 550 MPa/12 min/20°C [34]	
	HPL	Reduction of 80% at 650 MPa/12 min/20 °C [34]	
Orange juice	PME	Reduction of 50% at 400 MPa/90 min/30°C [2] Reduction of 93% at 450 MPa/30 min/50°C [35] Reduction of 100% at 650 MPa/3 min/25°C [36]	Reduction of 20% at 170 MPa/T inlet of 25°C [37] Reduction of 50% at 250 MPa/T inlet of 45°C [38] Reduction of 75% at 150 MPa at 68°C for 15 s [39] Reduction of 90% at 150 MPa at 68°C for 10 s [40] Reduction of 96% at 300 MPa/T inlet of 20°C [41]
Strawberry pulps/ purée	BGL	Activation of 16.6% at 400 MPa/20 min/25°C and reduction of 41.4% at 600 MPa/25 min/25°C [42]	Indicative of residual enzyme activity for color change [43]
	POD	Reduction of 74.6% at 500 MPa/25 min/25°C [42]	
	PPO	Reduction of 51.5% at 600 MPa/25 min/25°C [42]	
	POD	Reduction of 50% at 500 MPa/15 min/50°C [44]	
	PPO	Reduction of 72% at 500 MPa/15 min/50°C [44]	
Wild berry pulp	POD	Activation of ~182% at 200 MPa/5 min/20°C [45]	
	PPO	Activation of ~156% at 200 MPa/5 min/20°C [45]	
Apple juice	PPO	Reduction of 91% at 450 MPa/60 min/50°C [35] Reduction of ~51% at 750 MPa/50 min/50°C [46] Activation of ~120% at 200 MPa/1 min/25°C and reduction of ~95% at 900 MPa/1 min/25°C [47]	Reduction of 100% at 300 MPa/T inlet of 4 °C [48]
	POD	Activation of ~70% at 500 MPa/1 min/25°C and reduction of ~95% at 900 MPa/1 min/25°C [47]	

Matrix	Enzyme	Effect of HIP on enzymatic activity	Effect of HPH on enzymatic activity
	Amylase	Reduction of 90% at 400 MPa/30 min/22°C [49]	
	PME	Reduction of 99.7% at 400 MPa/180 min/25°C [50] Reduction of ~81% at 750 MPa/90 min/50°C [46]	Reduction of 100% at 300 MPa/T inlet of 4°C [48]
Banana juice	Pectinase	–	Activation of ~262% at 100 MPa/T inlet of 4°C [51]
Kiwifruit juice	POD	Activation of ~10% at 200 MPa/20 min/10°C and reduction of ~70% at 600 MPa/30 min/50°C [52]	
Litchi juice	POD	Activation of ~125% at 300 MPa/1 sec/25°C and reduction of ~30% at 600 MPa/45 min/25°C [29]	
	PPO	Activation of ~30% at 300 MPa/1 sec/25°C and reduction of ~30% at 600 MPa/45 min/25°C [29]	
Litchi-based mixed fruit beverage	POD	Reduction of ~80% at 480 MPa/10 min/70°C [53]	
	PPO	Reduction of ~80% at 550 MPa/10 min/70°C [53]	
	PME	Reduction of ~80% at 370 MPa/10 min/70°C [53]	
Pear	POD	Activation of ~23% at 600 MPa/3 min/40°C and reduction of ~92% at 600 MPa/5 min/100°C [54]	
	PPO	Reduction of ~90% at 600 MPa/3 min/100°C [54]	Activation of ~83% at 180 MPa/T inlet of 25°C [10]
	PME	Reduction of ~83% at 600 MPa/1 min/100°C [54]	
Mushroom	PPO		Activation of ~11% at 110 MPa/T inlet of 25°C [11]
Cocoyam	POD	Reduction of ~35% at 600 MPa/5 min/25°C [30]	
	PPO	Reduction of ~10% at 600 MPa/5 min/25°C [30]	
Peruvian carrot	POD	Reduction of ~40% at 600 MPa/5 min/25°C [30]	
	PPO	Reduction of ~70% at 600 MPa/5 min/25°C [30]	
Sweet potato	POD	Activation of ~15% at 600 MPa/5 min/25°C [30]	

Matrix	Enzyme	Effect of HIP on enzymatic activity	Effect of HPH on enzymatic activity
	PPO	Activation of ~48% at 600 MPa/5 min/25°C [30]	
		Meat and fish	
Rabbit muscles	m-Calpain	Reduction of 92% at 300 MPa/5 min/2°C [55]	
	μ-Calpain	Reduction of 98.4% at 300 MPa/5 min/2°C [55]	
	Calpastatin	Reduction of 95.7% at 300 MPa/5 min/2°C [55]	
Pork	Acid lipase	Activation of ~40% at 150 MPa/20 min/40°C [56] Reduction of ~100% at 600 MPa/20 min/50°C [56]	
	Neutral Lipase	Activation of ~10% at 150 MPa/20 min/<40°C [56] Reduction of ~100% at 450 MPa/20 min/55°C [56]	
	Phospholipase	Activation of ~10% at 300 MPa/20 min/40°C [56] Reduction of ~100% at 750 MPa/20 min/50°C [56]	
	Lipoxygenase	Activation of ~30% at 300 MPa/20 min/30°C [56] Reduction of ~100% at 450 MPa/20 min/55°C [56]	
Atlantic salmon	Acid phosphatase	Reduction of ~40% at 500 MPa/2 min/8-9 °C [57]	
Atlantic cod		Reduction of ~30% at 500 MPa/2 min/8–9°C [57]	
Mackerel		Reduction of ~20% at 500 MPa/2 min/8–9°C [57]	
Lean meat	Cathepsin D	Reduction of ~75% at 500 MPa/5 min/2°C [58]	
	Acid phosphatase	Reduction of ~15% at 500 MPa/5 min/2°C [58]	
	Cathepsin B	Reduction of ~17% at 500 MPa/5 min/2°C [58]	
	Cathepsin H	Reduction of ~85% at 500 MPa/5 min/2°C [58]	
	Cathepsin L	Reduction of ~20% at 500 MPa/5 min/2°C [58]	
	Aminopeptidase B	Reduction of ~78% at 500 MPa/5 min/2°C [58]	

Matrix	Enzyme	Effect of HIP on enzymatic activity	Effect of HPH on enzymatic activity
		Milk and dairy	
Milk	Alkaline phosphatase	Reduction of 100% at 800 MPa/8 min [59]	Activation from 100 to 150 MPa and inactivation above 175 MPa [60]
	Plasmin	Resistant at 400 MPa/30 min/25°C and reduction of 87% inactivation of plasmin at 400 MPa/15 min/60° C [61, 62]	Resistant up to 200 MPa [63]
	Lactoperoxidase	50% reduction after 800 MPa/4 hours/60°C [59]	Increased activity at 75 MPa [64]
	Lysozyme	Very resistant [65]	Activity increased at 75 MPa [64] Increase activity at 100 MPa [66]
	Lipases	350 and 400 MPa up to 100 min presented an increase up to 140% in the activity [67]	Increase in lipolysis during ripening (200 MPa, <58°C) or no effects (200 MPa, >71°C) [68, 69]
	Lactoferrin		Increase activity at 100 MPa [66]
Cheese	Proteases and lipases	Accelerating or delaying the ripening process depending on the conditions and microbial cultures [70, 71]	–
		Cereals and legumes	
Green beans (crude extract)	LOX	Reduction of ~50% of activity after 500 MPa/~20°C/10 min [72]	
Citrate and TRIS-HCl buffer (pH 4 -9)	LOX from soybeans	Reduction of >90% of activity after 600 MPa/45°C/2.5 min in all evaluated pH [73]	
Barley flours	Amylase	Increase of activity between 300 and 600 MPa/10–20 min (starch gelatinization) Inactivation at pressure > 600 MPa/10 min/room temperature [74]	
Wheat flour	Amylase	Increase of activity between 300 and 600 MPa/10–20 min (starch gelatinization) Inactivation at pressure >600 MPa/10 min/room temperature [74]	

PME, pectin methylesterase; PG, polygalacturonase; POD, peroxidase; PPO, polyphenoloxidase; BGL, β-glucosidase; LOX, lipoxygenase; HPL, hydroperoxide lyase.

Table 1. Enzymatic activity of the enzymes present in foods processed by HIP and HPH.

3.2. Enzymes structure

Relatively few works focused on the evaluation of the impact of HIP and HPH on enzymes structure, aiming to explain the alterations induced by these non-thermal processes on enzyme activity [4, 5], which includes activation, inactivation, or no change. These changes can be affected by multifactorial effects, including enzyme, substrate, media characteristics and process conditions, as described in item 3.1. Therefore, considering that enzymes alteration by HIP and HPH is an emerging field of research and the complexity of the question (due to the phenomena be governed by multifactorial effects), there is not a conclusive theory about the relationship between structural changes induced by HIP and its consequent activity changes.

The hypotheses, which discuss the impact of HIP or HPH on enzymes, consider two possible alterations: (i) changes on enzyme structure—mainly on quaternary and tertiary ones, with consequent exposure or entrapment of active site [75] and (ii) split of latent isoenzymes due to pressurization [6]. Additionally, HIP processing is necessary to consider the increase of enzyme reactivity under pressure, when enzyme and substrate are processed together [6].

The change on enzyme structure was evidenced by several studies [4–6, 10, 11, 76] that showed that pressure processing: (i) increase the exposure of hydrophobic amino acids, increasing the hydrophobicity of enzymes surface [11, 66, 76]; (ii) increase the exposure of SH groups due to unfolding of the protein and reduced the total SH content, due to new disulfide bonds formation [11, 76]; and (iii) slightly changes the content of α-helix, β-turn, β-sheet and random coil of processed enzymes, indicating that in some cases the processes can alter the secondary structure of enzymes [11]. On the other hand, no changes on primary structure were evidenced, which was expected since HIP is not considered to break covalent bounds.

At lower pressures (50–400 MPa), these slight molecular alterations were commonly linked with increase of enzyme conformational flexibility due to increase of its polar groups hydration [4], which facilitates the enzyme activity. Additionally, for enzymes processed at lower pressure, an increase of stability was attributed to higher enzyme intramolecular interactions and hydration of charged groups [4]. Conversely, the majority of work showed that there is a limit pressure/temperature/time conditions (specific for each enzyme/product) in which the enzyme energy gain is higher enough to induce sufficient structural changes that negatively affects the enzyme activity, being these effect reversible or not, depending on the applied pressure.

The split of latent isoenzymes is also an important factor to be considered for HIP and HPH applications in products that have endogenous enzymes, especially fruits and vegetables as carrots, peach, apples and others [6, 10, 11, 54]. Many reports showed that latent enzymes probably had activity induced by pressurization [6] and these enzymes normally have higher stability under pressure, being inactivated just at extreme conditions of HIP and just inactivated by HPH when the process is associated with mild heating [39]. This effect is particularly important for samples that contain polyphenoloxidase and peroxidases because of the occurrence of native resistance and occurrence of latent isoenzymes with consequent difficulty of inactivation [6, 46, 48].

Taking into account that HIP processing depends on a residence time at higher pressure, it is necessary to consider the reactivity of enzymes under pressure for samples processed (food) with endogenous enzymes and when enzyme is intentionally processed with the substrate. The changes on enzyme reactivity can be induced by several factors: (i) impact of the *Le Chatelier* principle (that postulates that enzyme reaction will be favored if the product formed by enzyme reaction had lower volume then the substrate or inhibited if the formed product had higher volume) [6], (ii) relative increase of substrate concentration due to solvent compression upon high pressure application [4], (iii) higher physical interaction between enzyme and substrate, with consequent increase of reactivity rate [30], (iv) higher concentration polar charged groups in the Michaelis complex and in the transitional state [77], favoring the equilibrium toward enzyme release against inhibition from their conjugated complex [4] and (v) changes induced on the substrate or media of reaction, modifying the substrates availability or enzyme:substrate contact [55].

Therefore, it can be concluded that both processes are able to transform the structure of enzymes and that the observed effects are similar for HIP and HPH, altering mainly the hydrophobicity and disulfide bonds. However, from the data available in literature, it is not possible to establish a sequence of transformations and the level of pressure required in each process to the changes occur, because these data are scarce and the effects observed can be attributed to a multiple reason. Further researches need to be performed using the same enzyme and matrix to allow the adequate comparison between HIP and HPH effects on enzyme structure.

3.2.1. Effect of HIP and HPH on fruit and vegetable enzymes

Endogenous enzymes can deteriorate the color, flavor, structure and nutritional value of fruits and vegetables [6]. Therefore, considering the intention of using HIP and HPH as emerging processes to be commercially used for stabilization of fruits and vegetables products, the inactivation of these degrading enzymes is very important and must be reached with or without the assistance of mild temperature. The effect of HIP and HPH has been studied for enzymes present in fruit and vegetable purees/juices like tomato, strawberry, apple, litchi, pear, coconut water and orange and different tubers.

The fruits and vegetables processed by HIP showed different changes on enzyme activity, being affected by the kind and source of enzyme, presence of isoenzymes, integrity of vegetable (cubes, purees/pulps and juices), processing conditions and activity measurement. For HIP, the enzymes activity normally increase or did not change up to 400 MPa [4] and, above to this pressure, enzyme inactivation occurs, being the effect enhanced by increase of the time (1–180 min) or temperature (20–100°C) (**Table 1**). Among the evaluated enzymes, drastic reduction/inactivation for polygalacturonase (PG), lipoxygenase (LOX) and hydroperoxide lyase (HPL) in tomato juice [33, 34], for pectin methylesterase (PME) in pear [54], orange [35, 36, 78], apple [46, 50] and tomato juice [33], for amylase in apple juice [49] and for peroxidase (POD) and polyphenoloxidase (PPO) in strawberry pulps/purée [42, 44] and pear [54] was highlighted. On the other hand, POD and PPO showed high resistance depending on the sample, being reduced just at extreme conditions (600 MPa associated with high temperature and/or long time) in cocoyam, Peruvian carrot, sweet potatoes [32], pear [54], apple [47], litchi [53] and strawberry juices [42]. Therefore, in general, it was observed that heat-resistant enzymes could also be baroresistant. Additionally, enzyme activation was also observed in sweet potato (up to 368% for PPO and 27% for POD in puree) [32] and in *Lonicera caerulea berry* [45], kiwi fruit [52] and banana juice [51], being associated to the split of isoenzymes and/or enzymes stabilization caused by pressure processing. The dependency of vegetable matrix could be observed for several works, however, it was not possible to establish a general rule to describe the effects, since the activation/inactivation was higher for cubes or puree in several matrix and in juices in other ones.

The impact of HPH on the inactivation of PME, POD and PPO was observed for process above 150 MPa [37–41], whereas in other ones, no significant inactivation was observed, causing negative effects (mainly browning and phase separation) in vegetable juices [43]. Comparing both processes, it was observed that the pressure required for enzyme inactivation in HPH is lower than HIP, indicating that high shear and the impact of abrupt pressure reduction in flow system are more important than the maintenance of the product at high pressure. In addition, the higher temperatures reached in HPH (increase of ~18°C for each 100 MPa) compared with temperature increase in HIP (3 °C for each 100 MPa) can partially explain the observed effect in homogenizers, indicating an association of homogenization and thermal effects.

Regarding to commercial application of HPH and HIP to process vegetable products with baroresistant enzymes, three possibilities are available: (i) the use of higher pressures (>600–1000 MPa for HIP and 400 MPa for HPH) associated with mild temperatures (50–80°C) aiming to inactivate the enzymes with better nutritional and sensory retention when compared with thermal processing alone [6], (ii) the use of mild pressure condition (up to 600 MPa for HIP or 300 MPa for HPH) and temperatures for inactivation of sensible fractions of enzymes, reducing the initial activity associated with other technological artifice to control the residual activity during product shelf life (e.g. package with good barrier to oxygen, use of gums to avoid phase separation and antioxidants) [46] and (iii) consumers education about the alterations induced by enzymes, explaining that these changes naturally occur and did not represent an unsafe product. A coconut water manufacturer adopted the strategy of explaining that the pink color occurs due to the activity of PPO that is not inactivated by mild process condition [79]. This seems to be a good alternative considering the growing consumer demand for unprocessed or low processed food and the comprehension that the use of mild processes impact the product's characteristics.

3.2.2. *Effect of HIP and HPH on cereals and legumes enzymes*

Enzymes are endogenously found in beans and cereals, being the lipases, lipoxygenases (LOX) and amylases are the classes of enzymes with major relevance. Amylases are able to differently hydrolyse starch, reducing the viscosity of products and it can be undesirable in several products due to the structural loss induced by the enzyme activity or desirable, when starch is used as raw material to obtain fermentable sugar by enzymatic saccharification process. On the other hand, lipoxygenases activity is always deleterious, due to the action as an antinutritional factor that affects in nutritional and sensory perception of the products, reducing its shelf life.

Few studies evaluated the impact of HIP on native cereals and legume enzymes [80]. Additionally, no work has evaluated the effect of HPH on crude extracts of cereals and legumes. Although several work focused on the impact of HPH on legume and cereals extract, it was previously thermally treated, inactivating the enzymes. In general, cereal and legume enzymes exposed to HIP at mild conditions demonstrated an increase in activity; however, when pressure is above to a critical energy limit (specific for each enzyme and influenced by matrix and processes conditions), enzymes lose their activity as consequence of modifications on its structure and active sites [80].

LOX is considered a relative-resistant enzyme in cereals due to the existence of isoenzymes present in this kind of food (types I, II, III). Studies showed that pressure around 600 MPa is interesting to inactivate all the isoenzymes [73], being this resistance lower than the observed for thermal processing. Therefore, HIP splits as an interesting alternative to promote LOX inactivation with minimum damage of the processed vegetable [72]. For amylases, a huge increase of the activity of starch-degrading enzymes α and β-amylases was observed at pressures of 400–600 MPa due to starch gelatinization that starts at 400 MPa and is almost complete at 600 MPa. Above this pressure, the higher intensity of the process induces negative and permanent unfolding and/or modifications of the enzymes activity site [74]. No studies evaluated the amylases enzyme activity after HIP process, but it is possible that pressures lower than 400 MPa are able to activate the amylases (as normally occurs with other enzymes). Future studies need to be performed to better explain the impact of both technologies in enzymes found in cereals and legumes.

3.2.3. *Effect of HIP and HPH on milk enzymes*

Enzymes present in dairy products include those endogenous of milk (as phosphatase, lactoperoxidase, lysozyme, latoferrin and plasmin), proteases and lipases from microbial origin, being these microorganisms from a contamination (especially present in milk stored for long time before processing) or intentionally added in dairy products, such as yogurt, fermented milk and cheese [59–71].

Enzymes have low ability to change the characteristics of raw or pasteurized milk due to the relative short shelf life and low storage temperature of milk, which decreases enzyme activity. To the contrary, commercial sterile milk (obtained by ultra high temperature—UHT—processing) are degraded by enzyme activity (endogenous proteases and especially from contaminants), which induces the age gelation phenomena, responsible for the end of UHT milk shelf life [61]. In dairy products, the action of these enzymes can be mostly desirable (enzymes that act as antimicrobials, preventing dairy product contamination) or eventually desirable depending on the product (e.g. lipase and protease are important for cheese maturation, but elevated activity can produce bitter peptides and rancid flavor) [70, 71]. Therefore, enzyme activation or inactivation can be desirable in different dairy products and in different time of product shelf life.

Considering the HIP effects, generally it is possible to say that enzymes are more resistant to the HIP or HPH processes than to thermal processing with similar microbial inactivation [65]. This is interesting considering the higher retention of enzymes, such as lactoperoxidase, lactoferrin and lysozyme and its antimicrobial activity [65]. On the other hand, the high stability of proteases and lipases can negatively affect the products obtained by using milk processed by HPH or HIP, since the product shelf life can be reduced especially due to changes on texture and flavor resulting from enzymes activity [67]. Therefore, the use of HPH and

HIP need to be evaluated carefully to substitute the pasteurization in milk process prior to the manufacturing of dairy products not only due to the enzyme action but also due to the effect of both processes on milk constituents.

Conversely, several works highlight the use of pressure processing as unitary operation to induce enzyme production of microorganisms due to baric stress and to lyse microbial cells, allowing the extravasation of cellular content with high amount of enzymes [81, 82]. These uses are normally interesting aiming to increase the enzyme activity on cheeses during ripening (lower pressures) and reducing the time to reach the desirable flavor and texture [82, 83]. In addition, higher pressures are highlighted as an interesting tool to prevent over ripening of cheeses due to partial enzymes inactivation [70], with consequent increment on the product shelf life. Additionally, when enzymes inactivation is desirable, mild/high temperature process can be associated with HIP (temperature during pressurization) or with HPH (high inlet temperature or temperature retention after homogenization valve) to improve the effectiveness of the process [61, 62].

3.2.4. *Effect of HIP on meat enzymes*

The tenderization is one of the most important effect in respect to meat quality. It occurs due to meat tissue changes induced by enzyme and biochemical reactions, being enzymes such as cathepsin (B, H, L e D), calpastatin, m-calpain and μ-calpain involved in this process. In this context, the HIP process can be a useful tool either by increasing the enzyme activity, greater exposure of enzyme/substrate, or by changes in proteins [84]. The phenomena specifically involved in these processes are as follows: (i) release of cathepsin present in lysosomes [58]; (ii) activation of calpains and cathepsin up to 150 MPa [85]; and (iii) inactivation of calpastatin, which are inhibitors of calpains [84]. Furthermore, as the calpains are calcium dependent, the HIP process can promote the increase of the calcium release and thereby increase the reaction efficiency [86]. Moreover, these enzymes can be inactivated at higher pressures probably due to denaturation [57, 84]. In this case, cathepsins are more resistant (pressures above 500 MPa [87]) compared with the calpains (activity reduced at 250 MPa for 10 min at room temperature [88]).

In contrast, the HIP process can increase lipid rancidity in meat. For example, in pork, lipases and lipoxygenases are activated up to 300 MPa and this increase is related to the level of release of these enzymes present in the lysosome [56]. This demonstrates that there is a greater influence of the increase of contact enzyme/substrate compared to the increase in enzyme activity. For HPH, there are no studies with meat because the process is carried out only for fluid.

3.2.5. *Effect of HIP and HPH on commercial enzymes*

The use of HIP and HPH technologies in the processing of commercial enzymes is growing over time aiming to improve its performance. Among the advantages, it highlights the increase of the activity in optimal and nonoptimal conditions and improvement/modification of the specificity.

The results obtained for enzymes processed by HIP showed that the aminopeptidases such as PepN, PepX and PepA from *Lactobacillus delbrueckii* ssp. bulgaricus ACA-DC 0105 (used as a starter culture in cheese production for ripening acceleration) were activated at pressures of up to 200 MPa and at temperatures of up to 40°C [27]. On the other hand, PepY and PepC were more sensitive to pressure and temperature, resulting in inactivation at pressures above 100 and 200 MPa, respectively [27]. For the glycolytic enzymes, a slight increase in activity was observed at pressures of 100 and 200 MPa, but no further changes were observed in the activity of these enzymes above 300 MPa [27]. The HIP process on commercial milk-clotting enzymes (recombinant Camel chymosin, calf rennet, bovine rennet, porcine pepsin, protease from *R. miehei*) was able to increase up to 25% the proteolytic activity (up to 300 MPa) or completely inactivate at high pressures (above 550 MPa) [26]. In other study, the leucine aminopeptidase obtained from *Aspergillus* (*A.*) *oryzae* used in the reduction of soy immunoreactivity was activated at 100–200 MPa/15 min/50°C [89].

The results obtained for HPH showed that this process was able to change the optimum temperature of neutral protease from 55 to 20°C after HPH at 200 MPa [28], improve glucose oxidase activity at 75°C after HPH at 150 MPa and increase between 100% and 400% its stability under storage [9]. In addition, it was observed an increase of amyloglucosidase activity at 80°C after HPH at 100 MPa [8] and an increase in the milk-clotting activity of coagulant enzymes, especially those that have chymosin in its constituents.

4. Future challenges

The studies of applying HIP and especially HPH, are recent. Therefore, there are some gaps of knowledge that must be fulfilled, allowing complete understanding about the effect and potential of both technologies. Firstly, from the data available in the literature, it is not possible to differentiate the effects of HIP and HPH since few enzymes were evaluated in both processes and, when evaluated, normally is not using the same diluting media, concentration, or method for activity measurement. Therefore, new studies need to be performed in a comparative way (varying just process conditions) to better establish the differences and equivalences of both processes. Ideally, a refined evaluation of molecular structure of enzymes must be determined to quantify the intensity of observed alterations and to describe the sequence of alterations caused specifically by HIP and HPH, helping to explain the enzymes transformation induced and consequently try to establish a mechanistic explanation of the alteration level that induces activation and inactivation of enzymes. From these explanations, it might be possible to predict the type of alteration expected for different enzymes, making these physical methods more interesting for industrial applications. Obtaining these results is mandatory to stimulate application of HIP and HPH, considering the industrial purpose of enzymes activation/modification/stabilization (mainly for enzyme manufactures) or the requirement of deleterious enzymes inactivation (especially for food processors).

5. Conclusions

High isostatic pressure and high pressure homogenization can be considered interesting unitary operations to be applied for inducing enzyme changes. Both processes are able to activate and/or stabilize the majority of enzymes when mild process conditions (low pressure and temperature) are applied and this is highly desirable since the use of physical methods are considered cleaner way to change molecules, minimizing the risks of transformed ones. The changes induced by HPH and HIP allow some enzymes overcome usage limitations, as inability to react at desirable process conditions (low or high temperature, pH and salt concentration) and as high costs since the increase of activity induced by the process reduces the amount of required enzyme in industrial reactions and, consequently, the costs of enzyme usage. Considering the multifactorial effects that governed the physical transformations induced by HIP and HPH, it is not possible to predict if is HIP or HPH the better method to obtain the desired activation and, possibly, for several enzymes, different conditions of each process can induce similar activation.

In respect to inactivation, a similar behavior with thermal treatment was observed, being the heat resistant also baroresistant enzymes (e.g. peroxidase, polyphenoloxidase), making necessary the association of pressure and mild temperatures to reach adequate levels of inactivation. In this case, HIP tends to be more effective than HPH to induce irreversible changes on enzymes.

The activation and inactivation effects can be related to the molecular energy input caused by HPH or HIP processing. Differently from the thermal effects, both processes involving pressurization can deliver energy enough to induce changes (enhancement of hydrophobic surface and disulfide bonds formation) that positively affect the enzymes activity due to higher flexibility and exposure sites. However, at extreme conditions, the higher energy delivered induces to drastic changes, with irreversible enzyme changes and consequent permanent activity loss. Therefore, it is concluded that HIP and HPH are versatile technologies to alter enzymes, being possible to reach activation or inactivation, depending on the process conditions chosen.

Acknowledgements

The authors like to thank the São Paulo Research Foundation (FAPESP) for the financial support.

Author details

Bruno Ricardo de Castro Leite Júnior[1]*, Alline Artigiani Lima Tribst[2] and Marcelo Cristianini[1]

*Address all correspondence to: brunorclj@gmail.com

1 Department of Food Technology (DTA), School of Food Engineering (FEA), University of Campinas (UNICAMP), Campinas, Brazil

2 Center for Food Studies and Research (NEPA), University of Campinas (UNICAMP), Campinas, Brazil

References

[1] Ramos, O S, Malcata, F X. Food-Grade Enzymes. In: Murray M-Y editor. Comprehensive Biotechnology 2nd ed., vol. 3, Elsevier; 2011, p. 555–569, DOI: 10.1016/B978-0-12-809633-8.09173-1

[2] Stryer, L. Enzymes: Basics concepts and kinetics. In: Stryer, L. Biochemistry. New York: W H Freeman & Company, 1995.

[3] Whitaker, J.R. Principles of enzymology for the food science. New York: Marcel Dekker, 1994.

[4] Eisenmenger, M.J.; Reyes-De-Corcuera, J.I. High pressure enhancement of enzymes: a review. Enzymes Microb Technol 2009; 45 (5): 331–347.

[5] Tribst AAL, Cota J, Murakami MT, et al. Effects of High Pressure Homogenization on the Activity, Stability, Kinetics and Three-Dimensional Conformation of a Glucose Oxidase Produced by *Aspergillus niger*. PLoS One 2014; **9**: e103410. DOI: 10.1371/journal.pone.0103410

[6] Chakraborty S, Kaushik N, Rao PS, et al. High-pressure inactivation of enzymes: A review on its recent applications on fruit purees and juices. Compr Rev Food Sci Food Saf 2014; **13**: 578–596. DOI: 10.1111/1541-4337.12071

[7] Leite Júnior BR de C, Tribst AAL, Bonafe CFS, et al. Determination of the influence of high pressure processing on calf rennet using response surface methodology: Effects on milk coagulation. LWT - Food Sci Technol 2016; **65**: 10–17. DOI: 10.1016/j.lwt.2015.07.063

[8] Tribst AAL, Cristianini M. Increasing fungi amyloglucosidase activity by high pressure homogenization. Innov Food Sci Emerg Technol 2012; **16**: 21–25. DOI: 10.1016/j.ifset.2012.03.002

[9] Tribst AAL, Cristianini M. Changes in commercial glucose oxidase activity by high pressure homogenization. Innov Food Sci Emerg Technol 2012; **16**: 355–360. DOI: 10.1016/j.ifset.2012.08.002

[10] Liu WEI, Jianhua LIU, Mingyong XIE, et al. Characterization and high-pressure microfluidization-induced activation of polyphenoloxidase from Chinese pear (*Pyrus pyrifolia Nakai*). J Agric Food Chem 2009; **57**: 5376–5380. DOI: 10.1021/jf9006642

[11] Liu W, Liu J, Liu C, et al. Activation and conformational changes of mushroom polyphenoloxidase by high pressure microfluidization treatment. Innov Food Sci Emerg Technol 2009; **10**: 142–147. DOI:10.1016/j.ifset.2008.11.009

[12] Diels AMJ, Michiels CW. High-pressure homogenization as a non-thermal technique for the inactivation of microorganisms. Crit Rev Microbiol 2006; **32**: 201–216. DOI: 10.1080/10408410601023516

[13] Dumay E, Chevalier-Lucia D, Picart-Palmade L, et al. Technological aspects and potential applications of (ultra) high-pressure homogenisation. Trends Food Sci Technol 2013; **31**: 13–26. DOI: 10.1016/j.tifs.2012.03.005

[14] Bermúdez-Aguirre D, Barbosa-Cánovas G V. An update on high hydrostatic pressure, from the laboratory to industrial applications. Food Eng Rev 2011; **3**: 44–61. DOI: 10.1007/s12393-010-9030-4

[15] Avure Technologies, Inc. [Internet]. 2016. Available from: http://www.avure-hpp-foods.com/hpp-equipment/ [Accessed: 2016-09-15]

[16] Hiperbaric High pressure processing. [Internet]. 2016. Available from: http://www.hiperbaric.com/en/equipment [Accessed: 2016-09-15]

[17] Stansted Fluid Power Ltd [Internet]. 2016. Available from: http://www.homogenizersystems.com/production_high_pressure_homogenizers.html [Accessed: 2016-09-12]

[18] Balasubramaniam V M, Martínez-Monteagudo S I, Gupta R. Principles and application of high pressure-based technologies in the food industry. Annu Rev Food Sci Technol 2015; 6:435-462. DOI: 10.1146/annurev-food-022814-015539

[19] Rendueles E, Omer MK, Alvseike O, et al. Microbiological food safety assessment of high hydrostatic pressure processing: A review. LWT - Food Sci Technol 2011; **44**: 1251–1260. DOI: 10.1016/j.lwt.2010.11.001

[20] Knorr D. Novel approaches in food-processing technology: new technologies for preserving foods and modifying function. Curr Opin Biotechnol 1999; **10**: 485–491. DOI: 10.1016/S0958-1669(99)00015-4

[21] Campos FP, Cristianini M. Inactivation of *Saccharomyces cerevisiae* and *Lactobacillus plantarum* in orange juice using ultra high-pressure homogenisation. Innov Food Sci Emerg Technol 2007; **8**: 226–229. DOI: 10.1016/j.ifset.2006.12.002

[22] Floury J, Bellettre J, Legrand J, et al. Analysis of a new type of high pressure homogeniser. A study of the flow pattern. Chem Eng Sci 2004; **59**: 843–853. DOI: 10.1016/j.ces.2003.11.017

[23] GEA Process Engineering Ltd [Internet]. 2016. Available from: http://www.gea.com/en/products/homogenizers-ariete-series.jsp [Accessed: 2016-09-12]

[24] Mozhaev VV, Lange R, Kudryashova EV, et al. Application of high hydrostatic pressure for increasing activity and stability of enzymes. Biotechnol Bioeng. 1996. Epub ahead of print 1996. DOI: 10.1002/(SICI)1097-0290(19961020)52:2<320::AID-BIT12>3.0.CO;2-N

[25] Leite Júnior BR de C, Tribst AAL, Cristianini M. Comparative effects of high isostatic pressure and thermal processing on the inactivation of *Rhizomucor miehei* protease. LWT - Food Sci Technol 2016; **65**: 1050–1053. DOI: 10.1016/j.lwt.2015.09.042

[26] Leite Júnior BR de C, Tribst AAL, Cristianini M. The effect of high pressure processing on recombinant chymosin, bovine rennet and porcine pepsin: Influence on the proteolytic

and milk-clotting activities and on milk-clotting characteristics. LWT - Food Sci Technol 2016; 1–10. DOI: 10.1016/j.lwt.2016.04.018

[27] Malone AS, Wick C, Shellhammer TH, et al. High pressure effects on proteolytic and glycolytic enzymes involved in cheese manufacturing. J Dairy Sci 2003; **86**: 1139–1146. DOI: 10.3168/jds.S0022-0302(03)73696-0

[28] Tribst AAL, Augusto PED, Cristianini M. Multi-pass high pressure homogenization of commercial enzymes: Effect on the activities of glucose oxidase, neutral protease and amyloglucosidase at different temperatures. Innov Food Sci Emerg Technol 2013; **18**: 83–88. DOI: 10.1016/j.ifset.2013.01.002

[29] Kaushik N, Kaur BP, Rao PS. Inactivation of polyphenol oxidase and peroxidase enzymes during pulsed, static and cyclic pressurization of litchi (*Litchi chinensis*) juice. Institution of Chemical Engineers. 2016; 100:412–423. DOI:10.1016/j.fbp.2016.07.015.

[30] Tribst AAL, Leite Júnior BR de C, De Oliveira MM, et al. High pressure processing of cocoyam, Peruvian carrot and sweet potato: Effect on oxidative enzymes and impact in the tuber color. Innov Food Sci Emerg Technol 2016; **34**: 302–309. DOI: 10.1016/j.ifset.2016.02.010

[31] Leite Júnior BRC, Tribst AAL, Cristianini M. Influence of high pressure homogenization on commercial protease from *Rhizomucor miehei*: Effects on proteolytic and milk-clotting activities. LWT - Food Sci Technol 2015; **63**: 739–744. DOI: 10.1016/j.lwt.2015.03.020

[32] Tribst AAL, Augusto PED, Cristianini M. The effect of the high pressure homogenisation on the activity and stability of a commercial neutral protease from *Bacillus subtilis*. Int J Food Sci Technol 2012; **47**: 716–722. DOI: 10.1111/j.1365-2621.2011.02898.x

[33] Andreou V, Dimopoulos G, Katsaros G, et al. Comparison of the application of high pressure and pulsed electric fields technologies on the selective inactivation of endogenous enzymes in tomato products. Innov Food Sci Emerg Technol. Epub ahead of print 2016. DOI: 10.1016/j.ifset.2016.07.026

[34] Rodrigo D, Jolie R, Loey A Van, et al. Thermal and high pressure stability of tomato lipoxygenase and hydroperoxide lyase. J Food Eng 2007; **79**: 423–429. DOI: 10.1016/j.jfoodeng.2006.02.005

[35] Bayındırlı A, Alpas H, Bozoğlu F, et al. Efficiency of high pressure treatment on inactivation of pathogenic microorganisms and enzymes in apple, orange, apricot and sour cherry juices. Food Control 2006; **17**: 52–58. DOI: 10.1016/j.foodcont.2004.09.002.

[36] Nienaber U, Shellhammer TH. High-pressure processing of orange juice : Kinetics of pectin methylesterase inactivation food engineering and physical properties. J Food Sci 2001; **66**: 328–331. DOI: 10.1111/j.1365-2621.2001.tb11341.x

[37] Lacroix N, Fliss I, Makhlouf J. Inactivation of pectin methylesterase and stabilization of opalescence in orange juice by dynamic high pressure. Food Res Int 2005; **38**: 569–576. DOI: 10.1016/j.foodres.2004.11.010

[38] Welti-Chanes J, Ochoa-Velasco CE, Guerrero-Beltrán JÁ. High-pressure homogenization of orange juice to inactivate pectin methylesterase. Innov Food Sci Emerg Technol 2009; **10**: 457–462. DOI: 10.1016/j.ifset.2009.05.012

[39] Carbonell J V., Navarro JL, Izquierdo L, et al. Influence of high pressure homogenization and pulp reduction on residual pectin methylesterase activity, cloud stability and acceptability of Lane Late orange juice: A study to obtain high quality orange juice with extended shelf life. J Food Eng 2013; **119**: 696–700. DOI: 10.1016/j.jfoodeng.2013.06.041

[40] Navarro JL, Izquierdo L, Carbonell J V., et al. Effect of pH, temperature and maturity on pectin methylesterase inactivation of citrus juices treated by high-pressure homogenization. LWT - Food Sci Technol 2014; **57**: 785–788. DOI: 10.1016/j.lwt.2014.01.033

[41] Velázquez-Estrada RM, Hernández-Herrero MM, Guamis-López B, et al. Impact of ultra high pressure homogenization on pectin methylesterase activity and microbial characteristics of orange juice: A comparative study against conventional heat pasteurization. Innov Food Sci Emerg Technol 2012; **13**: 100–106. DOI: 10.1016/j.ifset.2011.09.001

[42] Cao X, Zhang Y, Zhang F, et al. Effects of high hydrostatic pressure on enzymes, phenolic compounds, anthocyanins, polymeric color and color of strawberry pulps. J Sci Food Agric 2011; **91**: 877–885. DOI: 10.1002/jsfa.4260

[43] Karacam CH, Sahin S, Oztop MH. Effect of high pressure homogenization (microfluidization) on the quality of Ottoman Strawberry (*F. Ananassa*) juice. LWT - Food Sci Technol 2015; **64**: 932–937. DOI: 10.1016/j.lwt.2015.06.064

[44] Marszałek K, Mitek M, Skąpska S. The effect of thermal pasteurization and high pressure processing at cold and mild temperatures on the chemical composition, microbial and enzyme activity in strawberry purée. Innov Food Sci Emerg Technol 2015; **27**: 48–56. DOI: 10.1016/j.ifset.2014.10.009

[45] Liu S, Xu Q, Li X, et al. Effects of high hydrostatic pressure on physicochemical properties, enzymes activity and antioxidant capacities of anthocyanins extracts of wild *Lonicera caerulea* berry. Innov Food Sci Emerg Technol 2016; **36**: 48–58. 10.1016/j.ifset.2016.06.001

[46] Valdramidis VP, Graham WD, Beattie A, et al. Defining the stability interfaces of apple juice: Implications on the optimisation and design of high hydrostatic pressure treatment. Innov Food Sci Emerg Technol 2009; **10**: 396–404. DOI: 10.1016/j.ifset.2009.02.006

[47] Anese M, Nicoli MC, Dallaglio G, et al. Effect of high-pressure treatments on peroxidase and polyphenoloxidase activities. J Food Biochem 1995; **18**: 285–293. DOI: 10.1111/j.1745-4514.1994.tb00503.x

[48] Suárez-Jacobo Á, Saldo J, Rüfer CE, et al. Aseptically packaged UHPH-Treated apple juice: Safety and quality parameters during storage. J Food Eng 2012; **109**: 291–300. DOI: 10.1016/j.jfoodeng.2011.09.007

[49] Riahi E, Ramaswamy HS. High pressure inactivation kinetics of amylase in apple juice. J Food Eng 2004; **64**: 151–160. DOI: 10.1016/j.jfoodeng.2003.09.025

[50] Riahi E, Ramaswamy HS. High-pressure processing of apple juice: kinetics of pectin methyl esterase inactivation. Biotechnol Prog 2003. **19**: 908–14. DOI: 10.1021/bp025667z

[51] Calligaris S, Foschia M, Bartolomeoli I, et al. Study on the applicability of high-pressure homogenization for the production of banana juices. LWT - Food Sci Technol 2012; **45**: 117–121. DOI: 10.1016/j.lwt.2011.07.026

[52] Fang L, Jiang B, Zhang T. Effect of combined high pressure and thermal treatment on kiwifruit peroxidase. Food Chem 2008; **109**: 802–807 DOI: 10.1016/j.foodchem.2008.01.017

[53] Jayachandran LE, Chakraborty S, Rao PS. Inactivation kinetics of the most baro-resistant enzyme in high pressure processed litchi-based mixed fruit beverage. Food Bioprocess Technol 2016; **9**: 1135–1147. DOI: 10.1007/s11947-016-1702-5

[54] Terefe NS, Tepper P, Ullman A, et al. High pressure thermal processing of pears: Effect on endogenous enzyme activity and related quality attributes. Innov Food Sci Emerg Technol 2016; **33**: 56–66. DOI: 10.1016/j.ifset.2015.12.001

[55] Homma N, Ikeuchi Y, Suzuki A. Levels of calpain and calpastatin in meat subjected to high-pressure. Meat Sci 1995; **41**: 251–260. DOI: 10.1016/0309-1740(95)00005-6

[56] Huang Y, Wang Y, Wu Z, et al. Combined effects of high-pressure and thermal treatments on lipid oxidation and enzymes in pork. Food Sci Biotechnol 2016; **25**: 261–266. DOI: 10.1007/s10068-016-0038-2

[57] Rode TM, Hovda MB. High pressure processing extend the shelf life of fresh salmon, cod and mackerel. Food Control 2016; **70**: 242–248. DOI: 10.1016/j.foodcont.2016.05.045

[58] Homma N, Ikeuchi Y, Suzuki A. Effects of high pressure treatment on the proteolytic enzymes in meat. Meat Sci 1994; **38**: 219–228. DOI: 10.1016/0309-1740(94)90111-2

[59] Rademacher B, Pfeiffer B, Kessler HG. Inactivation of microorganisms and enzymes in pressure treated raw milk. In: Isaacs NS, editors. High pressure food science, bioscience and chemistry. Cambridge: The Royal Society of Chemistry; 1998. p. 145–151.

[60] Picart L, Thiebaud M, René M, et al. Effects of high pressure homogenisation of raw bovine milk on alkaline phosphatase and microbial inactivation. A comparison with continuous short-time thermal treatments. J Dairy Res 2006; **73**: 454–463. DOI: 10.1017/S0022029906001853

[61] Huppertz T., Fox PF, Kelly, AL. Plasmin activity and proteolysis in high pressure-treated bovine milk. Lait 2004; **84**: 297–304. DOI: 10.1051/lait:2004003

[62] Garía-Risco MR, Olano A, Ramos M, et al. Micelar changes induced by high pressure. Influence in the proteolytic activity and organoleptic properties of milk. J Dairy Sci 2000; **83**: 2184–2189. DOI: 10.3168/jds.S0022-0302(00)75101-0

[63] Iucci L, Lanciotti R, Kelly A, et al. Plasmin activity in high-pressure-homogenised bovine milk. Milchwissenschaft 2008; **63**: 68–70.

[64] Vannini L, Lanciotti R, Baldi D, et al. Interactions between high pressure homogenization and antimicrobial activity of lysozyme and lactoperoxidase. Int J Food Microbiol 2004; **94**: 123–135. DOI: 10.1016/j.ijfoodmicro.2004.01.005

[65] Viazis S, Farkas BE, Allen JC. Effects of high-pressure processing on immunoglobulin A and lysozyme activity in human milk. J Hum Lact; 2007; **23**: 253–261. DOI: 10.1177/0890334407303945

[66] Iucci L, Patrignani F, Vallicelli M, et al. Effects of high pressure homogenization on the activity of lysozyme and lactoferrin against *Listeria monocytogenes*. Food Control 2007; **18**: 558–565. DOI: 10.1016/j.foodcont.2006.01.005

[67] Pandey PK, Ramaswamy HS. Effect of high-pressure treatment of milk on lipase and γ-glutamyl transferase activity. J Food Biochem. 2004; **28**: 449–462. DOI: 10.1111/j.1745-4514.2004.02603.x

[68] Datta N, Hayes MG, Deeth HC, et al. Significance of frictional heating for effects of high pressure homogenisation on milk. J Dairy Res 2005; **72**: 393–399. DOI: 10.1017/S0022029905001056

[69] Juan B, Quevedo JM, Zamora A, et al. Lipolysis of cheeses made from goat milk treated by ultra-high pressure homogenization. LWT - Food Sci Technol 2015; **60**: 1034–1038. DOI: 10.1016/j.lwt.2014.10.003

[70] Delgado FJ, González-Crespo J, Cava R, et al. Changes in microbiology, proteolysis, texture and sensory characteristics of raw goat milk cheeses treated by high-pressure at different stages of maturation. LWT - Food Sci Technol 2012; **48**: 268–275. DOI: 10.1016/j.lwt.2012.03.025

[71] Costabel LM, Bergamini C, Vaudagna SR, et al. Effect of high-pressure treatment on hard cheese proteolysis. J Dairy Sci 2016; 1–13. DOI: 10.3168/jds.2015-9907

[72] Indrawati, Van Loey AM, Ludikhuyze LR, Hendrickx ME. Elevated Temperatures. 1999; **47**: 2468–2474. DOI: 10.1021/jf9811875

[73] Tangwongchai R, Ledward DA, Ames JM. Effect of high-pressure treatment on lipoxygenase activity. J Agric Food Chem 2000; 48: 2896–2902. DOI: 10.1021/jf9913460

[74] Gomes MRA, Clark R, Ledward DA. Effects of high pressure on amylases and starch in wheat and barley flours. Food Chem 1998; **63**: 363–372. DOI: 10.1016/S0308-8146(98)00014-4

[75] Hendrickx M, Ludikhuyze L, Van den Broeck I, et al. Effects of high pressure on enzymes related to food quality. Trends Food Sci Technol 1998; **9**: 197–203. DOI: 10.1016/S0924-2244(98)00039-9

[76] Liu W, Zhang ZQ, Liu CM, et al. The effect of dynamic high-pressure microfluidization on the activity, stability and conformation of trypsin. Food Chem 2010; **123**: 616–621. DOI: 10.1016/j.foodchem.2010.04.079

[77] Kudryashova EV, Mozhaev VV, Balny C. Catalytic activity of thermolysin under extremes of pressure and temperature: modulation by metal ions. Biochim Biophys Acta - Protein Struct Mol Enzymol 1998; **1386**: 199–210. DOI: 10.1016/S0167-4838(98)00055-7

[78] Basak S, Ramaswamy HS, Simpson BK. High pressure inactivation of pectin methylesterase in orange juice using combination treatments. J Food Biochem 2001; **25**: 509–526. DOI: 10.1111/j.1745-4514.2001.tb00811.x

[79] Harmless Harvest [Internet]. 2016. Available from: http://www.harmlessharvest.com/original/ [Accessed: 2016-09-12]

[80] Estrada-Girón Y, Swanson BG, Barbosa-Cánovas G V. Advances in the use of high hydrostatic pressure for processing cereal grains and legumes. Trends Food Sci Technol 2005; **16**: 194–203. DOI: 10.1016/j.tifs.2004.10.005

[81] Hayes MG, Kelly AL. High pressure homogenisation of milk (b) effects on indigenous enzymatic activity. J Dairy Res 2003; **70**: 307–313. DOI: 10.1017/S0022029903006319

[82] Vannini L, Patrignani F, Iucci L, et al. Effect of a pre-treatment of milk with high pressure homogenization on yield as well as on microbiological, lipolytic and proteolytic patterns of 'Pecorino' cheese. Int J Food Microbiol 2008; **128**: 329–335. DOI: 10.1016/j.ijfoodmicro.2008.09.018

[83] Lanciotti R, Chaves-López C, Patrignani F, et al. Effects of milk treatment with dynamic high pressure on microbial populations and lipolytic and proteolytic profiles of Crescenza cheese. Int J Dairy Technol 2004; **57**: 19–25. DOI: 10.1111/j.1471-0307.2004.00121.x

[84] Ma H, Ledward D. High pressure processing of fresh meat--is it worth it? Meat Sci 2013; **95**: 897–903. DOI: 10.1016/j.meatsci.2013.03.025

[85] Sikes A, Tornberg E, Tume R. A proposed mechanism of tenderising post-rigor beef using high pressure–heat treatment. Meat Sci 2010; **84**: 390–399. DOI: 10.1016/j.meatsci.2009.09.007

[86] Cheftel JC, Culioli J. Effects of high pressure on meat: A review. Meat Sci 1997; **46**: 211–236. DOI: 10.1016/S0309-1740(97)00017-x

[87] Jung S, Ghoul M, de Lamballerie-Anton M. Changes in lysosomal enzyme activities and shear values of high pressure treated meat during ageing. Meat Sci 2000; **56**: 239–246. DOI: 10.1016/S0309-1740(00)00048-6

[88] Jin Y, Nan Q, Che R. Effects of high pressure on the activity of major enzymes in beef. Meat Research (China) 2001; **3**: 13–16.

[89] Meinlschmidt P, Brode V, Sevenich R, et al. High pressure processing assisted enzymatic hydrolysis – An innovative approach for the reduction of soy immunoreactivity. Innov Food Sci Emerg Technol. Epub ahead of print 2016. DOI: 10.1016/j.ifset.2016.06.022

Chapter 4

Kinetic Modelling of Enzyme Catalyzed Biotransformation Involving Activations and Inhibitions

Ganapati D. Yadav and Deepali B. Magadum

Additional information is available at the end of the chapter

http://dx.doi.org/10.5772/67692

Abstract

To achieve transition from lab scale enzyme studies to industrial applications, understanding of enzyme kinetics plays a critical role. The widely applied Michaelis Menten equation of the single substrate kinetics, sequential and double replacement mechanism of bisubstrate reaction and the relevant kinetics, inhibition and activation of enzyme are all integral parts of this discussion. In this chapter, we have discussed different types of inhibition and kinetic modelling. Systematic approach to generate data and its interpretation as well as designing of inhibitors is also explained.

Keywords: enzyme catalysis, mechanism, activation, deactivation, assay methods, inhibition, kinetic modelling, microwave irradiation

1. Introduction

The chapter gives brief introduction of enzyme, its classification and kinetics. It also discusses about the different types of inhibitions and activations involved in the enzymatic reactions. The classification of different types of enzyme inhibitors and activators and their mechanism of action will also be explained with appropriate examples. It will also give gist of different methods to identify inhibition or activation of specified reaction by given molecule. Different types of kinetic models used for the enzyme inhibition and activation will be systematically discussed with case studies. The applications of bioinformatics and computational modelling for *in-vitro* designing of the modulators and the study of transition states of enzyme-inhibitor complex will also be brought into light. The new developments and future challenges are summarized in this chapter. Enzyme catalysis, classification and effect of physiological conditions are also discussed.

Continuous rigorous efforts have been put into the understanding of enzymatic kinetics in order to use them for reactor design and industrial applications. The systematic study of the reactivity of enzymes under different controlled external conditions reveals the reaction mechanism, activation and deactivation of enzymes and their reactivity towards a given set of substrates [1–4]. The crystallographic data supporting three-dimensional structural properties and mechanism in combination with the kinetic data serve as a powerful tool to understand working principles of enzyme in cellular fluids and thereby provide an opportunity to design the therapeutic molecules [1, 2]. International Commission on Enzymes was instituted by the joint efforts of International Union of Pure and Applied Chemistry (IUPAC) and International Union of Biochemistry (IUB) which have classified and assigned an EC number to each enzyme. The respective four-digit enzyme commission (EC) number associated with the enzyme signifies the details about the reactions catalysed by the enzyme. The readers can refer the biochemistry literature present in the books written by Stryer, Voet and Voet [3], Nelson and Cox [4] and the official website of the enzyme classification [5, 6]. Enzymes as the biological catalyst are mainly classified in six classes, namely, oxidoreductase, ligase, lyase, transferases, isomerase and hydrolases [3–5, 7, 8]. More importantly, enzymes do not alter the equilibrium constants of chemical reaction as they also have the ability to catalyse backward and forward reactions. Hence, they only affect the rate at which the equilibrium is established. However, as almost all the biological systems in human body work at the steady state, the net effect of enzyme to convert the substrate into the product occurs rapidly as the products are removed instantaneously [4, 8, 9]. Various factors and physiological conditions affect the activity and efficiency of enzymes to the various extents. Therefore, it is essential to understand their effects on the enzymes [10, 11].

1.1. Effect of temperature

The fluctuations in temperature can affect the activity of enzymes in either way, i.e. they may cause activation or deactivation, depending on the nature of enzymes and properties of the solution. In most of the catalytic reactions, the average kinetic energy and velocity of molecules rise with increase in temperature, which results in higher probability of effective collisions and the cumulative changes lead to increase the rate of reaction with temperature [12–14]. Below the denaturation temperature, on an average, rate increases with per 10°C increase in temperature. Ratio by which rate changes for 10°C rise is known Q_{10} (approximately from 1.7 to 2.5). Generally, as the enzymes are evolved to work with optimum efficiency at physiological conditions of the source microorganism, the same conditions are observed to be the optimal conditions for most of the enzymes at which the rate of reaction is maximum. [18] concluded that in the case of macro-heterogeneous systems such as lipase, the rate of reaction of enzymes is partly regulated by diffusion and partly by temperature. Inactivation by heat expends on stability of the enzyme up to certain temperature range at different experimental conditions. However, purified enzymes have higher probability of deactivation than crude enzymes. This is because the liability is often subsidized by the protective protein and their colloids associated with the enzymes of interest. Therefore, the critical temperature and temperature coefficient depend upon properties and composition of the enzyme preparation along with the properties of enzyme itself. The critical temperature is at which the activity of its enzyme is destroyed to half of the reactions. If the enzymes are used for the reaction at higher

http://dx.doi.org/10.5772/67692

temperature than optimum, it catalyses the reaction and will get deactivated over the period of time. If the temperature is lowered to optimum, reaction will be completed earlier. When the optimum temperature is maintained slightly above the critical temperature of inactivation, reaction will reach to the completion at a faster rate. Practically, optimum temperature gives rough idea of heat stability of enzymes and there are no other significant data available at this temperature. Hence, it does not serve as a definite character even for the given enzyme system. In some cases, increase in catalase activity in a yeast cell suspension occurs by 49 times after heating for an hour. This is attributed to the destruction of an inhibitory substance at higher temperatures. Similar properties are observed for the blood catalase due to hemolysis as corpuscular enzyme differs from the intra-corpuscular enzyme in its properties [15].

1.2. Action of radiations on enzyme

The exposure of enzyme solution to the various radiations generally leads to an irreversible destruction, occasionally an activation of the enzymes [16]. The rate of destruction is practically independent of temperature during the treatment of enzyme preparation with radiations. Perhaps, it is affected by pH and chemical conditions like presence of substrate, optical sensitizers and oxygen [17, 18]. While enzymes are very stable to heat inactivation near their optimal pH, the reverse is often the case with regard to ultraviolet (UV) radiations. Generally, diluted solutions are more affected by the UV radiations than the concentrated ones due to absence of protective colloids and ions. The deactivating effect of X- and γ-rays may be due to the electrons ejected by them which lead to the inactivation of enzymes [19]. In case of radioactive sources, the amount of deactivation can be calculated. If Q is the amount of enzyme present, then

$$\frac{I}{Q}\frac{dQ}{dW} = c \tag{1}$$

where W is an average activity of radioactive source and c is a constant multiplied by time over which it acts.

1.3. Effect of microwave radiations

Since last five decades, microwave irradiations have been established as an efficient source of heat energy for chemical reactions, wherein rate of reactions has been enhanced by a multiple fold than conventional heating and has led to the reduction in overall reaction time drastically with high selectivity. The electromagnetic components of microwave irradiation lead to the vibration of bonds to chemical reactants and catalysts resulting into the lowering of activation energy and increase in collision frequency [20, 21]. Microwave irradiation triggers the effect that cannot be created with the help of thermal heating. Especially, in case of polar solvent molecules, it acts as a bridge between two molecules through absorption energy [22–24]. The enzyme hydration layer can be used for the improvement of the process or modification of the selectivity as it interacts with the microwave radiations [25–27]. Hence, the microwave irradiation affects synergistically to the enzymes.

1.4. Effect of pH

The optimal pH of the enzyme is generally the same as that of fluid at which enzyme functions *in vivo*. Enzyme activity profile is bell shaped curve signifying that before and after the optimal pH, the activity of the enzyme decreases drastically. The presence of charged amino acids in enzyme structure, chemical nature of substrate and enzyme-substrate complexes is the reason for the pH dependence of enzyme [28]. Hence, pH causes dissociation or undissociation of the charged amino acids that affect catalysis. Ionization of these groups depends upon pKa values. The pH and nature determine the pKa value of the medium. The change in pH alters interaction of the substrate with an active site and rate of breakdown of enzyme-substrate complex into product [29–31]. The effect of pH on the biocatalysed reaction can be represented with the change in V_{max} and K_m which is summarized in Scheme 2.1 and Expression 2.1.

$$\begin{array}{ccccc} E- & & ES- & & \\ \uparrow\downarrow K_{E2} & & \uparrow\downarrow K_{ES2} & & \\ EH+S & \underset{k_{-1}}{\overset{k_1}{\rightleftarrows}} & ESH & \overset{k_2}{\rightarrow} & EH+P \\ \uparrow\downarrow K_{E1} & & \uparrow\downarrow K_{ES1} & & \\ EH2+ & & ESH2+ & & \end{array} \tag{2}$$

Scheme 2.1: Mathematical equations modelling pH effects on enzyme catalysed reactions.

$$v = \frac{V_{mapp}[S]}{K_{mapp} + [S]} \tag{3}$$

$$V_{mapp} = \frac{V_m}{1 + \frac{[H+]}{K_{es1}} + \frac{K_{es2}}{[H+]}} \tag{4}$$

$$K_{mapp} = \frac{K_m(1 + \frac{[H+]}{K_{e1}} + \frac{K_{e2}}{[H+]})}{1 + \frac{[H+]}{K_{es1}} + \frac{K_{es2}}{[H+]}} \tag{5}$$

Expression 2.1: Kinetic expression for the effect of pH on enzyme catalysed reactions.

1.5. Effect of immobilization

With the purpose of reusability, enzymes are immobilized on porous and non-porous solid medium using various methods. This immobilization along with the stabilization of the enzyme brings the extrinsic and intrinsic diffusional limitations and increased probability of substrate or product inhibition. The kinetic constants (e.g. K_m, V_{max}) of immobilized enzymes may be altered by the process of immobilization, intrinsic specificity of enzymes, properties of the solution and molecular diffusion within the local environment [32–34]. The relationship between these intrinsic and apparent parameters is shown below.

$$\text{Damkohler number } (D_a) = \text{Maximum rate of reaction/maximum rate of diffusion} = \frac{V'_{max}}{K_L[S_b]} \tag{6}$$

http://dx.doi.org/10.5772/67692

If $D_a >> 1$, diffusion rate is limiting the observed rate.

If $D_a << 1$, reaction rate is limiting the observed rate.

The Michealis Menten equation for the immobilized enzyme is represented in equation (7).

$$v_s = \eta \frac{V'_{max}[S_b]}{K_m + [S_b]} \quad (7)$$

Expression 2.2: Modified Michealis Menten equation for immobilized enzymes.

The substances are partitioned in the two environments, internal solution and surrounding of the immobilized enzyme. Substrate molecule (S_b) diffuses to the micro-environment of catalytic centre from the macro-environment to convert into product (*P*) [35].

2. Study of enzyme mechanism and kinetics

2.1. Study of enzyme mechanism

To study the mechanism of enzyme catalysis, one should be able to estimate the inferences of the experimentally observed effects with respect to the change in molecular events. The basic mechanism can be predicted and validated for the enzymatic reaction under investigation by analysing the data available from a number of techniques. Combination of the basic physico-chemical properties with steric structure and quantum mechanical calculations gives solid foundation for the prediction of mechanism [36]. The kinetic data and graphical representation validate the data and play a crucial role in establishing the mechanism of the enzyme. Additionally, the study of enzyme inhibition not only unveils the mode of action of enzyme but also opens up other crucial information which is boon to the other fields of science. Some of the methods that are used for the prediction of the mechanism of the enzymatic catalysis are isotope exchange, irreversible inhibition, pH dependence, fluorescence labelled substrates, etc. [37].

2.2. Study of enzyme kinetics

To upscale the laboratory level research to an industrial scale with improved productivity and reduction in cost of the process, it is very necessary to study the kinetic models of the desired biocatalytic reactions, which facilitate process designing and intensification [38, 39]. The understanding of enzyme kinetics is also essential to design modulator that can help in designing various drugs. Transient-state kinetics, steady-state kinetics and rapid-equilibrium kinetics are the three major classes of kinetic studies [40–42]. Transient-state kinetics is applied to the very fast reactions of which mechanisms are dependent on the enzyme structure [43, 44]. The steady state of enzyme kinetics is derived with the hypothesis that each enzymatic step is in steady state and remains in steady state, even though the outer environment continuously changes in a given catalytic system [36, 45]. For rapid equilibrium kinetics, the reaction components of the step, preceding to the rate determining step, are in equilibrium with various enzyme forms like the enzyme, enzyme-substrate complex and substrate [46–48]. The initial velocity of the enzyme catalysed reaction gets affected by the modifiers [49]. The addition of single type of modifier changes the kinetics and yields two rate constants, whereas addition of two types of modifiers in two different reactions; five independent equilibrium conditions and three routes for synthesizing products are observed [50]. The chemical equilibrium is significantly affected by the external conditions such as drugs, activators, metals, toxins and pH, some of which were discussed earlier in Section 2. However, the enzyme kinetics can differ marginally in a cellular environment because of the dynamic nature of system [41].

2.2.1. *Single substrate kinetics*

In 1902, Brown found that at high concentration of sucrose, reaction follows zero-order reaction. Therefore, he proposed that the reaction is composed of two elementary steps. (1) Pre-steady-state: the formation of ES complex (2) At steady state: formation of product. This initiated the study of enzyme kinetics and derivation of the relevant rate expression (6 and 7). The next significant efforts were put forth by Michaelis and Menten in 1913 which were further improved by Briggs and Haldane [3, 7]. The reaction rate is directly proportional to $[E]$ if excess of free substrate concentration is present. At low substrate concentration, reaction follows first-order kinetics. Enzyme substrate interaction obeys the mass action law. For given $[E]$, reaction velocity increases initially with increasing substrate concentration up to certain maxima and further addition does not change the velocity anymore. Plotting and shaping of rectangular hyperbola characterize the shape of non-allosteric enzymes [8, 33]. The progress curve for the simple enzyme for catalysed reaction is represented in **Figure 1**.

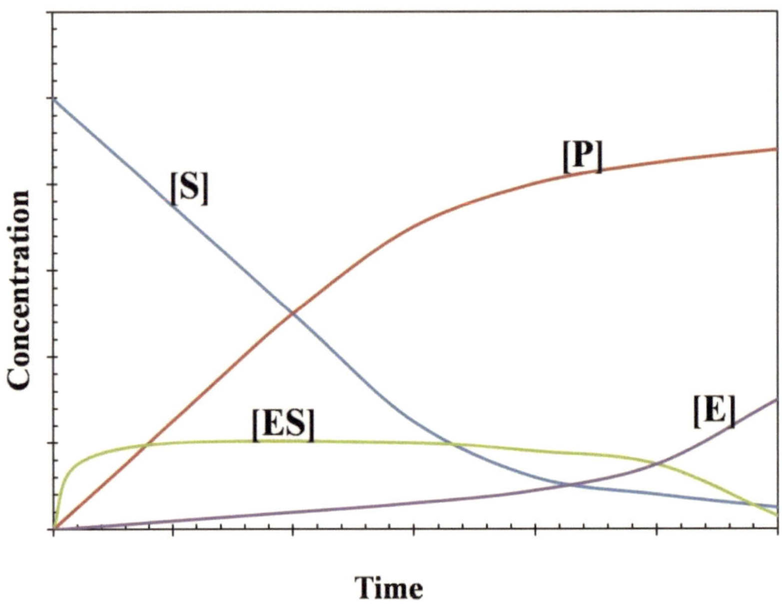

Figure 1. Progress curve for simple enzyme for catalysed reaction [7].

Unfortunately, steady-state kinetics measures are incapable of revealing the number of intermediates. It is referred as 'black box'. Perhaps, it provides phenomenal description of enzymatic behaviours. The nature of intermediate remains indeterminate. Hence, even though steady-state kinetics is not helpful for predicting the mechanism, it is useful to eliminate the proposed mechanisms [8, 51]. This developed expression relates initial rate of reaction with the substrate concentration and rate constants. The representation of reaction is as follows:

$$S + E \underset{k_{-1}}{\overset{k_1}{\rightleftarrows}} ES \overset{k_2}{\rightarrow} E + P \tag{8}$$

In the given rate expression, the first step is reversible, whereas second step became irreversible. The second step is not reversible as the sufficient amount of energy is not available to cross the barrier as the energy liberated during the reaction. To derive the rate expression for the desired single substrate enzymatic reaction, following assumptions are made [3, 4].

1. Concentration of substrate ($[S]$) is greater than the concentration of enzyme ($[E]$): concentration of substrate is not so large that all enzyme become enzyme-substrate (ES) complex form. But, it must be sufficiently large so that the concentration of substrate does not rapidly become so small that concentration of substrate becomes larger than the concentration of enzyme. Under these conditions, concentration of ES complex rapidly becomes constant, so that the rate of formation of ES complex is equal to the rate of breakdown of ES complex wherein steady-state kinetics can be applied.

2. Only initial velocities are measured so that the concentration of substrate is not depleted during the measurement (velocity measured before more than ~10% of conversion). Initial velocity measured below ~10% minimizes interference by parameters like reversible nature of reactions, inhibition of the enzymes by its product and progressive inactivation of the enzymes.

3. Concentration of the ES complex is static throughout the process which infer that:

   a. Sufficient time has passed, which is needed to build the concentration of the ES complex after mixing of E and S.

   b. Adequate time has been given to the reaction for the rate of formation of ES complex resulting into the depletion of substrate.

   c. If it is $k_2 >> k_{-1}$, ES complex breakdown occurs as rapidly as it is produced to form product and thus steady state can never be achieved. When $k_2 << k_{-1}$, the concentration of ES can build up inside the system and then,

$$\text{Rate of formation of ES complex} = \frac{d[ES]}{dt} = k_1[E][S] \tag{9}$$

$$= k_1([E_T - [ES])[S] \tag{10}$$

$$\text{Rate of breakdown of ES complex} = \frac{d[ES]}{dt} = k_{-1}[ES] + k_2[ES] \tag{11}$$

Under steady state conditions,

Rate of formation of ES complex = rate of breakdown of ES complex

$$k_1([E_T] - [ES])[S] = (k_{-1} + k_2)[ES] \tag{12}$$

Rearranging the Eq. (12)

$$\frac{([ET] - [ES])[S]}{[ES]} = \frac{k_{-1} + k_2}{k_1} = K_m \tag{13}$$

Wherein, K_m = Michealis Menten constant which is a ratio of rate constant and not an equilibrium constant.

$$[S][E_T] - [S][ES] = K_m[ES] \tag{14}$$

$$[S][ET] = (K_m + [S])[ES] \tag{15}$$

$$[S] = \frac{[ES]}{[E_T]}(K_m + [S]) \tag{16}$$

At any point of time, it is very difficult to measure the concentration of enzyme-substrate complex in the system. Hence, Eq. (16) is not useful experimentally to deduce any quantitative results. On the contrary, various experimental methods can be used for quantify the velocity (v) and maximum velocity (V_{max}). Therefore, V_{max} is the limiting value, if $[S] \to \infty$. In this case, all enzymes are bound to the substrate and $[E] = 0$, $[E_T] = [ES]$. Thus,

$$v = k_2[ES] \tag{17}$$

$$V_{max} = k_2[E_T] \tag{18}$$

$$k_2 = \frac{v}{[ES]} \tag{19}$$

$$k_2 = \frac{V_{max}}{[E_T]} \tag{20}$$

$$\frac{v}{V_{max}} = \frac{[ES]}{[E_T]} \tag{21}$$

So, after putting these substitutes, Eq. (16) becomes,

$$[S] = \frac{v}{V_{max}}(K_m + [S]) \tag{22}$$

This expression is known as Michealis Menten equation for the prediction of enzyme kinetics. This is the most widely accepted and applied enzyme kinetics to the various systems *in vitro* and *in vivo*.

Deductions derived from the Michealis Menten expression:

i. Michealis Menten constant (K_m) does not change for desired enzyme and is independent of the concentration of substrate and enzyme.

ii. Maximum velocity (V_{max}) depends on concentration of enzyme and at saturating concentration it is independent of substrate concentration.

iii. Both K_m and V_{max} may be affected by temperature, pH, immobilization and other physical constraints.

iv. A graph of velocity (v) versus concentration of substrate $[S]$ fits hyperbolic rectangular function.

v. If enzyme can catalyse more than one reaction, then K_m values can be used to measure the relative affinity towards particular substrate. If K_m increases, affinity of substrate towards enzyme decreases. In metabolic pathway studies, rate limiting step from the sequential pathway is determined with the values of K_m for that enzyme for different substrate.

vi. When $k_{-1} >> k_2$, concentration of ES complex is implied that it reaches to equilibrium with concentration of enzyme and substrate and ES is dissociated to yield the product. Under these conditions, the dissociation constant of enzyme substrate complex (K_s) is,

$$K_S = \frac{[E][S]}{[ES]} = \frac{k_{-1}}{k_1} \tag{23}$$

$$K_m = \frac{k_{-1} + k_2}{k_1} \tag{24}$$

$$K_m = \frac{k_{-1}}{k_1} = K_S \quad k_{-1} \gg k_2 \tag{25}$$

vii. When $k_2 >> k_{-1}$, the rate of dissociation is low so that the rate of product formation is high, reaction sequence becomes irreversible at both steps.

$$E + S \xrightarrow{k_1} ES \xrightarrow{k_2} E + P \tag{26}$$

As overall rate of reaction is determined by the concentration of ES complex,

- At low substrate concentration, formation of ES complex follows second-order reaction. The rate of reaction is proportional to the concentration of enzyme and substrate.

- At high substrate concentration, $[ES]$ complex remains constant and reaction rate is proportional to the $[ES]$ complex. The reaction follows the first-order reaction rate.

viii. The turnover number is defined as the rate at which an enzyme can accomplish its catalytic activity per unit time or it can also be defined as a frequency at which the ES complex leads to the formation of product. When $[S] >> K_m$, the turnover number for the enzyme is calculated. It can be determined under saturation conditions as,

$$\frac{dP}{dt} = V_{\max} = k_2[ES] = k_2[E_T] \tag{27}$$

If $[E_T]$ increases x-times, while concentration of substrate is established at the saturating levels as compared to the total concentration of enzyme, then $V_{\max}$ increases x-times. The rate of reaction is proportional to the concentration of total enzyme concentration. The reaction tends to become first order.

In saturation condition, the rate of reaction is maximum. The rate constant k_2 is denoted as k_{cat} at saturation levels and gives the value of turnover number. It ranges from 1 to 10^4 per second. Sometimes, it reaches to 10^5. Upper limit to the value of k_{cat}/K_m cannot be greater than k_2 i.e. decomposition of ES complex to E and P. The most efficient catalyst has values of k_{cat}/k_m near the controlled diffusion limits of 10^{-8}–10^{-9} m^{-1} s^{-1}. These enzymes catalyse reaction almost every time they colloid with substrate and achieve virtual catalytic perfection.

$$k_{cat} = \frac{V_{\max}}{[E_T]} \tag{28}$$

The ability of enzyme to produce a given amount of product to use in given time is changes proportionally with turnover number and total amount of enzyme present in cell. However, turn over number is measured for purified enzymes. Hence, the enzyme activity is measured as a specific activity (μmol of substrate converted per minute per mg of enzyme to form product). That is also denoted as Katal (kat). It is defined as concentration of enzyme that transforms 1 mol of substrate into product per second. As for most of the clinical disorders, activity is measured in biological fluids. It is redefined as 1 μmol of substrate to product per minute at optimal conditions (IU).

ix. When concentration of substrate is far less that the value of K_m, the product formation rate rises linearly with rise in substrate concentration and reaction becomes first order with reference to the concentration of substrate.

At low substrate concentrations, reaction rate is proportional to the total enzyme concentration. Reaction velocity at low concentration of substrate can be represented as,

$$v = k'[E][S] \tag{29}$$

$$k' = \frac{k_{cat}}{K_m} \tag{30}$$

Therefore, the rate equation must specify a second-order dependence on the concentration of substrate and total enzyme. When concentration of substrate is small there is first order dependence in concentration of total enzyme alone. When concentration of substrate is large, rate constant is k_{cat} for first order and k_{cat}/K_m for second order in which

http://dx.doi.org/10.5772/67692

K_m is the value of the concentration of substrate. When $v = \frac{1}{2}V_{max}$, k_{cat}/K_m measures total activity of the enzyme, which includes the ability of enzyme to bind with a particular substrate.

x. When $K_m = [S]$, results in the velocity that is $V_{max}/2$,

$$\frac{V_{max}}{2} = \frac{V_{max}[S]}{K_m + [S]} \tag{31}$$

$$K_m + [S] = 2[S] \tag{32}$$

Initial velocity increases with increase in V_{max} at constant concentration of the substrate and K_m and Initial velocity decreases with increase in K_m at constant concentration of the substrate and V_{max}.

- Limitation of Michealis Menten relation:

It is very tedious to quantify the velocity of the enzyme which can catalyse the multiple reactions. In other cases, the enzyme may be inactivated by its substrate or impurities in it. The enzyme that consists of several components and velocity-substrate relationship depend upon their ratios (e.g. urease). In these cases, the use of Michealis Menten equation cannot be accurately predicted.

- Representation of Michealis Menten Data:

For enzyme catalysis the data mostly fit into a rectangular hyperbola wherein the initial velocity of reaction is plotted against the concentration of substrate in the system [52, 53]. It can be represented as follows (**Figure 2**):

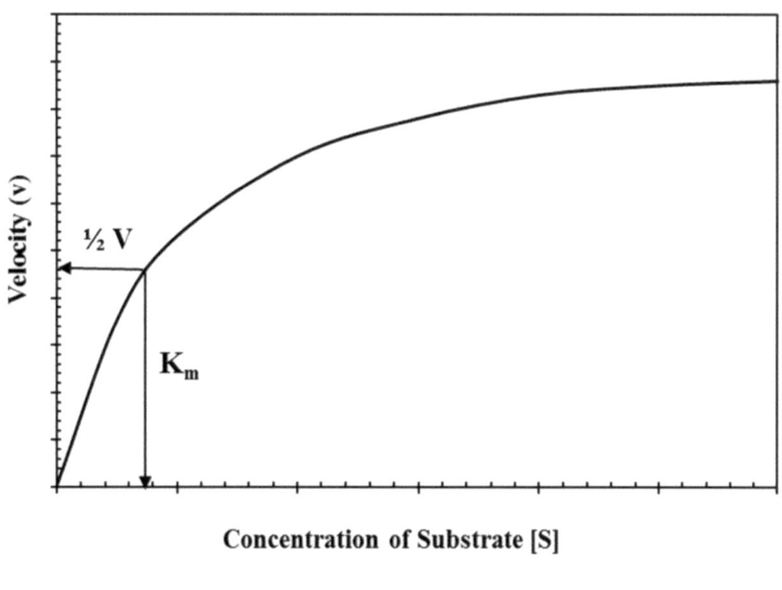

Figure 2. Initial velocity of reaction versus the concentration of substrate [4].

In the direct diagram, V_{max} and K_m are calculated by extrapolating the graph. As from the given data points, there are numerous ways to represent the data. Secondly, as the experiment is conducted at the saturation level, it is often that velocity is overlooked and therefore K_m. Also, the inhibitory effect of the excessive substrate or the limited solubility of the substrate causes the misinterpretation of the values of kinetic parameters. Practical prediction of this error prone data becomes more tedious as it becomes very difficult to predict and evaluate the experimental error from data. Hence, the need of accurate prediction of the data resulted into development of various types of reciprocal and logarithmic plots derived on the basis of Michealis Menten equation. Because straight lines are easier to evaluate, the efforts have been taken to interpret that hyperbolic data in the form of a straight line by various ways like Lineweaver Burk plot, Eadie Hofstee plot, Hanes plot, Dixon plot, etc. [47, 53].

- Lineweaver Burk plot (Double reciprocal plot)

It is based on the reciprocal of the Michealis Menten Equation and represented as,

$$\frac{1}{v} = \frac{1}{V_{max}} + \frac{K_m}{V_{max}} \frac{1}{[S]} \tag{33}$$

The plot is developed by plotting the inverse of velocity of reaction against the inverse of concentration of substrate with abscissa at $1/K_m$. This method is the most frequently used method for the prediction of kinetic parameters. The major disadvantage is the irregular distribution of data. The reciprocal action causes distortion of equally distributed substrate concentrations either in the compressed form towards the co-ordinates or over-extended in the other direction. To overcome the limitation, if the analysis is carried out at small range for uniform dispersal of data, the results do not cover the range of study at satisfactory levels. The tangible advantage of the linear graph is that the variables (v and $[S]$) are displayed on the separate co-ordinators from each other. The external disturbances and change in mechanism pattern cause the deviations from straight lines nature of the plot. This plot serves as primary step to distinguish the inhibition in the system as well as it is useful to understand the mechanism followed by the reactions [7, 8, 54].

- Eddie-Hofstee plot

Eddie-Hofstee plot is more equally placed in comparison with Lineweaver Burk plot. The equation can be obtained as,

$$v(\frac{k_m + [S]}{[S]}) = V_{max} \tag{34}$$

$$v\left(\frac{k_m}{[S]}\right) + v = V_{max} \tag{35}$$

$$v = V_{max} - K_m\left(\frac{v}{[S]}\right) \tag{36}$$

Plotting v against $v/[S]$ gives V_{max} from the ordinate intercept and $-K_m$ from the slope. This method of linearization is not only associated with distortion of error limits at higher concentrations of substrates but also there is no separation of variables from each other [3, 55].

- Hanes plot

The expression is derived by simple multiplication of the concentration of the substrate with the reciprocal Michaelis-Menten equation. It is represented as,

$$\frac{[S]}{v} = \frac{K_m}{V_{max}} + \frac{[S]}{V_{max}} \tag{37}$$

Plotting of $[S]/v$ against $[S]$ gives K_m/V_{max} (ordinate intercept) and $1/V_{max}$ (slope). The error limits are only slightly deflected at low substrate concentrations when simple linear regressions can be applied. However, substrate concentrations variable is represented on both the co-ordinates [7, 33].

- Eisenthal and Cornish-Bowden plot

Eisenthal and Cornish-Bowden have derived over a period of time in various forms. The final equation of this type of direct linearization is represented as,

$$\frac{1}{K_m} = \frac{V_{max}}{K_m v} - \frac{1}{[S]} \tag{38}$$

By entering $1/v$ against $1/[S]$, $1/K_m$ is quantified from the intercept and V_{max}/K_m from the slope of the graph. The reciprocal transformation of the equation leads to the distortion of scale. The deviation of the graph from the ideal behaviour is difficult to trace as it is overlapped by the distortion error and gives the false values of kinetic constants [52, 56, 57].

- Dixon plot

In a Dixon analysis, two types of graphs are plotted to evaluate the type of inhibition that is caused by the addition of the inhibitor in the reaction system. In 1953, the first Dixon plot of $1/v$ versus concentration of inhibitor at static substrate level was studied. Based on the nature of lines, this plot can be used to differentiate between the partial and complete inhibition (**Table 1**).

In some cases, the plots create confusion in the prediction of non-competitive and competitive inhibition. This led to the development of second Dixon plot ($[S]\nu{-}1$ is plotted versus [I]) (1972), which requires change in the turnover rate with respect to the inhibitor concentration at a static saturating concentration of the substrate. In this case, the enzyme activity follows hyperbolic decrease reaching to zero at complete inhibition. In case of the competitive inhibition, the lines plotted remain parallel to each other while in case of uncompetitive inhibition, the Dixon plot showed the presence of intersecting lines [33, 55].

Type of inhibition	Nature of Dixon plot I
Competitive inhibition	Intersect with each other at I = −Ki
Non-competitive inhibition	Intersect on abscissa at I =−Ki
Uncompetitive inhibition	Remain parallel with each other

Table 1. Analysis of Dixon plots.

There are some generalized methods that are used for determining various rate and inhibition constants along with the prediction of mechanism that is followed by the reaction under particular set of conditions.

2.2.2. *Kinetics of two substrate reaction*

The study of enzyme kinetics was initiated with the single substrate molecule which can be further applied to study the bisubstrate reactions which constitute 60% of biochemical bisubstrate reactions in nature [8].Two substrate reactions are more complex than single substrate reactions. For example, most of dehydrogenase and aminotransferase follow bisubstrate enzyme kinetics. The Cleland notation is used for the representation of higher order enzymatic reactions. According to which, the substrates are assigned with A, B, C and D letters based on the sequence of binding to the enzyme, products are assigned by letters P, Q, R and S on the basis of release sequence. Enzyme is designated with letter E, and sometimes with F, which is a slightly modified form of a stable enzyme observed in double-displacement reaction [58–60]. Hence forward, this notation is used for the representation of bisubstrate reaction. To study reaction kinetics by applying Michaelis Menten kinetic to bisubstrate enzyme reactions, concentration of one of the substrates (*B*) is kept constant, whereas another one (*A*) is varied and *vice versa*. As concentration of substrate *B* is kept arbitrarily constant, the obtained values of the kinetic parameters for concentration of substrate *A* can be erroneous. On the other hand, varying the concentration of both the parameters at a single point of time intricates large complexity in data acquisition and analysis. Bisubstrate reactions can be broadly classified into two groups, namely, sequential and double-displacement reactions. In sequential reaction mechanism, both substrates bind to the enzyme which leads to the development of transition state complex followed by the product. The binding of substrate decides further classification of sequential reaction mechanism. When binding of one substrate *A* becomes obligatory prior to other substrate *B*, then the reaction follows ordered sequential mechanism (**Figure 3**). On the other hand, the sequence of binding of the substrates to the enzyme has very less importance, the reaction undergoes random sequential reaction mechanism (**Figure 4**) [8, 33, 61].

These types of sequential mechanisms are further bifurcated into different sub-groups based on the location rate limiting step. In case of a rapid equilibrium mechanism, the rate-limiting step is chemistry of the reaction; on the other hand, step except chemistry of reaction controls the rate of reaction for a steady-state ordered mechanism. In Theorell-Chance mechanism, every step is fast except the release of the second product, *Q*, in a reaction pathway. For steady-state random mechanism, the substrate binds with the enzyme in any order [55, 62].

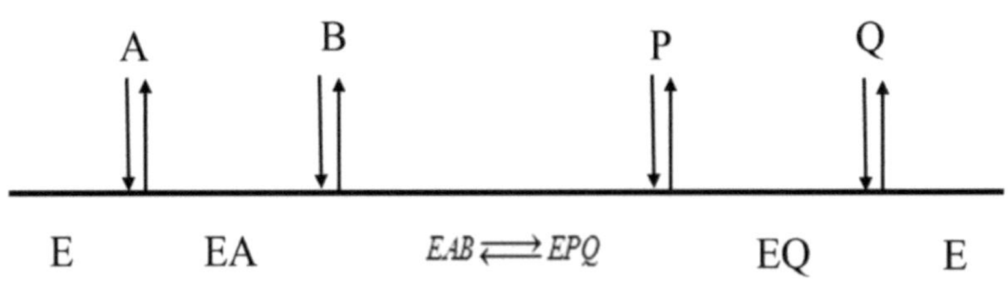

Figure 3. Ordered sequential mechanism.

As the name suggests, a double-displacement mechanism occurs when a catalytic process can proceed with binding of one of the two substrates to the enzyme (**Figure 5**). On completion of first catalytic event, first product (*P*) leaves from the active site leaving some of its portion inside the active site. After the release of first product, some chemical group of substrate *A* left behind in the catalytic site of the enzyme creating new form of stable enzyme (*F*). After binding of second substrate, the catalysis proceeds further to produce product *Q* with the regeneration of an original enzyme *E*. The plot of initial velocity and substrate concentration does not signify any noticeable change in different types of mechanisms by visual inspection and hence, not helpful in assigning any proper rate equation. However, the double-reciprocal plot, i.e. $1/v_0$ versus $1/[A]$ at the different concentrations of *B*, significantly distinct between a sequential or ping-pong mechanism [13, 63, 64].

In case of sequential mechanism (ternary complex mechanism) of reaction, the lines on the double reciprocal plot intersect with each other on the left of the *x*-axis (**Figure 6**), while that of the double-displacement mechanism (ping-pong mechanism), the double reciprocal plot is represented by the parallel lines (**Figure 7**) [58, 65, 66].

Thus, the double reciprocal plot gives clear indication of mechanism of reaction with some exception [67]. The different rate equations and plot information are presented in **Table 2**.

The rapid-equilibrium random, steady-state ordered and Theorell-Chance mechanisms are represented with the same rate expression and the primary plot or the double reciprocal plot cannot differentiate between these mechanisms from each other. They only differ at the location of rate limiting step and can be differentiated on the basis of kinetics studies such as dead-end inhibition studies, isotope effects and pre-steady-state kinetics [55, 59, 68]. The bisubstrate

Figure 4. Random sequential mechanism.

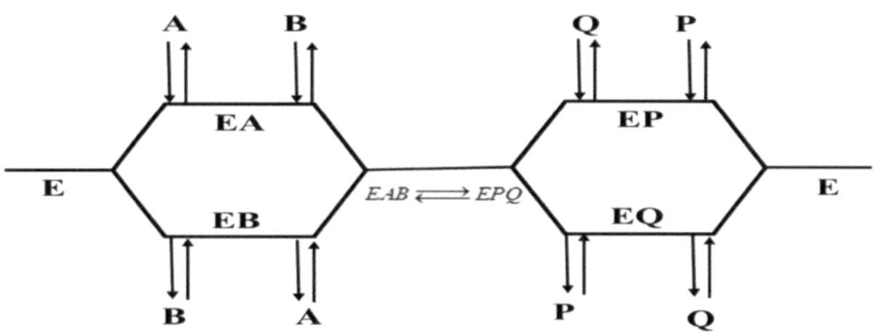

Figure 5. Double-displacement reaction mechanism.

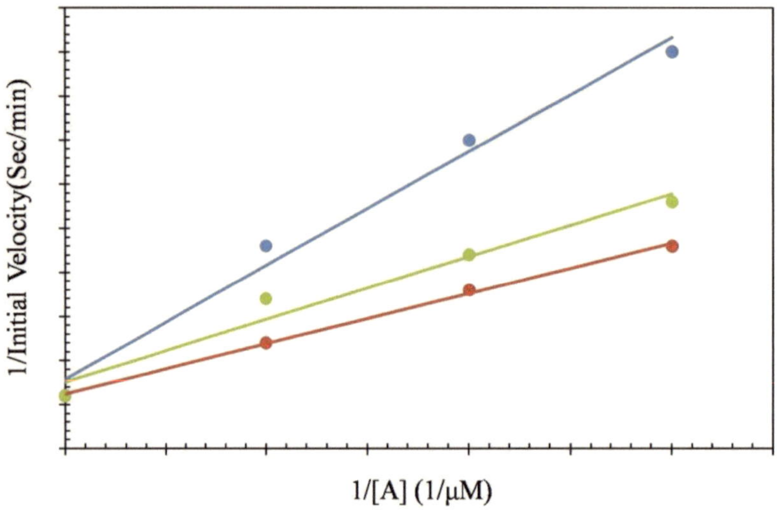

Figure 6. Double reciprocal plot for double-displacement mechanism (at different levels of *B*).

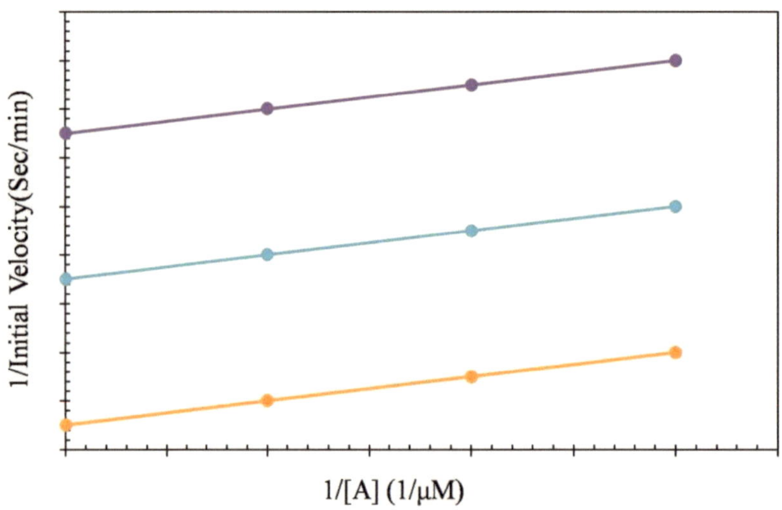

Figure 7. Double reciprocal plot for double-displacement mechanism (at different levels of *B*).

Mechanism	Rate equation	Double reciprocal plots
Rapid equilibrium random sequential mechanism	$v = \frac{VAB}{(K_{ia}+K_{b}A+K_{a}B+AB)}$	Intersecting lines
Steady state ordered sequential mechanism	$v = \frac{VAB}{(K_{ia}K_{b}+K_{b}A+K_{a}B+AB)}$	Intersecting lines
Theorell-Chance mechanism	$v = \frac{VAB}{(K_{ia}K_{b}+K_{b}A+K_{a}B+AB)}$	Intersecting lines
Rapid equilibrium ordered sequential mechanism	$v = \frac{VAB}{(K_{ia}K_{b}+K_{b}A+AB)}$	Intersecting lines
Double-displacement mechanism	$v = \frac{VAB}{(K_{a}B+K_{b}A+AB)}$	Parallel lines

Table 2. Rate expressions for different type of bisubstrate mechanisms.

kinetic study also reveals which form of enzyme can exist and which cannot, which has significant implications in the study of dead-end inhibition and therapeutic application [68, 69]. The rapid equilibrium ordered mechanism is a unique sequential mechanism as the chemistry step is the slowest step along the reaction pathway. This results in the absence of kinetic term $K_a[B]$ from the rate of expression.

$$v_0 = \frac{V_{\max}[A][B]}{K_{ia}K_b + K_b[A] + [A][B]} \tag{39}$$

This mechanism cannot be differentiated from other on the basis of double-reciprocal patterns.

$$\frac{1}{v_0} = \frac{1}{[A]}\frac{K_{ia}K_b}{V_{\max}[B]} + \frac{K_b}{V_{\max}[A]} + \frac{1}{V_{\max}} \tag{40}$$

As discussed previously, both the slope ($K_{ia}K_b/V_{max}[B]$) and intercept ($K_b/V_{max}[B]$ + $1/V_{max}$) terms reveal a dependency on the concentration of substrate B. In this instance, the graph of intercept will not differ much. By contrast, the slope value ($y = 1/[B](K_{ia}K_b/V_{max})$) when plotted as the slope versus $1/[B]$, a line that passes directly through the origin as K_a/V_{max} does not exist. However, the slope-of-the-slope value reflects free enzyme and is equivalent to $K_{ia}K_b/V_{max}$. Similarly, the implications are valid when bisubstrate reaction is studied by varying concentration B at various levels of the concentration of A, are kept constant.

This distinctive character works as a diagnostic tool to distinguish rapid-equilibrium ordered mechanism from all the other double-reciprocal plots. For various rate expressions for other mechanism, please refer the above section. The double-displacement or ping-pong mechanism, which represents the other type of bireactant mechanism, has symmetrical equation as it can be broken into two separate equations representing each half of the complete reaction [70]. For example, at saturating concentrations of substrate, the equation can be simplified as,

$$v_0 = \frac{V_{\max}[A]}{K_a + [A]} \quad \text{at saturating concentration of } B \tag{41}$$

$$v_0 = \frac{V_{\max}[B]}{K_a + [B]} \text{at saturating concentration of } A \tag{42}$$

The diagnostic double-reciprocal pattern for a double-displacement mechanism is a series of parallel lines. When varying the concentration of A, the double-reciprocal equation becomes,

$$\frac{1}{v_0} = \frac{1}{[A]}\frac{K_a}{V_{\max}} + (\frac{K_b}{[B]} + 1)\frac{1}{V_{\max}} \tag{43}$$

When the intercept of intercept is plotted, it provides the values of K_b/V_{max} and $1/V_{max}$. By contrast, the slope value from the primary plot is defined as $y = (K_a/V_{max})$.

When reaction mechanism is predicted based on the initial velocity data, one has to be sceptical as value of K_{ia} is much greater than that of K_a, rapid-equilibrium random mechanism gets degenerated into a rapid-equilibrium ordered mechanism. If the value of K_{ia} is very less, then the sequential mechanism degenerates into a ping-pong mechanism [11, 45, 55, 61].

Examples

— Ternary complex mechanism

Cinnamyl acetate is a major ingredient of food and cosmetic products as a flavouring or fragrance agent [71]. Yadav and Devendran [71] discuss transesterification of cinnamyl alcohol with vinyl acetate to produce cinnamyl acetate. The parameters were optimized to minimize the errors in the prediction of kinetic constants due to change in temperature and solvent system. The mass transfer resistance was removed to accurately estimate the kinetic constant. Hence, the reaction was conducted in toluene with 10 mg novozym 435 as a catalyst. The mole ratio was maintained at 1:2 of cinnamyl alcohol to vinyl acetate. Under these optimized conditions, the reaction kinetics was studied. The detailed experimental process is explained in the section below. The kinetics was predicted by systematically changing the concentrations of reactants on a wide range of the concentrations keeping others constant. The Lineweaver Burk plot was plotted with the initial velocities of reaction (**Figure 8**). It was observed that for given system under given set of conditions, reaction followed ternary complex mechanism as the plot showed the lines intersecting with each other [71].

The formation of dead-end complex with alcohol at saturating level of cinnamyl alcohols is observed in double reciprocal graph with linear increase in slope and intercept with rise of concentration of cinnamyl alcohol (**Figure 8**). The equation obtained with this mechanism is represented as,

$$v = \frac{V_{\max}[A][B]}{(K_{\mathrm{ia}}K_{\mathrm{mb}}(1 + ([B]/K_{\mathrm{ib}})) + K_{\mathrm{mb}}[A] + K_{ma}[B](1 + ([B]/K_{ib})) + [A][B])} \tag{44}$$

Where v is the velocity of reaction, $V_{\max}$ is the maximum velocity of reaction and $[A]$ and $[B]$ represent the initial concentration of vinyl acetate and cinnamyl alcohol, respectively. K_{ma} and

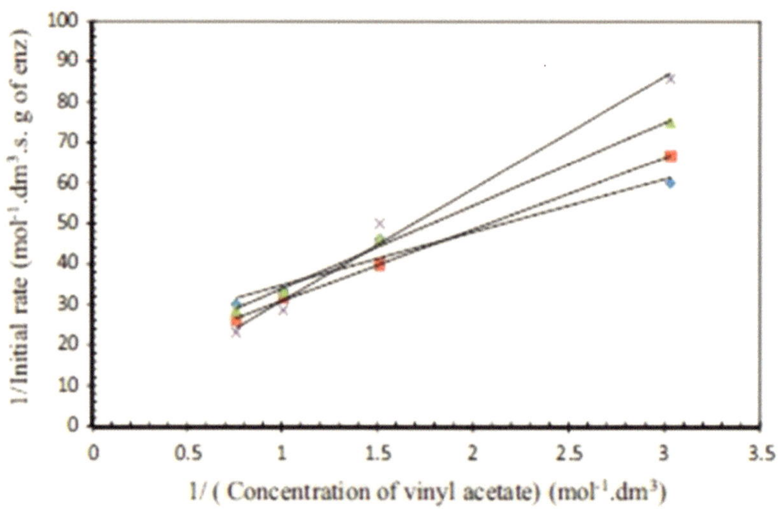

Figure 8. Lineweaver Burk plot of initial velocity versus concentration of vinyl acetate.

K_{mb} are the Michaelis constants for vinyl acetate and cinnamyl alcohol, respectively; K_{ia} and K_{ib} are the inhibition constants for vinyl acetate and cinnamyl alcohol, respectively. The kinetic constants were predicted using Polymath 5.1 software. The model was validated by plotting the simulate versus experimental rate of reactions for proposed reaction mechanism.

Ping-pong Bi-Bi mechanism

The reaction between ethyl-3-phenylpropanoate and n-butanol was studied using novozyme 435. The reaction conditions of transesterification are optimized for maximum conversion and initial rates to understand the kinetics and mechanism. There is synergism between enzyme catalysis and microwave irradiation. The reaction kinetics was studied after eliminating the external mass transfer limitation. The Lineweaver Burk plot showed that the lines are parallel to each other. The reaction follows the ping-pong bi-bi mechanism with inhibition by n-butanol as the double reciprocal graph represents the data in the form of parallel lines (**Figure 9**) [72].

With experimental data analysis and Lineweaver Burk plot, the reaction rate equation was predicted as,

$$v_0 = \frac{V_{max}[A][B]}{K_{mb}[A] + K_{ma}[B](1 + \frac{[B]}{K_i}) + [A][B]} \tag{45}$$

where K_{ma} and K_{mb} are the respective Michaelis constants for ethyl-3-phenylpropanoate and n-butanol. v_0 and V_{max} represent the respective initial and maximum velocity of the reaction. The data are validated with a parity plot.

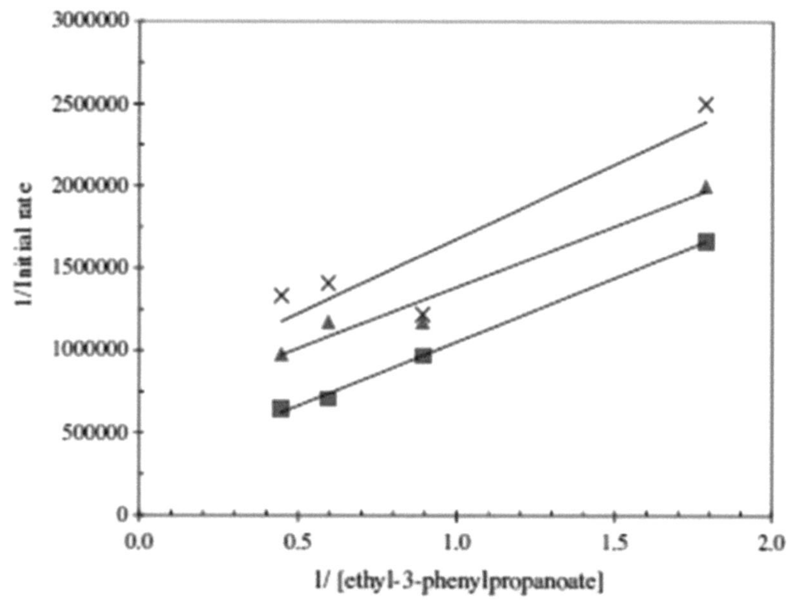

Figure 9. Lineweaver Burk plot of initial velocity versus concentration of ethyl-3-phenylpropionate.

Ordered Bi-Bi mechanism

For the production of perlauric acid, Novozyme 435 was used as catalyst in toluene solvent. The conversion and initial velocity are optimized with reference to the different parameter of reaction. Lineweaver-Burk plots indicated the formation of a ternary complex. The reaction undergoes ordered bi-bi mechanism based on which the kinetic parameters were calculated (**Figure 10**) [73].

Reusability studies indicated that there was enzyme deactivation and from the preliminary study of the deactivation, it was observed that the deactivation obeys a pseudo first-order model. The deactivation was observed at higher levels of hydrogen peroxide wherein the methionine and cysteine amino acid got oxidized due to the hydrogen peroxide. The ordered Bi-Bi mechanism showed the best fit for the given data and it is represented as,

$$v = \frac{V_{\max}[A][B]}{(K_{\text{ia}}K_{\text{mb}}(1 + ([B]/K_{\text{ib}})) + K_{\text{mb}}[A] + K_{ma}[B](1 + ([B]/K_{ib})) + [A][B])} \quad (46)$$

Some of the examples for the bisubstrate reaction are presented in **Table 3**.

2.2.3. Termolecular reaction

Termolecular reactions are the reactions wherein three reactants react simultaneously to form a desired product. They are unusual because the simultaneous collision of three molecules is a rare event. Fourth and higher-order reactions are unknown. These types of reaction are given in literature [74].

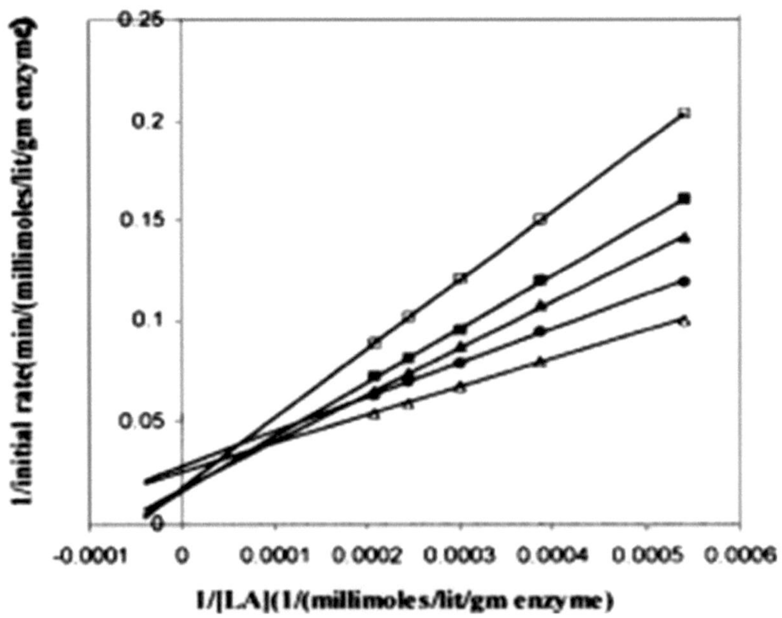

Figure 10. Lineweaver Burk plot of initial velocity versus concentration of lauric acid.

Sr. no.	Enzyme	Reaction	Kinetic model	Inhibiting molecule	Reference
1.	Mucor Miehei lipase	Esterification of pentanol with caprylic acid	Ping-pong Bi-Bi mechanism	n-Pentanol	[66]
2.	Candida antarctica lipase B	Reaction of citronelly alcohol and vinyl acetate	Ping-pong Bi-Bi mechanism	Both citronelly alcohol and vinyl acetate	[28]
3.	Candida antarctica lipase B	Reaction of n-butyl amine with benzyl acetate	Ternary complex mechanism	n-butyl amine	[77]
4.	Candida antarctica lipase B	Esterification of cinnamyl alcohol with lauric acid	Ternary complex mechanism	Cinnamyl alcohol	[78]
5.	Candida antarctica lipase B	Resolution of (RS)-methyl mandelate	Ordered Bi-Bi mechanism	S-methyl mandelate	[79]
6.	Candida antarctica lipase B	Esterification of butyl alcohol and levulinic acid	Ternary complex mechanism	Levulinic acid	[80]
7.	Candida antarctica lipase B	Reaction of Styrene with perlauric acid	Ternary complex mechanism	Lauric acid	[81]
8.	Candida antarctica lipase B	Transesterification of n-octanol with vinyl acetate	Ordered bi-bi mechanism	n-octanol	[17]
9.	Candida antarctica lipase B	Reaction of (DL)-(+/−)-3-phenyl lactic acid with vinyl acetate	Ordered bi-bi mechanism	(DL)-(+/−)-3-phenyl lactic acid	[27]
10.	Candida antarctica lipase B	Reaction of (+/−)-1-(1-naphthyl) ethanol with vinyl acetate	Ping-pong Bi-Bi mechanism	Inhibition by both substrate	[82]
11.	Candida antarctica lipase B	Hydrolysis of methyl mandelate	Ordered bi-bi mechanism	(RS)-methyl mandelate	[83]

Table 3. List of bisubstrate reaction with their mechanism and inhibition.

3. Enzyme inhibition

The inhibition of enzymes is a naturally occurring phenomenon that control and regulate body's defence and repair system along with various other essential functionalities. It also helps in regulating the optimal use of limited resources available within the cell. Naturally occurring inhibition processes are blood coagulation, blood clot dissolution, complement activation, connective tissue turnover, inflammatory reaction, etc. [57]. Hence, it is very essential to study the nature of inhibitor, mode of action, quantitative estimation of the inhibitory effect of the inhibitor on the enzyme. Such mechanistic and kinetic observations provide the information for the designing of various inhibitors. The inhibitors are usually classified into two groups namely, reversible and irreversible. In case of reversible, inhibitors bind with the enzyme with non-covalent interaction which can be reversed at any point of time by dilution or dialysis. In second class of inhibitors, i.e. irreversible inhibitors, the inhibitors bind covalently or tightly with the enzyme which cannot revert back causing permanent damage to the catalytic site [68, 75, 76]. The detailed classification of the inhibition is shown in **Figure 11**.

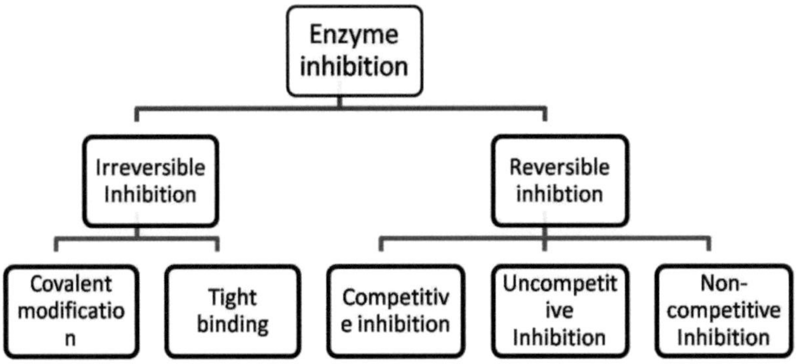

Figure 11. The classification of enzyme inhibition.

3.1. Reversible inhibition

The reversible inhibition is the class of inhibitors which bind with the enzymes and this binding can be reversed with the complete regain of enzyme activity. The reversible inhibition further is mainly classified in three categories competitive, uncompetitive and non-competitive inhibition based on its binding with direct enzyme, enzyme-substrate complex or both [77].

3.1.1. Competitive inhibition

The competitive inhibition occurs due to either binding of substrate or inhibitor to the active site. Both substrate and inhibiting molecule compete for the same active site on the enzyme leading to the reduction in the velocity of the reaction. The diverse set of substrate analogues studied for their activity in the presence of different kinds of inhibitors reveals immense information useful to understand the interaction of the active site and catalytic mechanism of the enzyme. The basic assumption in the study of this type of inhibition is that the inhibitor can only bind to the active site of the free enzyme [3, 74, 78].The Cleland notation for the competitive inhibition can be written as,

$$\begin{array}{l} E + A \underset{k2}{\overset{k1}{\rightleftarrows}} EA \overset{k2}{\rightarrow} E + P \\ + \\ I \\ \uparrow\downarrow \\ EI \end{array} \tag{47}$$

which can also be represented as,

http://dx.doi.org/10.5772/67692

$$A + E \underset{k_{-1}}{\overset{k_1}{\rightleftarrows}} EA \overset{k_2}{\rightarrow} E + P \tag{48}$$

$$E + I \overset{k_i}{\rightleftarrows} EI \tag{49}$$

The Michealis Menten equation for the competitive inhibition can be written as,

$$v = \frac{V_{\max}[S]}{[S] + K_m(1 + \frac{[I]}{K_i})} \tag{50}$$

Hence, the Lineweaver Burk equation can be derived as,

$$\frac{1}{v} = \frac{(1 + \frac{[I]}{K_i})K_m}{V_{\max}[S]} + \frac{1}{V_{\max}} \tag{51}$$

After plotting $1/v$ versus $1/[S]$, the slope and intercept will give the values of apparent K_m and V_{max} (**Figure 12**).

In this case, K_m changes with V_{max} unchanged. Increase in K_M depends upon the concentration of I. The competitive inhibitors can be applied as targeted blockers of enzyme in the pharmaceutical industry. Product can also act as a competitive inhibitor in various regulation pathways in the cells which increases the efficiency of cellular processes by eliminating the accumulation product and diverting the substrate to another pathway [79]. Both structural analogue and in some cases, unrelated compounds act as competitor inhibitors. For example, alkaline phosphatases are inhibited by the inorganic inhibitors where both substrate and inhibitor have similar affinities. Because of high selectivity, it provides many *in vivo* application opportunities, e.g. penicillinase, prostaglandin cyclooxygenase.

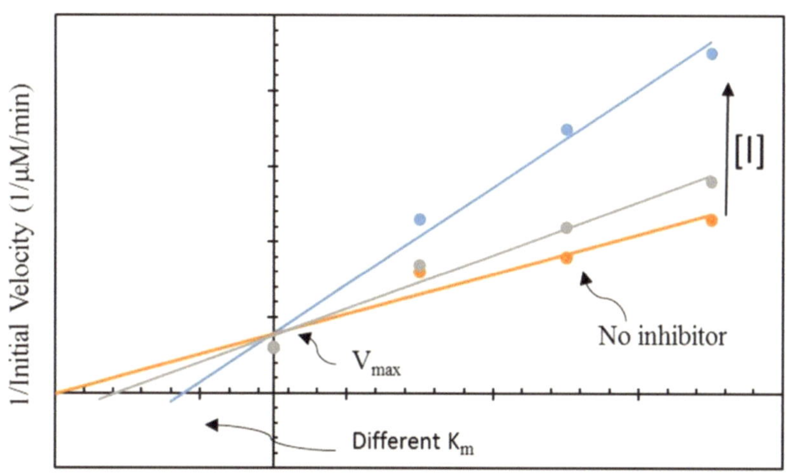

Figure 12. Graphical representation of competitive inhibition.

The reactions requiring the presence of metal ion as co-factors compete with similar ones for the catalytic site on the enzyme, e.g. Ca ions compete with Mg requiring enzyme. Similarly, Na requiring enzymes are inhibited by the Li and K ions. In double-displacement reaction mechanism, high concentration of the second substrate acts as a competitive inhibitor with reference to binding of first substrate, e.g. aminotransferase.

Adulteration of ethanol with methanol makes it unsuitable for human consumption, commonly known as denatured alcohol. Methanol is oxidized in liver and kidney to form formaldehyde and formic acid. This causes damage to retinal cells that may cause blindness which is followed by severe acidosis which lead to death. This may also lead to depression of CNS. Retardation of first step in oxidation of methanol can be achieved by administration of ethanol. The removal of methanol is done by gastric lavage, haemodialysis and administration of exogenous bicarbonate. Ethylene glycol is an anti-freezing agent used in automobiles. Ingestion in the body leads to depression of CNS and causes metabolic acidosis with severe renal damage after oxidation by alcohol dehydrogenase which is inhibited by ethanol or 4-methyl pyrazole. Kidney damage is resulted due to the deposition of oxalate crystal into convulsed tubules. Elevated anion-gap metabolic acidosis is caused by glycolic acid and lactic acid. The shift in redox potential causes the production of lactic acid instead of pyruvate. The treatment is same as that of methanol adult injection. Fomepizole drug (4-methylpyrazole) can be used in the treatment without any side effect that is caused by the ethanol. Isopropanol is major constituent in rubbing alcohols such as hand lotion and anti-freezing preparations. If ingested accidentally is oxidized and converted into acetone, a toxic non-metabolized product by alcohol dehydrogenase. It also causes depression in CNS, coma, gastritis, vomiting and haemorrhage. This can be treated by haemodialysis.

Toxicity by these substrates is done by evaluation of following serum components: Na, K, Cl, HCO_3, glucose, urea nitrogen, blood osmolality, blood gap, anion gap and metabolic acidosis along with pertinent medical history. Serum osmolal gap is difference between measured osmolality and calculated osmolality.

Serum osmolal gap = Measured osmolality-calculated osmolality

Calculated osmolality=2 $\times$ Na^+ (mM/L) + glucose (mM/L) + Urea nitrogen (mM/L)

Small rise in toxic substances increase osmolality significantly. Hence, it becomes very easy to detect the toxicity in the body based on this test.

3.1.2. Uncompetitive inhibition

Uncompetitive inhibition occurs when instead of enzyme, inhibitor binds with enzyme substrate complex to inhibit the reaction. The binding of inhibitor is possible only after the binding of substrate to the enzyme. The binding site of the inhibitor forms when the substrate binds with enzyme. This type of inhibition is rare where inhibitor binds with the enzyme substrate complex [7, 52]. This can be represented in the form of Cleland notation as,

http://dx.doi.org/10.5772/67692

$$\begin{array}{c} E + A \underset{k2}{\overset{k1}{\rightleftarrows}} EA \overset{k2}{\rightarrow} E + P \\ + \\ I \\ \uparrow\downarrow \\ EAI \end{array} \tag{52}$$

can also be represented as,

$$A + E \underset{k_{-1}}{\overset{k_1}{\rightleftarrows}} EA \overset{k_2}{\rightarrow} E + P \tag{53}$$

$$EA + I \overset{k_i}{\rightleftarrows} EAI \tag{54}$$

The Michealis Menten equation for the uncompetitive inhibition becomes,

$$v = \frac{V_{\max}[S]}{K_{\mathrm{m}} + [S](1 + \frac{[S]}{K_{\mathrm{i}}})} \tag{55}$$

The Lineweaver Burk equation can be written as,

$$\frac{1}{v} = \frac{K_{\mathrm{m}}}{V_{\max}[S]} + \frac{(1 + \frac{[I]}{K_{\mathrm{i}}})}{V_{\max}} \tag{56}$$

The plot is shown in **Figure 13**.

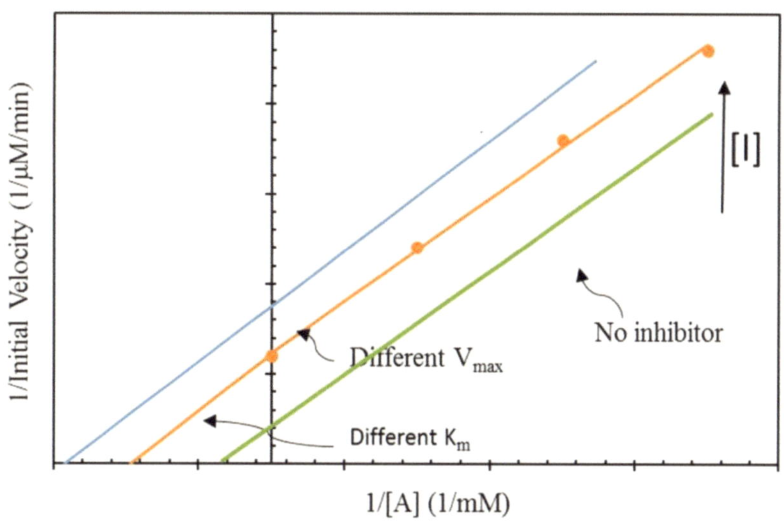

Figure 13. Graphical representation of uncompetitive inhibition.

The binding with ES complex yields parallel lines wherein both slope and intercept change. The apparent K_m and V_{max} include division by (1+ [*I*]/K_i). This type of inhibition is rarely observed in single substrate reactions. The classical example of this type of inhibition is an intestinal alkaline phosphatase which is inhibited by l-phenylalanine.

3.1.3. Non-competitive inhibition

Non-competitive inhibition is observed when inhibitor can bind with both enzymes and enzyme-substrate complex. The inhibitor bears no structural resemblance to substrate and bind to distinct site than the substrate. There is no competition between substrate and inhibitor for the active site of the enzymes. This type of inhibition cannot be overcome by increasing the substrate concentration. This may bind to enzyme or enzyme-substrate complex making both of them catalytically inactive [8, 80]. This can be denoted as,

$$A + E \underset{k_{-1}}{\overset{k_1}{\rightleftarrows}} EA \overset{k_2}{\rightarrow} E + P \tag{57}$$

$$E + I \overset{k_i}{\rightleftarrows} EI \tag{58}$$

$$EA + I \overset{k_i}{\rightleftarrows} EAI \tag{59}$$

The Cleland notation is denoted for the non-competitive inhibition.

$$E + A \underset{k2}{\overset{k1}{\rightleftarrows}} EA \overset{k2}{\rightarrow} E + P + I \rightleftarrows EI + I \rightleftarrows EAI \tag{60}$$

The double reciprocal plot for the non-competitive inhibition is represented in **Figure 14**.

Examples:

• Enzymes with sulfhydryl group that participate in maintenance of three-dimensional confirmation of the molecules are non-competitively inhibited by heavy metal ions such as silver (Ag), lead (Pb) and mercury (Hg).

$$ESH + Hg^{+2} \rightleftarrows ESHg^{+} + H^{+} \tag{61}$$

Heavy metal ions react with S-, O- and N-containing ligand. Hence, they can inhibit enzymes in the metabolic pathway (see **Table 4**).

3.2. Irreversible inhibition

Irreversible inhibition occurs, when inhibitor molecule bind with enzyme so strongly that it does not dissociate from the enzyme. This kind of inhibitor binds rapidly with the enzyme and deactivated the enzyme completely. The activity decreases exponentially with binding of the inhibitor, at saturating levels of inhibitor concentration. At lower concentration, the rate of reaction decreases linearly. The covalent modification and tight binding (K_d< 10^{-8} M) are two

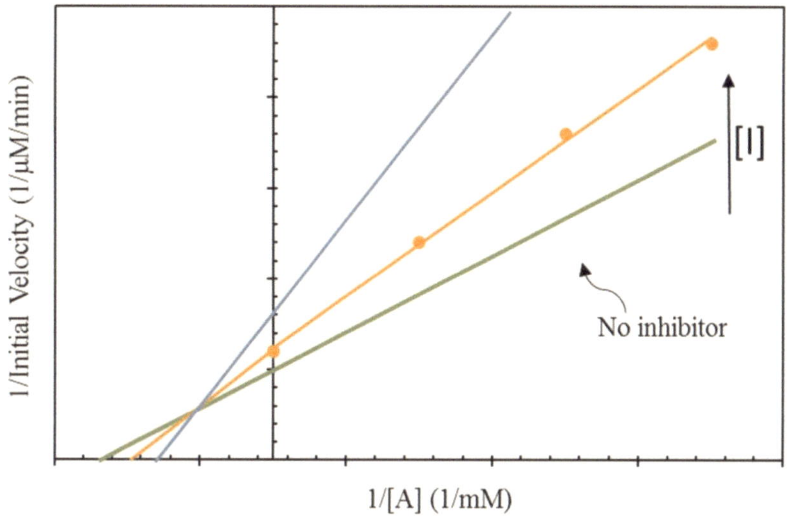

Figure 14. Graphical representation of noncompetitive inhibition.

types of irreversible inhibitions; for practical purposes, there is no dissociation of E and I. Thus, physical separation processes are ineffective in removing the irreversible inhibitor from the enzyme [33, 75, 81]. Reaction is written as,

$$E + I \underset{k_{-1}}{\overset{k_1}{\rightleftarrows}} EI \overset{k_2}{\rightarrow} EI^* \tag{62}$$

3.2.1. *Transition state analogues*

Transition state is a state in which substrate is strongly bound and interacting with enzyme for short period (in picosecond). Therefore, the transition state analogues bind tightly with transition state enzyme and cannot be easily dissociated from the enzyme [82–84, 94, 102, 103, 104, 105]. The rate expression for such type of inhibition is represented as,

$$E_t = Ee^{-\frac{k_2[I]}{[I]+k_i}} \tag{63}$$

3.2.2. *Suicide inhibition*

The suicide substrate binds covalently with an active site of the enzyme and blocks the enzyme completely. The mechanism of binding of suicide substrate with the active site of the enzyme gives the understanding of enzyme mechanism [84–86].

$$\begin{array}{c} E + I \underset{k_{-1}}{\overset{k_1}{\rightleftarrows}} EI \overset{k_2}{\rightarrow} EI^* \overset{k_3}{\rightarrow} E + I^* \\ \downarrow k_{+4} \\ E - I \end{array} \tag{64}$$

Sr. no.	Enzyme	Substrate	Inhibitor	Mechanism of action	Reference
Competitive inhibition					
1.	Cytochrome c oxidase	Oxygen	Cyanides	Competes with active site	[92]
2.	Succinate Dehydrogenase	Succinate	Malonate, oxaloacetate, oxaloacetate	Competes with active site	[93, 94]
3.	HMG-CoA reductase	HMG-CoA	Lovastatin	Competes with active site	[54]
4.	Sucrase	Sucrose	Acarbose, nojirimycin, and deoxynojirimycin	Competes with active site	[95]
5.	Influenza neuraminidase	Neuraminic acid	Oseltamivir carboxylate	Competes with active site and prevent escape of viral flu particles (Influenza)	[96, 97]
6.	Dihydropteroate synthetase	p-Aminobenzic acid	Sulfonamides	Competes with active site (bacterial infection)	[98]
7.	Dihydrofolate reductase	Folate	Methotrexate	Competes with active site and prevent rapid DNA synthesis (HIV)	[98]
8.	Xanthine oxidase	Hypoxanthine	Allopurinol	Competes with active site to block formation of uric acid (Gout, hemotological disorders, antineoplastic therapy)	[99, 100]
9.	Cytochrome P450 enzyme	Range of substrate	Biapigenin (hypericum perforatum extracts)	Competes with active site	[101]
Non-competitive inhibition					
1.	Porphobilinogen synthase and ferrocene lactase	Heme group synthesis pathway	Lead (Pb)	Bind with sulfuryl group of enzyme	[102, 103]
2.	Enzyme requiring divalent ions	Magnesium and calcium	Ethylenediamine tetracetate (EDTA)	Chelates the metallic ion	[104]
3.	Enolase (magnesium and manganese complexes	2-phosphoglycerate	Fluorides	Competes with magnesium and manganese complexes and prevent glycolysis pathway.	[105, 106]
4.	Acetylcholinesterase	Acetylcholine	cis-2-dimethylaminocyclohexanol, (hydroxyphenyl)-trimethylammonium derivatives	Binds with esterase active site	[107]
Uncompetitive inhibition					
	Inositol monophosphatase	Inositol	Lithium	Interfere with polyphosphoinositide metabolism in brain	[108, 109]

Table 4. List of examples of different types of inhibitions.

Examples:

- Enzymes containing free sulfhydryl group at active site of the enzymes. These groups react with alkylating agent such as iodoacetic acid. This cause inactivation of enzymes. Imidazole ring also showed the reactivity with the iodoacetic acid, e.g. Ribonuclease, two residue Histidine 12 and Histidine 19 loses activity by reacting with iodoacetic acid at pH 5.5 [8].
- Seryl containing hydroxyl group at active site modified by organophosphorus diisopropyl phosphate fluoridate (DFP) and it inactivates serine hydrolase. Acetylcholine esterase has two types of active sites: (1) esteric activity: site contains Seryl OH sites whose nucleophilicity has been enhanced by properly placed imidazolium group. Generally, function as base catalyst (2) site containing negative charges. The seryl hydroxyl site reacts with DFP and inactivates its activity. The inhibition of acetylcholine esterase can be reversed and enzyme can be reactivated. The reactivation by hydrolysis is very slow. Therefore, nucleophilic reagents such as hydroxyl amine, hydroximic acid and oximes can reactivate enzyme more rapidly [87].
- In activation and deactivation of cytochrome oxidase, cyanides are most rapidly acting toxic substances. It inhibits intracellular respiration and causes tissue hypoxia by binding to terminal component of mitochondrial electron transport chain. This chain uses molecular oxygen to generate energy. Cyanides severely impede mitochondrial respiration. It causes cell death and affects CNS and respiratory system. It is an oxidized form of cytochrome oxidase has high affinity for CN^- forms and loses complex with $Fe^{+2}$of porphyria which forms stable complex with negative cyanides ions. It prevents oxygen uptake, e.g. methemoglobin [88].
- Proteinase inhibitor and their applications: Mostly, proteinases are proteinaceous in nature. They are present in intracellular and extracellular fluids in the form of enzymes and peptides. Other protein inhibitors are very rare. Within the cells, it contributes in the control of blood clot, activation of complement cascades, formation and destruction of peptide hormones. Protein inhibitor combines with target proteinase. It converts then into non-dissociable form. It binds to reactive or/and active site. Specific amino acid is recognized as primary binding site. For example, elastase is inhibited by alpha-1 protein inhibitors, which inhibit serine-containing proteinase. Alpha-2 macroglobins are present in plasma of mammals. It can combine with the variety of proteinase. It only inhibits proteolytic enzymes acting towards large protein. It is very useful in metastasis of cancer. It requires remodelling of extracellular matrix (ECM). Viral proteinase offers unique target HIV protease inhibitors. Reverse transcription of RNA to dsDNA is inhibited by nucleoside analogue such as zidovudine, didanosine, zalcitabine, stavudine and lamivudine. After integration, it produces protease, i.e. aspartly protease, which is a homodimer cleaves at phenylalanine-proline bonds and is observed in mammals [88] (see **Table 5**).

3.3. Inhibition in combination with each other

The kinetics expression for mutually exclusive inhibitors is studied for single substrate system. The overall reaction velocity for the system in presence of types of inhibition is given by,

Sr. no.	Enzyme	Substrate	Inhibitor	Mechanism of action	Reference
1.	ATPase	Phosphoenolpyruvate, ATP	Rutamycin, bongkrekic acid	Tight binding inhibitions	[116]
2.	Human mast-cell tryptase	Benzamidine	CRA-001390 inhibitor	Tight binding inhibitions	[113]
3.	Adenosine deaminase	Adenosine (purine metabolism)	Deoxycoformycin, 1,6-dihydro-6-hydroxymethyl purine ribonucleoside	Tight binding inhibitions	[117]
4.	Human caspases	Proteins	Peptide based (4 amino acid sequence)	Tight binding inhibitions	[118]
5.	Serine proteases	Proteins	Phenylmethanesulfonyl fluoride	Suicide inhibition	[36]
6.	Thymidylate synthase and Cofactor methylene tetrahydrofolate	Deoxyuridine monophosphate	Fluorouracil	Suicide inhibition	[36]
7.	Chymotrypsin	Protein	Tosyl phenylalanyl chloromethylketone	Suicide inhibition	[119]
8.	Monoamine oxidase (MAO)	Norepinephrine, serotonin	Clorgyline, deprenyl, pargyline	Suicide inhibition	[120]
9.	Cyclooxygenase 2	Arachidonic acid	Vioxx, celebrex	Suicide inhibition	[121–123]

Table 5. Examples of irreversible inhibitions.

$$\frac{1}{v_{1,2,\ldots n}} = \sum_{i=1}^{n} \frac{1}{v_i} - \frac{n-1}{v_0} \tag{65}$$

Wherein, v_0 is the velocity in the absence of inhibition and 'n' is the number of inhibitors used in combination. Similarly, the kinetic expression is derived for the two substrate systems. For example, two substrate reaction with two inhibitors: ping-pong Bi-Bi mechanism.

Two inhibitors: I_1 is competitive with respect to the substrate A (binds to E) and uncompetitive with respect to the substrate B; I_2 is competitive with respect to the substrate A (binds to E) and uncompetitive with respect to the substrate B.

$$\frac{1}{v_0} = \frac{(AB + K_b A + K_m B)}{VAB} \tag{66}$$

$$\frac{1}{v_1} = \frac{(AB + K_b A + K_m B(1 + \frac{I_1}{K_{I_1}}))}{VAB} \tag{67}$$

$$\frac{1}{v_2} = \frac{(AB + K_b A + K_m B(1 + \frac{I_2}{K_{I_2}}))}{VAB} \tag{68}$$

$$\frac{1}{v_{1,2}} = \frac{(AB + K_b A + K_m B(1 + \frac{I_1}{K_{I_1}} + \frac{I_2}{K_{I_2}}))}{VAB} \tag{69}$$

Overall velocity of the reaction is given as,

$$\frac{1}{v_{1,2}} = \frac{1}{v_1} + \frac{1}{v_2} - \frac{1}{v_0} \tag{70}$$

Synergism:

$$\frac{1}{v_{1,2}} \rangle \frac{1}{v_1} + \frac{1}{v_2} - \frac{1}{v_0} \tag{71}$$

Antagonism:

$$\frac{1}{v_{1,2}} \langle \frac{1}{v_1} + \frac{1}{v_2} - \frac{1}{v_0} \tag{72}$$

The expression is useful to understand the effect of inhibitors on each other's activity when used in combination. The overall velocity of the reaction is independent of mechanism of reaction, type of inhibition and number substrate used in the study [89].

4. Analytical aspects of enzyme inhibition

4.1. Assay conditions for single substrate enzyme kinetics

To estimate the reaction rate, it becomes very essential to quantify accurately the change in the concentration of either substrate or product or both during the progress of reaction. If the substrate and product generate different signals, the spectroscopic techniques like UV-visible spectroscopy and fluorescence spectroscopy can be applied to estimate these parameters [90]. If there is no significant difference between the products and substrate signal, then the better alternative is a discontinuous assay which measures the rates by intermittent sampling. This sampling can be done with removal of aliquots of a reaction mixture or sampling from the batch reactor. These reaction samples represent the time points of a reaction, and their respective concentration of product formed and substrate consumed. In some cases, labelling the molecules with chromophore or the fluorescent dye can serve as a better option. The estimation of amount of an ATP can be quantified by determining the amount of Pi present by the Malachite Green assay with absorbance at 660 nm [91]. Chromatography techniques can separate the substrate from the product which is then estimated with the spectroscopic method in continuous or discontinuous manner. In depth understanding, the enzyme mechanism is studied by radiolabelling the substrate, which is tracked for the transition of substrate through intermediate state to the product during the reaction progress [92]. The crystallographic structures of separated intermediate complexes and the various states of enzymes during the reaction progress reveal most of the mechanistic aspect of the reaction. With the advent of instrumentation technology, the product can be separated or analysed by thin-layer chromatography, separation techniques such as high-performance liquid chromatography, electrophoresis and gel filtration. The high-performance liquid chromatography is the most applied system as it gives the separation and quantification of substrate as well as product in real time.

Recent studies showed that the mass spectrometric (MS) techniques are applied to study the conversion of substrate into product and their behaviour in the presence of modulators. Matrix-assisted laser desorption ionization-time of flight-mass spectroscopy (MALDI-TOF-MS) is proved as more effective technique when compared to other mass spectrometric techniques, such as electronspray ionization MS, as it showed negligible interference due to the presence of buffers and reagents. MALDI-TOF-MS can quantify the ratio of substrate to product form. Greis et al. studied the phosphorylation catalysed by kinases using this technique for accurate prediction of the kinetics of reaction [93]. This technique has been used in combination with various chromatographic techniques such as capillary isoelectric focusing, frontal affinity chromatography and size exclusion chromatography to analyse the inhibition of the enzyme in direct or indirect manner [93–96]. This method is free of laborious work of labelling or derivatization and can be done in short period of time.

To estimate the enzyme inhibition and relative kinetic parameters, the stepwise process has to follow with the aim of minimum human error, which is discussed in the following sections.

4.1.1. *Experimental measurement of* V_{max}, k_{cat}, *and* K_m

As discussed in the above sections, the quantification of the initial velocity, the region of the curve wherein velocity does not change with time (up to 10% of the conversion), is of prime importance to estimate the reaction kinetics under the influence of other factors such as product inhibition. This region varies with the type of enzyme, nature of substrate and medium conditions and type of modulator. Hence, for any given system, it cannot be pre-determined. The instrument capacity needs, therefore, has to be very precise and effective to detect the signal generated from the build-up of product and to plot it. As the measurement needs to be done at a very constrained range, it may severely compromise on measurements because of the poor instrument capacity.

The linear range for the product of an instrument can be estimated by measuring the response of product at various known concentration and preparing the standard curve. Before the start of experimentation for evaluating enzyme kinetics, it is very essential to optimize the conditions of the reactions so that the enzyme remains stable throughout the reaction progress. If the maximum plateau value of product formed does not reach the same for all levels of tested enzyme, it is likely due to the enzyme instability over time [10, 53].

Conditions required for measuring initial velocity of an enzyme reaction:

- Equilibrate and maintain all the reaction reagents to the optimum temperature.
- Keep the reaction medium condition static (previously optimized or derived from literature) for the optimum performance of enzyme.
- Replicate the same reaction for minimum three times to avoid the manual errors.
- Proper analytical technique is developed for the estimation of product formed or substrate consumed below 10%.
- Record the signal of the reaction mass at time zero for the background.

For kinase assays, generally, the background can be recorded without the enzyme or the substrate. The condition with highest background level should be used for further estimation of the parameters. In case of EDTA, background is taken with control without EDTA during validation of a kinase assay. Once the assay has been validated, if the background measured with EDTA is the same as that of no enzyme and no substrate control, then EDTA could be used.

Measurement of K_m and V_{max}

After primary experiments, wherein the initial velocity conditions have been established and change in velocity with respect to time has been estimated, the substrate concentration should be varied. These data will generate a saturation curve and can be used for the determination of K_m and V_{max} values. The Michaelis Menten kinetic model shows that the $K_m = [S]$ at $V_{max}/2$. In order for competitive inhibitors in which the substrate competes, the substrate concentration should be maintained around or below the K_m. At the substrate concentration higher than the K_m, the identification of competitive inhibitors itself becomes more difficult.

For kinase assays, two set of experiments are carried out. In first experiment, the K_m for ATP is determined at saturating levels of the substrate concentration. Subsequent reactions need to be conducted with optimum ATP concentration (around or below the K_m value). However, simultaneous determination of K_m for ATP and specific substrate gives more accurate estimation of kinetic constants with maximum information and any potential co-operativity between substrate and ATP [91]. To achieve steady state, ratio of substrate to enzyme is maintained in between 100 and one million.

How to measure K_m

- Minimum eight experiments should be conducted with various substrate concentrations from 0.2 to 5.0 K_m (from the literature, if available).
- The initial estimate is established by using the wide range of six substrate concentrations when no data are available in the literature. This is then followed by the first step.
- For bisubstrate reaction, the K_m of the substrate of interest is measured in replicates by keeping the other substrate at saturating concentrations and vice-versa.
- The data are fitted in rectangular hyperbola function using non-linear regression analysis.

The reaction product is measured at various times for eight different levels of substrate. The product generated (*Y*-axis) is plotted against the reaction time (*X*-axis) with each curve of different concentration of substrate. The slope of the line is the initial velocity (v) of the reaction curve.

These resulting initial velocities (*Y*-axis) are plotted against the concentration of substrate (*X*-axis) and fitted with a non-linear regression analysis in a rectangular hyperbola (**Figure 2**). The K_m is one-half the maximum velocity determined under saturating substrate concentrations.

The linear plots are generated to determine the kinetic constants besides fitting the data by non-linear regression such as Lineweaver–Burk plot wherein the reaction rates ($1/v_o$) are plotted against reciprocal substrate concentrations ($1/[S]$) with y-intercept equivalent to $1/V_{max}$; and the slope of K_m/V_{max}. Various linearization plots such as Eisenthal–Cornish–Bowden plot, Dixon plot and Hanes plot can also be used to estimate kinetic and inhibition constant (see Section of Enzyme Kinetics).

Optimization experiments

Published literature information should be used in selecting these factors. The following parameters should be optimized prior to kinetic study so that the enzyme remains stable throughout the experiments.

- Divalent cations, for example Ca^{2+}, Mg^{2+}, Mn^{2+}
- Salts, for example NaCl, KCl
- Reducing agents, such as β-ME, DTT and glutathione
- Bovine serum albumin
- Detergents, such as Triton, CHAPS and DMSO
- Buffer source, for example HEPES versus acetate buffer
- pH

The assay conditions should be validated to avoid the loss of enzyme activity due to the wrong selection of buffer, pH or temperature [97].

4.2. Assay conditions for bisubstrate reaction

The plotting of Lineweaver Burk plot is more important to distinguish between a sequential reaction mechanism (plotted lines intersect each other) and a double-displacement or 'ping-pong' kinetic mechanism (plotted lines remain parallel). In some cases, these graphs gave preliminary evidences in different types of sequential kinetic constants. In the process of predicting kinetic and inhibition constant for the bisubstrate reaction, it is compulsory to understand the mechanism of the reaction to predict the appropriate rate equations. The data were generated by plotting the reciprocals of the initial rate of product formation versus $[A]$ at the differing $[B]$ and vice-versa. Most importantly, the values of kinetic parameters such as K_m and V_{max} change for substrate A with the change in substrate B concentration [12, 41, 63]. In case of random sequential reaction, the double reciprocal plot of rate expressions gives the apparent value of slopes and intercept which include the term for concentration B.

$$\frac{1}{v_0} = \frac{1}{[A]}\left(\frac{K_{ia}K_b}{V_{max}[B]} + \frac{K_b}{V_{max}}\right) + \frac{K_a}{V_{max}[B]} + \frac{1}{V_{max}} \quad [B] = \text{constant} \tag{73}$$

The double reciprocal plot gives the confirmation of the mechanism followed by the reaction. But, as the slope $\left(\frac{K_{ia}K_b}{V_{max}[B]} + \frac{K_b}{V_{max}}\right)$ and intercept $\left(\frac{K_a}{V_{max}[B]} + \frac{1}{V_{max}}\right)$ terms involve the concentration term, it signifies the dependency on the concentration B. Also, it is noted that the slope and intercept are the linear equations and can be plotted against $1/[B]$. When the intercept is plotted against $1/[B]$, it gives the values of $1/V_{max}$ and K_a/V_{max} and this plot is called as intercept of intercept (**Figure 15**). At different concentration of substrates various different forms of

http://dx.doi.org/10.5772/67692

enzyme exist in the system. To obtain the intercept of intercept of the secondary plot, concentration of both A and B are maintained at higher concentrations which give rise to higher concentration of EAB Form.

To calculate the K_a/V_{max} of desired enzymatic reaction, the EA form of enzyme should be predominant. Hence, the slope of the intercept is calculated by maintaining the concentration of A at high levels while concentration of B is kept low. The plot of slope versus $1/[B]$ will give the values of K_b and K_{ia}, which are measured graphically (**Figure 16**).

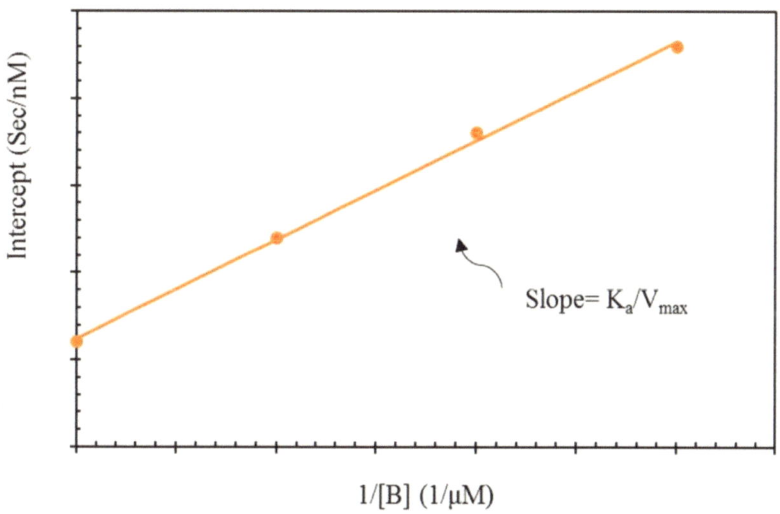

Figure 15. Secondary plot of intercept versus reciprocal of concentration of B.

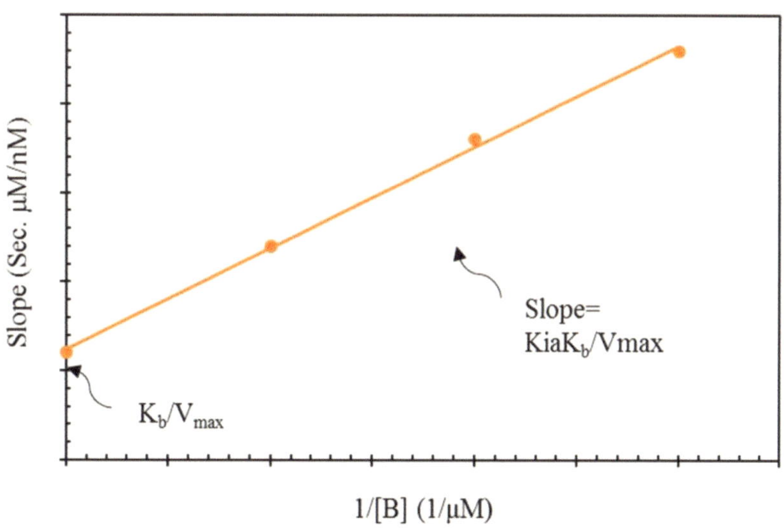

Figure 16. Secondary plot of slope versus reciprocal of concentration of B.

Similarly, same kind of analysis can be performed to find out the V_{max} and K_b/V_{max} by keeping high levels of *B* and *A*. The plot of intercept of intercept, i.e. intercept versus 1/[*A*], will give the values of $1/V_{max}$ and K_b/V_{max}. The slope plotted against the 1/[*B*] was studied for deriving the values of K_a and K_{ib} by analysing at high levels of *B* and low concentrations of *A* (building up the concentration of EB). These secondary plots elaborate various aspects of enzyme kinetics some of which are discussed in the kinetics section.

4.3. Software used in kinetic data analysis

The data-fitting process can be accomplished by using a software program that provides non-linear regression-fitting capability. Various programs such as Kaleidagraph are developed wherein the user put the experimental data along with probable fitting function for the prediction of rate constants. Some other software such as dynafit, mathematica, sigmaplot and prism are applied for the estimation of rate expression and effect of inhibition or activator on the enzyme [98]. Alternatively, some of the programs like Enzfitter have a predefined library of equations which can also be used for the prediction kinetics of enzyme. While using the software for data prediction, one should be able to differentiate in errors added due to the lack of fit and pure error as they contribute as a source of error [99]. Although both sources of error normally contribute to the sum of squares of deviations from a model, they can be separated. The inconsistencies between replicate observations are unaffected by the choice of model and thus allow calculations of how much of the total sum of squares is due to the pure error, and from this one can calculate the contribution of lack of fit [57, 80, 100].

The basic assumption during the development of related software is that the rate of reaction is zero in the absence of substrate. The buffer solution used for the reaction should be selected so that the *pKa* is not greater than 1 unit of pH that the desired one. Generally, the operational pH should be less than *pKa*. This will maintain the desired configuration required for the optimal activity of the enzyme. Also, the used buffer system should not react with the enzyme used for the catalysis or the buffer interfering with the analysis method [10, 52, 88]. Primary assumptions that any developed software have done are as follows:

1. Enzyme to substrate interaction ratio is maintained as 1:1.

2. Inhibitor compete with enzyme active site/or active groups at enzyme allosteric site to form enzyme-inhibitor-substrate and/or enzyme-inhibitor complex.

3. The loss of active free energy decides the formation of enzyme-inhibitor-substrate complex.

4. Reaction is reversible at every stage of the interaction between enzyme-substrate, enzyme inhibitor and enzyme-substrate-inhibitor interaction.

5. The binding of inhibitor or substrate with enzyme is expressed in terms of kinetic constants of a catalytic reaction.

6. The physiological conditions such as pH, temperature, concentration of reactants and reaction period are kept constant while determining the kinetic constants.

7. The overall enzyme reaction rates and mode of inhibition are depended upon the intermolecular forces between enzyme subunits, substrate or inhibitor.

While deriving the kinetics and mechanistic parameters in appropriate conditions at various points in iterative manner, various well established programs make some primary assumption during the development of the rate expression. For example, the popular program sigmaplot can fit Michaelis-Menten data very easily, but if used in its default state it incorporates assumptions that: (1) the errors in the observed rates are subjected to a normal (Gaussian) distribution and substrate concentration are exactly known; and (2) all of the rates have the same standard deviation and are independent of each other as magnitude of error in one rate measurement do not affect the measurement of any other rate.

Some of the assumptions like no deviation in standard deviation cause the problem that need to be eliminated from the model. Very less number of software databases allows doing so. Hence, the selection of the software to process the data has to be done very meticulously. It is always preferable to derive the rate expression and related constant manually with desired set of assumptions for error free fitting of data which may be crossed checked with the computerized data fitting. The computer-based data fitting serves as boon to researchers when more complex and complicated rate expression are observed (e.g. bisubstrate reaction with inhibition and termolecular reactions). In the gist, software-based studies of kinetics become essential part of the system [2].

5. Design of inhibitors and activators

The use of inhibitors usually has two outputs when used against the target enzyme as drug. First, the inhibition of enzyme leads to the accumulation of substrate that it meant to process and second wherein the concentration of metabolite decrease as the enzyme is inactivated by the presence of inhibitors. The inhibition of enzyme results into accumulation of the substrate or the metabolite which then can be detected by the various analytical methods [101]. The reduction of the metabolite due to inhibition of the selective enzyme from the pathway leads to the blocking of further activation sequence and therefore the consequences are resulted due to the same. For example, allopurinol inhibits the action of xanthine oxidase in the treatment of gout disease while inhibition of co-enzymes in the pathway leads to the prevention of undesired cancerous growth. The combination of two inhibitors may work in synergism or may not. It has to be studied thoroughly before using it for the human treatment. Most of the anti-cancerous and antibacterial drugs (e.g. cotrimoxazole) are used in synergism to fight against the targeted disease. Some of the inhibitor (e.g. 5-fluorouracil and doxorubicin) acts on the enzyme-co-enzyme complexes and DNA-enzyme complexes to form dead-end complex instead of binding to the enzyme [85, 102, 103]. Other specific inhibitors can be described in several forms including:

1. Coenzyme inhibitors: e.g. cyanide, hydrazine and hydroxylamine that inhibit pyridoxal phosphate, and, dicumarol that is a competitive antagonist for vitamin K.
2. Inhibitors of specific ion cofactor: e.g. fluoride that chelates Mg^{2+} of enolase enzyme.

3. Prosthetic group inhibitors: e.g. cyanide that inhibits the heme-prosthetic group of cytochrome oxidase.

4. Apoenzyme inhibitors that the molecules which attack the apoenzyme component of the holoenzyme and

5. Physiological modulators of reaction such as pH and temperature denature the enzyme and disturb the catalytic activity drastically.

5.1. Rational approach to the design of enzyme inhibitors

5.1.1. General discovery method for lead inhibitor

A lead inhibitor with low potency and selectivity acts as a base scaffold molecule for designing the highly potent and selective inhibitors. The pool of initial molecule (5000–10,000 related compounds) to screen as a lead compound can be decided by two ways. Either by studying the existing drug molecules creating side effects by reacting with proposed enzyme or from the compounds having low potency in pharmacological, anti-bacterial, anti-parasitic, or anti-viral effects in the screening experiments against the proposed target enzyme.

The selection of molecules to study for primary screening of lead compound can be decided on the basis of prior knowledge of (1) modification of the structural scaffold of a lead inhibitor, (2) information of substrate enzyme and inhibitor interactions and their catalytic mechanism and (3) use of molecular modelling software. Among this pool, the compounds are then studied for their inhibitory action. The molecules showing dissociation constant less than 1 μM against the desired enzyme are designated as lead compound. These molecules are then studied with various permutation and combination of approaches to enhance their activity against the selective target. Some of the approaches that can be used for the redesigning of the molecules are replacing existing functional group, enhancing the hydrophobic characters, removal of steric hindrances as well as stereogenic centres. While applying the approaches, it needs to understand the solubility of the drug molecule and its adsorption within the body. After modifications, if the selected molecules are designed with the intent of pharmaceutical applications, then they tested and studied on the animal model. After clearing the animal studies, it the molecules fail in the clinical trials, it requires further structural and manipulation of to improve in vivo profiling [55].

5.1.1.1. Process of development of lead inhibitor

The complementary structure of the molecule to the active site of the enzyme forms non-covalent bonds such as hydrogen bonds, ionic bonds and hydrophobic interaction to bind with the enzyme. The structure specific requirements of the molecules are maintained by replacing the isosteric replacement of groups. Isosteric replacement is done by replacing the atoms or the group with similar electronic and steric configurations which can be the elements within the same vertical group of the periodic table. The replacement of groups is done on the basis of the outer electronic configuration as well as by the ring equivalents without disturbing the spatial arrangement of the lead molecule. In the process of improving potency of the lead

http://dx.doi.org/10.5772/67692

molecule, the atoms or the groups can be replaced by the non-specific substitute which may increase the hydrophobicity or hydrophilicity of the molecule to enhance the membrane penetration or the solubility. It is necessary to understand the dissociation state of the functional group at the physiological pH during the process of modification of the lead compound [68, 102].

The strategy of replacement leads compound group with the analogous group (=F, =OH, = NH_2 and =CH_3) and is successfully applied to design the hypoglycemic agents, inhibitory compounds such as aminopterin, 6-mercaptopurine, 6-thioguanine, etc. The cis- and trans-configurational similarities can also serve as the way to design better lead compound. For example, replacement of ester functionality of the anaesthetics prolongs their action in body which led to the group of drugs such as anticholinergic spasmolytics and anti-depressant drugs. The pyridyl containing compound has more hydrophilicity as compared to the benzene because of the presence of p-electron deficiency which improve the reactivity of most of the drugs. This concept has already applied to the development of tricyclic, anti-histaminic and neuroleptic (major tranquilizing) drugs as well as in sulfide-containing groups [69, 104].

Ring replacement of =N< by =HC< and its subsequent modification to >C— have resulted in a variety of useful drugs. This is seen in the development of psychotherapeutics (chlorprothixene), the anti-inflammatory drug (sulindac). Several other modifications such as reversal of group, ring opening and closure and groups with similar polar effect can lead to the development of more potent lead compound [97].

5.2. Design from a knowledge of the catalytic mechanism

In most of the examples, the prior data on the enzyme active site are not available. The new lead compounds are designed based on the knowledge of substrate and catalytic mechanism followed by the reaction. The design of the molecules and their interaction with predicted structure of enzymes can also serve as better option to design the lead compound wherein molecular modelling plays an essential role [105–107]. The enzymes about which no data are available, structure of substance acts as a guiding line. For example, before discovery of the structural data of the angiotensin-converting enzyme (ACE), the inhibitors are decided based on the knowledge of substrate and poison (snake venom) and its similarity with the carboxypeptidase A enzyme. The discovery that D-benzylsuccinic acid was a potent inhibitor of carboxypeptidase A was challenged to the molecule similar to snake venom with the attachment of terminal. Ala-Pro sequence leads to the development of inhibitor methyl glutamyl proline. For the further improvements to enhance similarity between the substrates, increasing the solubility of the compound and reduction of side effect caused the drug molecule, it undergoes a series of changes to develop highly potent and efficient drug [78, 94, 108, 109].

Nowadays, application of computer-based molecular modelling techniques has gained the relative importance over the conventional techniques. The highly resolved crystal structure of the enzyme and enzyme-inhibitor complex reveals the change structural configurations, the spatial arrangement of the inhibitor within the active site of the enzyme along with the binding site of the enzyme and inhibitors [110–112]. On the basis of analysis of these data, the molecular modelling software showed the best match for the given enzyme from the large pool for chemical molecules in its database. These might be the completely new structures which need to explore for their catalytic activity. It can also suggest the best modification that can serve for the developing potent inhibitor from the existing drugs. By using data from various sources such as structure of inhibitors from the various crystallographic databases and crystal structures of enzymes, enzyme-substrate and enzyme-inhibitor complex from the various PBD databases can be imported and analysed for the prediction of inhibition capacities of the various molecules [113]. Various *ab initio* and drawing softwares are used for the prediction of the structure of the inhibitors which are not present in the database. There spatial arrangement and group and chain positions are studied and model is minimized to confirm with lowest energy status. This is then docked inside the active site of the enzyme and positioned to have a maximum favourable interaction and tried for the energy minimization of the system. Thus, the finalized inhibitor then modified by various methods discussed earlier (Section 7.1) and analysed again in the software [77, 84, 114]. The use of high-resolution NMR studies or by homology modelling (minimum 30%) can give the relative desired protein structure. This can be used for the further docking experiments against the library of the inhibitors using computerized modelling software. The prediction of the active site on the surface of enzyme (pharmacophore) is done by docking the multiple inhibitor molecules. Because of the basic assumption of the rigid structure of the protein, the predicted inhibitor may differ in activity with the dynamic enzymes molecule [115, 116]. Thus, the developed lead compounds then can be applied in pharmaceutical, agrichemical industries and in the study of metabolic flux, genetic engineering and molecular biology for the betterment of life.

6. Enzyme activators

Activators are the molecules that enhance the rate of reaction by multi-fold which may or may not get utilized during the catalytic reaction progress. Depending upon their interaction with the enzymes, they are classified into various groups. The prosthetic groups generally bind covalently with the enzyme and act as a source of hydrogen or phosphorus group. Prosthetic

Sr. no.	Reaction type	Activators used
1	H-transfer reaction (vitamin-based activators)	(Vitamin-based activators) NAD, NADP, FMN, FAD
2	H-transfer (non-vitamin-based activators)	Lipoic acid, biopterin, coenzyme Q
3	Group transfer (vitamin based activators)	CoA, CoASH, TPP, pyridoxal phosphates, tetrahydrofolic acid (FH4), biotin, cobamide coenzyme, vitamin K, vitamins C
4	Group transfer	ATP, CDP, VDP, phosphoadenosine phosphosulfate
5	Transfer (non-vitamin-based activators)	(PAPs), heme-group, s-adenosylmethionine
6	Metals	Mg^{2+}—ATP utilizing enzymes Ca^{2+}—requires in functioning of muscle and blood clotting, essential for formation of cAMP Fe^{2+} and Fe^{3+}—heme-nonheme containing enzymes Cu^{1+} and Cu^{2+}—electron transport systems Zn^{2+}—present in dehydrogenase, polymerase Mo^{6+}—required for functioning of xanthanine oxidase

Table 6. Different types of enzyme activator and their examples.

group forms part of active group undergoes changes during reaction. The co-enzymes do not bind covalently with enzymes and very essential for the activity of the enzymes. Co-enzymes get chemically altered temporarily during the reactions which get reconverted into its original form by the same or another enzyme. The enzymes associated co-enzyme usually catalyses bisubstrate reactions [3, 4]. The co-factors are complex organic molecules or metal ions. Some of the activators are listed in **Table 6**.

7. New developments in enzyme inhibition

In the 1970s and 1980s, the stopped flow and continuous flow experiments brought about the major experimental advancement in enzyme kinetics field by allowing the process to reach steady state. The complicated enzyme kinetics and intermediates formed during the process are decoded using rapid sample mixing and high time resolution monitoring [50, 117, 118]. The application of single-molecule fluorescence imaging at room temperature for single-molecule manipulation has been applied to study the single molecule enzyme in 1990s [119, 120]. It gave a major breakthrough in the field of enzyme kinetics. Single-molecule enzymology has elaborated on molecular level insights on molecular motors and nucleic acid enzymes. In single molecule experiments, wherein the time pattern of enzymatic turnovers was analysed, it is observed that the catalytic activity is not consistent over time. The reaction follows parallel reaction pathway through different enzymatic conformation with different rate constants and rate constant for the rate-limiting step might become a function of time. This phenomenon is called dynamic disorder [120–123]. Single molecule experiments have shown that the stochastic waiting time of an enzymatic reaction exhibits a distribution of an exponential rise followed by an exponential decay [124, 125]. For a single molecule with slowly interconverting conformational states with different k_{cat} and K_m, it follows that

$$\frac{1}{\langle T \rangle} = \frac{k_{cat}[S]}{[S] + K_m} \tag{74}$$

Where $\langle T \rangle$ is the mean of the stochastic waiting time and the overbars denote the weighted averages of k_{cat} and K_m of different conformational state.

In case of reversible reactions, Hill showed that forward and backward reaction fluxes (J^+ and J^-, respectively) can be related to the chemical potential difference, $\Delta\mu$, between product and the substrate:

$$\Delta\mu = -\mathrm{k}T\ln\left(\frac{J+}{J-}\right) \tag{75}$$

where k is the Boltzmann constant and T is the absolute temperature. It also serves as bridge between the thermodynamic driving force and enzymatic kinetics.

A most commonly used fluorogenic substrate, resorufin â-D-galactopyranoside (RGP), is also hydrolysed by the enzyme, yielding a fluorescent product, resorufin. The fluorescent burst released during the process is used for the prediction of the single enzyme kinetics. The location, brightness and spectral properties of the fluorophore-labelled molecule affect the measurement of the signal significantly. This resolution directly yields kinetic information, for example, when measuring: (a) diffusion coefficients with respect to changes in the position of molecule (b) time-lapse experiments to analyse the systematic change in the brightness of fluorogenic substrate turnover with time or (c) temporal changes in molecule conformation or configuration when observing changes in FRET between a judiciously placed donor-acceptor pair.

The microscopic validity of chemical master equation (CME) has also been tested and verified by molecular dynamics for dilute chemical systems and using Brownian dynamics simulations for non-dilute crowded systems. The major assumption underlying the CME is that reactions are occurring in well-mixed environments. The characteristics of reactivity fluctuations in individual enzyme molecules relate to the second moment of turnover time statistics which can be defined by the randomness parameter R and time statistics. The randomness parameter R tends to become 1, when the waiting time distribution is a single exponential decay function. Any deviation of R from this predicted value is an indication of dynamic disorder [121]. When this is applied for the inhibition reaction, the expressions for the waiting time distributions for inhibition reactions became cumbersome. If more than one intermediate steps are present in the given set of reactions, then the chemical master equations become more and more complex. To avoid such complications, Chaudhary has used a simple analytical model based on the first passage time distribution between successive catalytic turnover events. This probability distribution function (PDF) is the moment of one monitored transition. If $\varphi(t)\mathrm{d}t$ is the probability of observing the next monitored transition between time t and (t+ dt) after this time moment. The assumption made is that there is formation of product molecules with regeneration of the free enzyme. Such events are actually detectable in single-molecule fluorescence experiments [126]. These experiments have given the final expression for competitive inhibition can be given as

$$\langle t_{\text{competitive}} \rangle = \frac{[S] + [I]\frac{k_3(k_{-1}+k_2)}{k_1 k_{-3}} + \frac{(k_{-1}+k_2)}{k_1}}{k_2[S]} \tag{76}$$

$$\langle t_{\text{uncompetitive}} \rangle = \frac{[S] + [I][S]\frac{k_3}{k_{-3}} + \frac{(k_{-1}+k_2)}{k_1}}{k_2[S]} \tag{77}$$

Various researchers are applying various approaches to derive rate expression for the enzymatic reaction, which are under continuous evaluation to finalize more generalized and simplistic equations.

8. Conclusion

Biocatalysis has evolved as a promising lead to make the chemical process more sustainable, green and environment friendly. The main challenge in biocatalysed reactions is their industrial viability and economical aspects. The advancement in genetic engineering, medium engineering and immobilization technology has made them more robust to sustain the industrial environment. To transform the lab scale process to industrial scale, the understanding of the

http://dx.doi.org/10.5772/67692

enzyme kinetics, inhibition and activation becomes very essential. The fragile nature of the enzymes makes them more sensitive towards the changes in medium and physical parameters of the surrounding environment. The various physical parameters such as temperature, pH, microwave and other radiations exposure, immobilization on different support affect the activity of the enzyme and in turn the kinetics of the reaction. The kinetics of the enzymes become more complex when the reaction becomes multi-substrate compared to single substrate reactions. The different kinds of methodologies and plots are used to predict the reaction kinetics of the multi-substrate reactions. The pattern observed during the analysis of the obtained experimental data is used to predict the model for the reactions such as ping-pong bi-bi or ternary complex model and mechanism of the reaction. The obtained model is then solved using various non-linear regression softwares available for the prediction of kinetic constants for the reaction. The inhibition or activation of the enzyme with the chemical entities or the reactant itself further modifies the reaction kinetics and in turn model of the reaction. To study these phenomena, a systematic research methodology needs to be followed to predict the exact reaction kinetics. The finding then can be used to design new enzyme inhibitors, activators which can be applied as possible drug molecules or for the prediction of the mechanism of the enzyme from the same family. This opens up the new area of the application such as pharmaceutical, molecular modelling, genetics studies and industrial production of fine and bulk chemicals. The use of the continuous and stopped flow experiments is becoming a new field to explore and understand enzyme kinetics and its inhibition to better extend.

Abbreviations

E	Enzyme
S, A	Substrate
H	Hydrogen
H+	Hydrogen ion
ES	Enzyme-substrate form
P	Product
ESH	Enzyme-substrate-hydrogen form
k_1	Rate constant for forward reaction to form ES
k_2	Rate constant for reaction to form P
k_{-1}	Rate constant for backward reaction to form ES
v	Initial velocity
V_{max}	Maximum velocity
K_m	Michaelis Menten constant
V_{mapp}	Apparent maximum velocity
K_{mapp}	Apparent Michaelis Menten constant
k_{e1}, k_{e2}, k_{es1}, k_{es2}	Equilibrium constant for hydrogen binding with various enzyme form
η	Effectiveness factor
S_b	Amount of substrate at boundary layer
V'_{max}	Maximum velocity within matrix
v_s	Reaction velocity within matrix

S_s	Amount of substrate in the solution
E_T	Total enzyme
K_a	Michaelis Menten constant for substrate A
K_{ia}	Intrinsic dissociation constant for A
K_{ib}	Intrinsic dissociation constant for B
k_i	Inhibition rate constant
B	Second substrate
P	Product 1
Q	Product 2
F	Stable modified form of enzyme
K_b	Michaelis Menten constant for substrate B

Author details

Ganapati D. Yadav* and Deepali B. Magadum

*Address all correspondence to: gdyadav@yahoo.com; gd.yadav@ictmumbai.edu.in

Institute of Chemical Technology, Mumbai, India

References

[1] Marti S, Roca M, Andres J, Moliner V, Silla E, Tunon I, et al. Theoretical insights in enzyme catalysis. Chem Soc Rev [Internet]. 2004;33(2):98–107. Available from: http://dx.doi.org/10.1039/B301875J

[2] Warshel A. Computer simulations of enzyme catalysis: methods, progress, and insights. Annu Rev Biophys Biomol Struct [Internet]. 2003 Jun 1;32(1):425–43. Available from: http://dx.doi.org/10.1146/annurev.biophys.32.110601.141807

[3] Voet D, Voet JG. Biochemistry [Internet]. 4th ed. Wiley; 2010. Available from: http://www.amazon.com/exec/obidos/redirect?tag=citeulike07-20&path=ASIN/0470570954

[4] Lehninger A, Nelson D, Cox M. Lehninger Principles of Biochemistry [Internet]. Book. New York: W. H. Freeman; 2008. Available from: citeulike-article-id:3823091

[5] Web refereence 1 [Internet]. Available from: 6. http://www.chem.qmul.ac.uk/iubmb/enzyme/rules.html

[6] Web reference 2. Available from: 8. http://goldbook.iupac.org/M03892.html

[7] Berg JM, Tymoczko JL, Stryer L. Biochemistry [Internet]. W. H. Freeman; 2010. Available from: https://books.google.co.in/books?id=O67WcAAACAAJ

[8] Bhagavan NV. Medical Biochemistry [Internet]. Amsterdam: Elsevier Science; 2001. Available from: https://books.google.co.in/books?id=eCOL8lW76F8C

[9] Noor E, Flamholz A, Liebermeister W, Bar-Even A, Milo R. A note on the kinetics of enzyme action: A decomposition that highlights thermodynamic effects. FEBS Lett [Internet]. 2013;587(17):2772–7. Available from: http://dx.doi.org/10.1016/j.febslet.2013.07.028

[10] Markus M, Hess B, Ottaway JH, Cornish-Bowden A. The analysis of kinetic data in biochemistry. A critical evaluation of methods. FEBS Lett [Internet]. 1976;63(2):225–30. Available from: http://dx.doi.org/10.1016/0014-5793(76)80100-7

[11] Badger MR, Collatz GJ. Studies on the kinetic mechanism of ribulose-1, 5-bisphosphate carboxylase and oxygenase reactions, with particular reference to the effect of temperature on kinetic parameters. Carnegie Inst Washingt Yearb. 1977;76:355–61.

[12] Krishna SH, Karanth NG. Lipase-catalyzed synthesis of isoamyl butyrate: a kinetic study. Biochim Biophys Acta (BBA)-Protein Struct Mol Enzymol. 2001;1547(2):262–7.

[13] Yadav GD, Devi KM. Immobilized lipase-catalysed esterification and transesterification reactions in non-aqueous media for the synthesis of tetrahydrofurfuryl butyrate: comparison and kinetic modeling. Chem Eng Sci. 2004;59(2):373–83.

[14] Yadav GD, Trivedi AH. Kinetic modeling of immobilized-lipase catalyzed transesterification of n-octanol with vinyl acetate in non-aqueous media. Enzyme Microb Technol. 2003;32(7):783–9.

[15] Kaplan JG. The alteration of intracellular enzymes: I. Yeast catalase and the Euler effect. Exp Cell Res. 1955;8(2):305–28.

[16] Gianfreda L, Scarfi MR. Enzyme stabilization: state of the art. Mol Cell Biochem. 1991;100(2):97–128.

[17] Quintiliani M. The oxygen effect in radiation inactivation of DNA and enzymes. Int J Radiat Biol Relat Stud Physics, Chem Med. 1986;50(4):573–94.

[18] Hutchinson F. Sulfhydryl groups and the oxygen effect on irradiated dilute solutions of enzymes and nucleic acids. Radiat Res. 1961;14(6):721–31.

[19] Mee LK. Effect of X-rays, ultraviolet light, and chemicals on the kinetic parameters of chymotrypsin. Radiat Res. 1964;21(4):501–12.

[20] Yadav GD, Lathi PS. Synergism between microwave and enzyme catalysis in intensification of reactions and selectivities: transesterification of methyl acetoacetate with alcohols. J Mol Catal A Chem. 2004;223(1):51–6.

[21] Yadav GD, Sajgure AD, Dhoot SB. Insight into microwave irradiation and enzyme catalysis in enantioselective resolution of RS-(±)-methyl mandelate. J Chem Technol Biotechnol. 2008;83(8):1145–53.

[22] Collins JM, Leadbeater NE. Microwave energy: a versatile tool for the biosciences. Org Biomol Chem. 2007;5(8):1141–50.

[23] Yadav GD, Pawar S V. Insight into microwave irradiation and enzyme catalysis in enantioselective resolution of DL-(±)-3-phenyllactic acid. Appl Microbiol Biotechnol. 2012;96(1):69–79.

[24] YaDav GD, DevenDrana S. Microwave assisted enzyme catalysis: practice and perspective. White Biotechnol Sustain Chem. 2015;52.

[25] Yadav GD, Thorat PA. Microwave assisted lipase catalyzed synthesis of isoamyl myristate in solvent-free system. J Mol Catal B Enzym. 2012;83:16–22.

[26] Rejasse B, Lamare S, Legoy M-D, Besson T. Influence of microwave irradiation on enzymatic properties: applications in enzyme chemistry. J Enzyme Inhib Med Chem. 2007;22(5):519–27.

[27] Yadav GD, Borkar IV. Kinetic and mechanistic investigation of microwave-assisted lipase catalyzed synthesis of citronellyl acetate. Ind Eng Chem Res. 2008;48(17):7915–22.

[28] Cleland WW. [22] The use of pH studies to determine chemical mechanisms of enzyme-catalyzed reactions. Methods Enzymol. 1982;87:390–405.

[29] Pörschke D, Eigen M. Co-operative non-enzymatic base recognition III. Kinetics of the helix—coil transition of the oligoribouridylic· oligoriboadenylic acid system and of oligoriboadenylic acid alone at acidic pH. J Mol Biol. 1971;62(2):361IN11365–364IN15381.

[30] Gill SC, Von Hippel PH. Calculation of protein extinction coefficients from amino acid sequence data. Anal Biochem. 1989;182(2):319–26.

[31] Dixon M. The effect of pH on the affinities of enzymes for substrates and inhibitors. Biochem J. 1953;55(1):161.

[32] Laurent N, Haddoub R, Flitsch SL. Enzyme catalysis on solid surfaces. Trends Biotechnol [Internet]. 2016 Sep 7;26(6):328–37.

[33] Bisswanger H. Practical Enzymology Enzymes in Industry Biocatalysts and Enzyme Technology. Enzyme Kinetics Principles and Methods. 2008. pp. 4–12 .

[34] Curtiss CF, Bird RB. Multicomponent diffusion. Ind Eng Chem Res. 1999;38(7):2515–22.

[35] Engasser J-M, Horvath C. Effect of internal diffusion in heterogeneous enzyme systems: evaluation of true kinetic parameters and substrate diffusivity. J Theor Biol. 1973;42 (1):137–55.

[36] Cook P, Cleland WW. Enzyme kinetics and mechanism. Garland Science; 2007.

[37] Mildvan AS. Mechanism of enzyme action. Annu Rev Biochem. 1974;43(1):357–99.

[38] Bornadel A, Åkerman CO, Adlercreutz P, Hatti-Kaul R, Borg N. Kinetic modeling of lipase-catalyzed esterification reaction between oleic acid and trimethylolpropane: A simplified model for multi-substrate multi-product ping–pong mechanisms. Biotechnol Prog. 2013;29(6):1422–9.

[39] Haki GD, Rakshit SK. Developments in industrially important thermostable enzymes: a review. Bioresour Technol. 2003;89(1):17–34.

[40] Marangoni AG. Enzyme kinetics: a modern approach. John Wiley & Sons; 2003.

[41] Cornish-Bowden A, Cornish-Bowden A. Fundamentals of enzyme kinetics. 2012;

[42] Lin CC, Segel LA. Mathematics Applied to Deterministic Problems. SIAM; 1974.

[43] Alberty RA. Biochemical thermodynamics and rapid-equilibrium enzyme kinetics. J Phys Chem B. 2010;114(51):17003–12.

[44] Vallières C, Fisher N, Meunier B. Reconstructing the Q o Site of Plasmodium falciparum bc 1 Complex in the Yeast Enzyme. PLoS One. 2013;8(8):e71726.

[45] Tu S-C, Becvar JE, Hastings JW. Kinetic studies on the mechanism of bacterial NAD (P) H: flavin oxidoreductase. Arch Biochem Biophys. 1979;193(1):110–6.

[46] Leskovac V. Comprehensive enzyme kinetics. Springer Science & Business Media; 2003.

[47] Cleland WW. 1 steady state kinetics. Enzym. 1970;2:1–65.

[48] Purich DL. Enzyme kinetics: catalysis & control: a reference of theory and best-practice methods. Elsevier; 2010.

[49] Fromm HJ. Initial rate enzyme kinetics. Springer Science & Business Media; 2013. 22.

[50] Tallarida RJ, Murray RB. Enzyme Kinetics I: Michaelis—Menten Equation. In: Manual of Pharmacologic Calculations: With Computer Programs. New York, NY: Springer New York; 1987. pp. 61–3.

[51] Cermak N. Fundamentals of enzyme kinetics: michaelis-menten and deviations michaelis-menten kinetics. Solutions. 2009;1–15.

[52] Eisenthal R, Cornish-Bowden A. The direct linear plot. A new graphical procedure for estimating enzyme kinetic parameters. Biochem J. 1974;139(3):715–20.

[53] Miller FP. Enzyme Kinetics: Enzyme Assay, Michaelis-Menten Kinetics, Lineweaver-Burk Plot, Enzyme Inhibitor, Eadie-Hofstee Diagram, Enzyme Catalysis. Alphascript Publishing; 2009. (Alphascript publishing).

[54] Menton M, Inhibition ES, Burk L. Two paths for E, less available for making P. 2009;1–10.

[55] Smith HJ, Simons C. Enzymes and Their Inhibitors: Drug Development. CRC Press; 2004. (CRC enzyme inhibitors series).

[56] Cornish-Bowden A. Principles of Enzyme Kinetics. Elsevier; 2014.

[57] Cornish-Bowden A. Analysis and interpretation of enzyme kinetic data. Perspect Sci. 2014;1(1–6):121–5.

[58] Cleland WW. The kinetics of enzyme-catalyzed reactions with two or more substrates or products: III. Prediction of initial velocity and inhibition patterns by inspection. Biochim Biophys Acta (BBA)-Specialized Sect Enzymol Subj. 1963;67:188–96.

[59] Cleland W. The kinetics of enzyme-catalyzed reactions with two or more substrates or products: I. Nomenclature and rate equations. Biochim Biophys Acta (BBA)-Specialized Sect Enzymol Subj. 1963;67:104–37.

[60] Yuzbasheva EY, Gotovtsev PM, Mostova EB, Perkovskaya NI, Lomonosova MA, Butylin V V, et al. Biodiesel production via enzymatic catalysis. Appl Biochem Microbiol. 2014;50(8):737–49.

[61] Anderson VE. Multiple alternative substrate kinetics. Biochim Biophys Acta - Proteins Proteomics. 2015;1854(11):1729–36.

[62] Hindson VJ, Shaw W V. Random-order ternary complex reaction mechanism of serine acetyltransferase from Escherichia coli. Biochemistry. 2003;42(10):3113–9.

[63] Yadav GD, Lathi PS. Intensification of enzymatic synthesis of propylene glycol monolaurate from 1, 2-propanediol and lauric acid under microwave irradiation: Kinetics of forward and reverse reactions. Enzyme Microb Technol. 2006;38(6):814–20.

[64] Da Silva AGM, Robles-Dutenhefner PA, Dias A, Fajardo H V., Lovón ASP, Lovón-Quintana JJ, et al. Gold, palladium and gold-palladium supported on silica catalysts prepared by sol–gel method: Synthesis, characterization and catalytic behavior in the ethanol steam reforming. J Sol–gel Sci Technol. 2013;67(2):273–81.

[65] Bousquet-Dubouch M-P, Graber M, Sousa N, Lamare S, Legoy M-D. Alcoholysis catalyzed by Candida antarctica lipase B in a gas/solid system obeys a Ping Pong Bi Bi mechanism with competitive inhibition by the alcohol substrate and water. Biochim Biophys Acta (BBA)-Protein Struct Mol Enzymol. 2001;1550(1):90–9.

[66] Romero MD, Calvo L, Alba C, Daneshfar A. A kinetic study of isoamyl acetate synthesis by immobilized lipase-catalyzed acetylation in n-hexane. J Biotechnol. 2007;127(2):269–77.

[67] Lonsdale R, Harvey JN, Mulholland AJ. A practical guide to modelling enzyme-catalysed reactions. Chem Soc Rev. 2012;41(8):3025–38.

[68] Chou TC, Talalay P. Quantitative analysis of dose-effect relationships: the combined effects of multiple drugs or enzyme inhibitors. Adv Enzyme Regul. 1984;22(C):27–55.

[69] Broom AD. Rational design of enzyme inhibitors: multisubstrate analog inhibitors. J Med Chem. 1989;32(1):2–7.

[70] Ivanetich KM, Goold RD, Sikakana CNT. Explanation of the non-hyperbolic kinetics of the glutathione S-transferases by the simplest steady-state random sequential Bi Bi mechanism. Biochem Pharmacol. 1990;39(12):1999–2004.

[71] Yadav GD, Devendran S. Lipase catalyzed synthesis of cinnamyl acetate via transesterification in non-aqueous medium. Process Biochem. 2012;47(3):496–502.

[72] Yadav GD, Pawar S V. Synergism between microwave irradiation and enzyme catalysis in transesterification of ethyl-3-phenylpropanoate with n-butanol. Bioresour Technol. 2012;109:1–6.

[73] Yadav GD, Devi KM. Enzymatic synthesis of perlauric acid using Novozym 435. Biochem Eng J. 2002;10(2):93–101.

[74] Fromm HJ. The use of competitive inhibitors in studying the mechanism of action of some enzyme systems utilizing three substrates. Biochim Biophys Acta (BBA)-Enzymology. 1967;139(2):221–30.

[75] Gray PJ, Duggleby RG. Analysis of kinetic data for irreversible enzyme inhibition. Biochem J. 1989;257(2):419–24.

[76] Fontes R, Ribeiro JM, Sillero A. Inhibition and activation of enzymes. The effect of a modifier on the reaction rate and on kinetic parameters. Acta Biochim Pol. 2000;47 (1):233–57.

[77] Laidler K, Bunting P. Enzyme kinetics, inhibition, and regulation. Chem Kinetics Enzyme Action. 1973; 147–57.

[78] Cushman DW, Cheung HS, Sabo EF, Ondetti MA. Design of potent competitive inhibitors of angiotensin-converting enzyme. Carboxyalkanoyl and mercaptoalkanoyl amino acids. Biochemistry. 1977;16(25):5484–91.

[79] Cornish-Bowden A. A simple graphical method for determining the inhibition constants of mixed, uncompetitive and non-competitive inhibitors. Biochem J. 1974;137(1):143–4.

[80] Cornish-Bowden A. The origins of enzyme kinetics. FEBS Lett. 2013;587(17):2725–30.

[81] Perdicakis B, Montgomery HJ, Guillemette JG, Jervis E. Analysis of slow-binding enzyme inhibitors at elevated enzyme concentrations. Anal Biochem. 2005;337(2):211–23.

[82] Murphy DJ. Determination of accurate KI values for tight-binding enzyme inhibitors: An in silico study of experimental error and assay design. Anal Biochem. 2004;327 (1):61–7.

[83] Kuzmic P, Hill C, Kirtley MP, Janc JW. Kinetic determination of tight-binding impurities in enzyme inhibitors. Anal Biochem. 2003;319(2):272–9.

[84] Kuzmič P, Sideris S, Cregar LM, Elrod KC, Rice KD, Janc JW. High-throughput screening of enzyme inhibitors: automatic determination of tight-binding inhibition constants. Anal Biochem. 2000;281(1):62–7.

[85] Zollner H. Handbook of Enzyme Inhibitors. Pharm Unserer Zeit. 1989;19(1):440.

[86] Chang TS. An updated review of tyrosinase inhibitors. Int J Mol Sci. 2009;10(6):2440–75.

[87] Gupta SP. QSAR studies on enzyme inhibitors. Chem Rev. 1987;87(5):1183–253.

[88] Sharma R. Enzyme inhibition. Edited by Rakesh Sharma.

[89] Chou T-C, Talalay P. A Simple Generalized Equation Inhibitions of Michaelis-Menten. J Biol Chem. 1977;252(18):6438–42.

[90] Freyer MW, Lewis EA. Isothermal titration calorimetry: experimental design, data analysis, and probing macromolecule/ligand binding and kinetic interactions. Methods Cell Biol. 2008;84:79–113.

[91] Mock H-P, Strack D. Energetics of the uridine 5′-diphosphoglucose: hydroxycinnamic acid acyl-glucosyltransferase reaction. Phytochemistry. 1993;32(3):575–9.

[92] Wolfenden R. Conformational aspects of inhibitor design: enzyme-substrate interactions in the transition state. Bioorganic Med Chem. 1999;7(5):647–52.

[93] Greis KD, Zhou S, Burt TM, Carr AN, Dolan E, Easwaran V, et al. MALDI-TOF MS as a label-free approach to rapid inhibitor screening. J Am Soc Mass Spectrom. 2006;17 (6):815–22.

[94] Hu L, Jiang G, Xu S, Pan C, Zou H. Monitoring enzyme reaction and screening enzyme inhibitor based on MALDI-TOF-MS platform with a matrix of oxidized carbon nanotubes. J Am Soc Mass Spectrom. 2006;17(11):1616–9.

[95] de Boer AR, Lingeman H, Niessen WMA, Irth H. Mass spectrometry-based biochemical assays for enzyme-inhibitor screening. TrAC - Trends Anal Chem. 2007;26(9):867–83.

[96] Xu Z, Yao S, Wei Y, Zhou J, Zhang L, Wang C, et al. Monitoring enzyme reaction and screening of inhibitors of acetylcholinesterase by quantitative matrix-assisted laser desorption/ionization fourier transform mass spectrometry. J Am Soc Mass Spectrom. 2008;19(12):1849–55.

[97] Brooks HB, Kahl SD, Smith MC, Weidner JR. Basics of Enzymatic Assays for HTS. 2012.

[98] Kuzmič P. Program DYNAFIT for the analysis of enzyme kinetic data: application to HIV proteinase. Anal Biochem. 1996;237(2):260–73.

[99] Moffitt JR, Bustamante C. Extracting signal from noise: kinetic mechanisms from a Michaelis-Menten-like expression for enzymatic fluctuations. FEBS J. 2014;281(2):498–517.

[100] Leone FA, Degreve L, Baranauskas JA. SIGRAF: A versatile computer program for fitting enzyme kinetic data. Biochem Educ. 1992;20(2):94–6.

[101] Acker MG, Auld DS. Considerations for the design and reporting of enzyme assays in high-throughput screening applications. Perspect Sci. 2014;1(1–6):56–73.

[102] Copeland RA. Evaluation of enzyme inhibitors in drug discovery: a guide for medicinal chemists and pharmacologists. John Wiley & Sons; 2013.

[103] Laskowski M, Qasim MA. What can the structures of enzyme-inhibitor complexes tell us about the structures of enzyme substrate complexes? Biochim Biophys Acta - Protein Struct Mol Enzymol. 2000;1477(1–2):324–37.

[104] Kubinyi H. Structure-based design of enzyme inhibitors and receptor ligands. Curr Opin Drug Discov Devel. 1998;1(1):4–15.

[105] Houston JB, Galetin A. Modelling atypical CYP3A4 kinetics: principles and pragmatism. Arch Biochem Biophys. 2005;433(2):351–60.

[106] Sculley MJ, Morrison JF, Cleland WW. Slow-binding inhibition: the general case. Biochim Biophys Acta - Protein Struct Mol Enzymol. 1996;1298(1):78–86.

[107] Galanakis CM, Patsioura A, Gekas V. Enzyme kinetics modeling as a tool to optimize food industry: a pragmatic approach based on amylolytic enzymes. Crit Rev Food Sci Nutr. 2014;8398(June 2014):37–41.

[108] Cushman DW, Ondetti MA. Design of angiotensin converting enzyme inhibitors. Nat Med. 1999;5(10):1110–2.

[109] Wyvratt MJ, Patchett AA. Recent developments in the design of angiotensin-converting enzyme inhibitors. Med Res Rev. 1985;5(4):483–531.

[110] Hanauske-Abel HM, Günzler V. A stereochemical concept for the catalytic mechanism of prolylhydroxylase: applicability to classification and design of inhibitors. J Theor Biol. 1982;94(2):421–55.

[111] Appelt K, Bacquet RJ, Bartlett CA, Booth CLJ, Freer ST, Fuhry MAM, et al. Design of enzyme inhibitors using iterative protein crystallographic analysis. J Med Chem. 1991;34 (7):1925–34.

[112] Hamada S, Suzuki T, Mino K, Koseki K, Oehme F, Flamme I, et al. Design, synthesis, enzyme-inhibitory activity, and effect on human cancer cells of a novel series of jumonji domain-containing protein 2 histone demethylase inhibitors. J Med Chem. 2010;53 (15):5629–38.

[113] Bode W, Huber R. Structural basis of the endoproteinase-protein inhibitor interaction. Biochim Biophys Acta - Protein Struct Mol Enzymol. 2000;1477(1–2):241–52.

[114] Mohan C, Long KD, Mutneja M. An Introduction to Inhibitors and Their Biological Applications. EMD Millipore Corp. 2013;3–13.

[115] Böhm H-J. The computer program LUDI: a new method for the de novo design of enzyme inhibitors. J Comput Aided Mol Des. 1992;6(1):61–78.

[116] Okamoto K, Eger BT, Nishino T, Kondo S, Pai EF, Nishino T. An extremely potent inhibitor of xanthine oxidoreductase crystal structure of the enzyme-inhibitor complex and mechanism of inhibition. J Biol Chem. 2003;278(3):1848–55.

[117] Loparo JJ, van Oijen A. Single-molecule enzymology. In: Handbook of Single-Molecule Biophysics. Springer; 2009. pp. 165–82.

[118] Xie XS, Lu HP. Single-molecule enzymology. J Biol Chem. 1999;274(23):15967–70.

[119] Grima R, Walter NG, Schnell S. Single-molecule enzymology à la Michaelis-Menten. FEBS J. 2014;281(2):518–30.

[120] Chaudhury S, Igoshin OA. Enzymatic reaction. 2009;31(3):13421–8.

[121] Qian H, Bishop LM. The chemical master equation approach to nonequilibrium Steady-state of open biochemical systems: Linear Single-molecule enzyme kinetics and nonlinear biochemical reaction networks. Int J Mol Sci. 2010;11(9):3472–500.

[122] Tuñón I, Laage D, Hynes JT. Are there dynamical effects in enzyme catalysis? Some thoughts concerning the enzymatic chemical step. Arch Biochem Biophys. 2015;582:42–55.

[123] Chaudhury S, Igoshin OA. Dynamic disorder in quasi-equilibrium enzymatic systems. PLoS One. 2010;5(8).

[124] Min W, English BP, Luo G, Cherayil BJ, Kou SC, Xie XS. Fluctuating enzymes: Lessons from single-molecule studies. Acc Chem Res. 2005;38(12):923–31.

[125] Terentyeva TG, Engelkamp H, Rowan AE, Komatsuzaki T, Hofkens J, Li CB, et al. Dynamic disorder in single-enzyme experiments: Facts and artifacts. ACS Nano. 2012;6 (1):346–54.

[126] Chaudhury S. Poisson indicator and fano factor for probing dynamic disorder in single-molecule enzyme inhibition kinetics. J Phys Chem B. 2014;118(35):10405–12.

Chapter 5

Telomerase Inhibitors and Activators: Pharmaceutical Importance

Ayse Gul Mutlu

Additional information is available at the end of the chapter

http://dx.doi.org/10.5772/65933

Abstract

Telomeres are specialized functional complexes that protect the ends of eukaryotic chromosomes. The telomeric DNA sequences are tandem repeats of a short hexameric sequence unit. The inability to DNA polymerase to replicate the end of the chromosome during lagging strand synthesis results in the loss of telomeric repeats when cell divides. Telomere shortening provides a barrier to cancer progression and the majority of the cancer cells depend on the activation of telomerase to gain proliferative immortality. Thus, telomerase is a molecular target for diseases since its discovery. Telomerase inhibition enables more specific ground for cancer therapy because the telomerase is not detected in most normal tissues. Some of the synthetic and natural telomerase inhibitors were tried on various cancer cells and there was a decrease in the number of cancer cells. But on the other hand, telomere shortening correlates with cellular aging. Some evidence suggests that the progressive loss of telomeric repeats of chromosomes may function as a molecular clock that triggers senescence. Telomerase-related gene mutations also result in some diseases. Because of this, telomerase activators are important for antiaging and telomerase-dependent disease treatments. This chapter summarizes the pharmaceutical importance of telomeres, telomerase structure, telomerase activators, and inhibitors.

Keywords: telomerase, telomerase inhibitors, telomerase activators

1. Introduction

1.1. Importance of telomeres

Telomeres are specialized functional complexes that protect the ends of eukaryotic chromosomes. The telomeric DNA sequences are, in most species, tandem repeats of a short hexameric sequence unit [1]. Overall, telomere sizes range from about 15 to 20 kbp at birth to sometimes less than 5 kbp in chronic disease states. Telomeric repeats help maintain chromosomal integrity [2].

Evolutionary conservation of this repetitive DNA sequence family might indicate that sequence is essential to the cellular function [3]. Telomeric DNA sequences and structure are similar among otherwise widely divergent eukaryotes. The telomeric repeat unit is TTAGGG for humans as well as other vertebrates [4]. The ends of telomeres are protected and regulated by telomere binding proteins and form a t loop structure [2, 5]. Mainly, the inability of DNA polymerase to replicate the end of the chromosome during lagging strand synthesis results in the loss of telomeric repeats when cell divides. This phenomenon eventually results in a growth arrest and telomeres become critically shortened when multiple chromosome end fusions occur, resulting in a loss of cell viability [2]. Telomere shortening provides a barrier to cancer progression by preventing immortalization and the majority of the cancer cells depend on the activation of telomerase to gain proliferative immortality [2, 6]. But on the other hand, telomere shortening correlates with cellular aging. Stem and progenitor cells express low levels of telomerase [6].

1.2. Telomerase structure and function

Greider and Blackburn identified a specialized DNA polymerase in extracts from the *Tetrahymena* that extends the chromosome ends in eukaryotes [7]. Telomerase adds multiple copies of certain DNA unit to the terminal portion of one strand of the repeat tract [1, 4]. This process is required for genomic stability and cell viability. Telomerase is a specialized reverse transcriptase. Telomerase subunit TER identified in the late 1980s and catalytic subunit TERT in 1997. Subsequent studies showed that the TER and TERT together form a tight complex that is sufficient for telomeric DNA repeat synthesis *in vitro* [6]. TER contains RNA template for reverse transcription [8–10]. TERT contains discrete domains that carry out the mechanically complicated reaction of nucleic acid and nucleotide binding and selectivity in a coordinated manner during telomerase replication [8]. Despite only TERT and TER are required for telomerase catalytic activity *in vitro*, the physiologically functional holoenzyme is a multisubunit ribonucleoprotein (RNP). *Tetrahymena* telomerase holoenzyme contains eight subunits, each of which is essential for telomere length maintenance [10].

Telomerase is a very important enzyme for the aging process and carcinogenesis. Primary human cells exhibit limited replicative potential but the cancer cell lines are immortal with passage in culture [11]. In embryonic stem cells telomerase is activated and maintains telomere length but the level of telomerase activity is low or absent in the majority of the stem cells. Thus, even in stem cells, except embryonic stem cells and cancer stem cells, telomere shortening occurs, possibly at a slower rate than that in normal somatic cells [12].

To grow indefinitely, human cancer cells must compensate the progressive loss of telomeric DNA by cell division [13]. This immortality is mainly a result of telomerase activity. Telomerase is expressed in more than 85% of cancer cells [14–17], but in some cells, the telomere length could be maintained in the absence of telomerase. It has been deduced that one or more alternative telomerase-independent mechanisms exist in human cells [13].

2. Telomerase inhibitors and pharmaceutical importance

Synthesis of DNA at chromosome ends by telomerase may be necessary for indefinite proliferation of human cells. According to results of Kim et al. in cultured cells representing

18 different human tissues, 98 of 100 were immortal and none of the 22 mortal populations were positive for telomerase. Similarly, 90 of 101 biopsies representing 12 human tumor types and none of the 50 normal somatic tissues were positive [14]. Tahara et al. observed that the positive telomerase activity in 28 hepatocellular carcinoma (HCC) tissues of 33. Interestingly, hepatitis B virus-positive patients were telomerase positive. Also, in 19 of 38 hepatitis tissues and 6 of 8 cirrhotic liver tissues, weak telomerase activity was detected. These results indicate that the expression of telomerase may play a crucial role in hepatocarcinogenesis [15]. Ferber et al. detected that the integrations of the hepatitis B virus and human papillomavirus into the hTERT gene in liver and cervical cancers [18]. Hiyama et al. showed that the 95% of the advanced stage breast cancer tissues have telomerase activity [16].

Telomerase is a molecular target since its discovery. The most important disadvantage of chemotherapy drugs used today is that they are not selective, they have effect on normal healthy cells together with cancer cells. Telomerase inhibition enables more specific ground because the telomerase is not detected in most normal tissues [2, 17]. Differences in telomerase expression, telomere length, and cell kinetics between normal and cancer tissues suggest that targeting telomerase for cancer therapy may be relatively safe [19]. Telomerase inhibitor effects on stem cells may thus be minor because these telomerase-competent cells only proliferate intermittently and typically have much longer telomeres than cancer cells [2].

Experimental and clinical studies for telomerase inhibition are currently carried out in many different ways, such as inhibition by nucleotides and nucleoside-type reverse transcriptase inhibitors; direct inhibition by nonnucleoside small molecules; oligonucleotide inhibitors of telomerase activity; gene therapy; telomerase-specific phosphorylation inhibitors; G quadruplex stabilizers; and TER directed hammer head ribozymes [2, 17, 20, 21].

In recent studies, some of the synthetic telomerase inhibitors were tried to bone marrow, prostate, brain, breast cancer, and pancreas cancer cells and there was a decrease in the number of cancer cells [17]. Telomerase inhibitor imetelstat (GRN163L) is the first telomerase inhibitor to advance to the clinical development (www.geron.com/imetelstat). Some izothiazolone derivatives show telomerase inhibition properties [20]. 2-[(E)-3-naphtalen-2-yl-but-2-enoylamino]-benzoic acid was reported as a selective telomerase inhibitor [20, 22]. According to the results of our previous studies (unpublished), some of Imidazo[1, 2-a]pyrazine derivatives can be used for telomerase inhibition. There are some studies which draw attention to various enzyme inhibition and anticancer activities of İmidazo(1,2-a)pyrazine derivatives [23–28].

In addition to synthetic compounds, various chemical compounds that occur naturally in plants like allicin, curcumin, and silbinin have been suggested as telomerase inhibitors (**Figures 1** and **2**) [29]. Allicin is organosulfur compound obtained from garlic that can inhibit telomerase activity and induce apoptosis of gastric cancer SGC-7901 cells [30]. Milk thistle's silymarin and silibinin have also been investigated by some researchers in terms of telomerase inhibition and activation. The treatment of the K562 human leukemia cell line with silymarin resulted in a significant inhibition of telomerase activity [31]. In Yurtcu et al.'s study, combination of silymarin and doxorubicin and silymarin alone inhibited telomerase activity in HepG2 hepatocellular carcinoma cell line [32]. But silymarin may activate the telomerase in noncancerous cells according to the results of Parzonko et al. In this study, silymarin increased telomerase activity in endothelial progenitor cells [33]. Thelen et al. showed that inhibition of telomerase

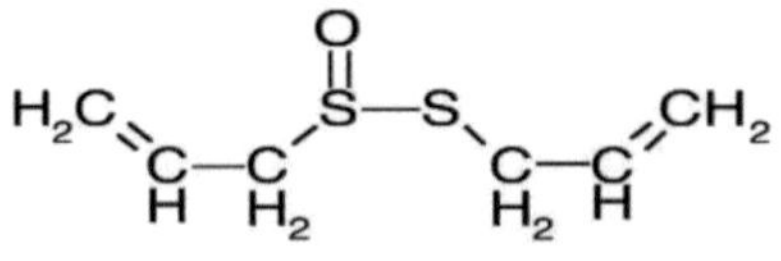

Allicin

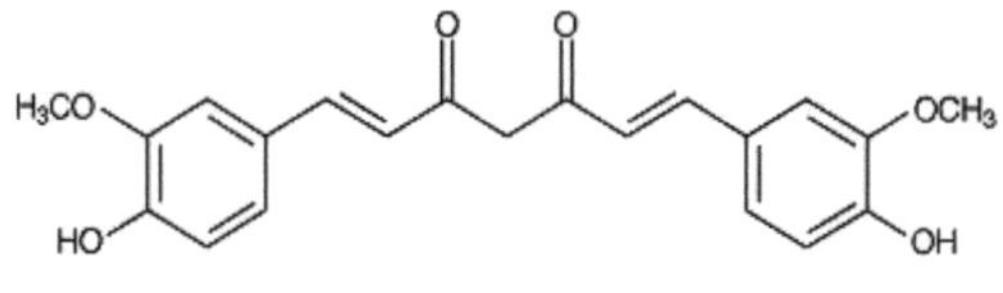

Curcumin

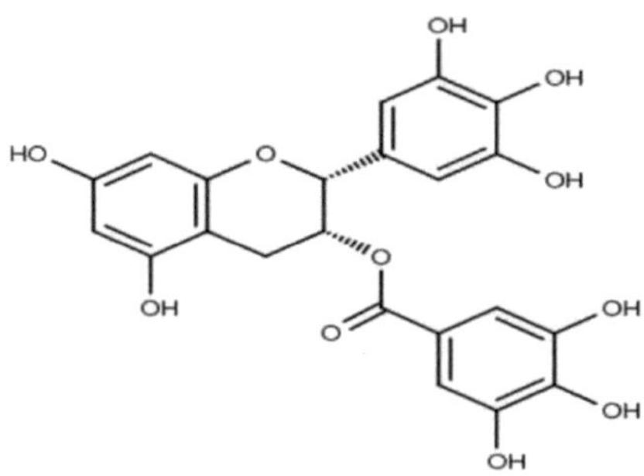

Epigallocatechin gallate

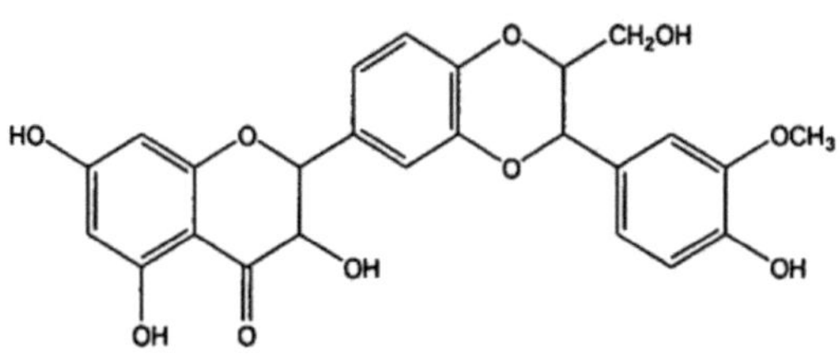

Silibinin

Figure 1. Chemical formula of some natural telomerase inhibitors.

activity by silibinin in prostate cancer cells [34]. Silibinin can also inhibit hTERT gene expression in T47D breast cancer cells. Silibinin and curcumin combination could be more effective in the way of inhibition of telomerase [35]. Curcumin is a phenolic compound isolated from the rhizome of the *Curcuma longa* L. Curcumin has antitumor, antiangiogenic, and apoptotic

http://dx.doi.org/10.5772/65933

Figure 2. Various chemical compounds that occur naturally in garlic, turmeric, grape, and milk thistle are telomerase activity modifiers.

properties. Chakraborty et al. suggested that telomerase inhibition is a main mechanism of curcumin-induced apoptosis in human leukemia cell line K-562 [36]. In Ramachandran et al.'s study, increasing concentrations of curcumin caused a steady decrease in the level of hTERT mRNA and inhibited telomerase activity in MCF-7 breast cancer cells [37].

Antibiotics oflaxacin and levofloxacin inhibit the telomerase in cell extracts. Helenalin (a natural sesquiterpene lactone), polyunsaturated fatty acids with cis-double bond, also inhibit the telomerase [20]. However, some dietary polyphenols have been suggested as telomerase inhibitors [38]. Major tea catechin epigallocatechin gallate strongly and directly inhibits telomerase. It is suggested that telomerase inhibition could be one of the major mechanisms underlying the anticancer effects of tea [39]. Naasani et al. proposed that the inhibition of telomerase is a key mechanism in cancer inhibition by epigallocatechin gallate [40]. Besides, epigallocatechin gallate inhibits telomerase and induces apoptosis in drug resistant lung cancer and cervical cancer cells [41, 42]. Epigallocatechin gallate and sulforaphane combination treatment induce apoptosis in paclitaxel-resistant ovarian cancer cells through hTERT and Bcl-2 downregulation [43]. Sulforaphane (SFN) is a dietary isothiocyanate. SFN decreases viability and telomerase activity in hepatocellular carcinoma Hep3B cells. Moon et al. suggested that the reactive oxygen species (ROS) are essential for the suppression of SFN-mediated telomerase regulation [44]. According to Meeran et al.'s results, SFN causes epigenetic downregulation of hTERT expression in human breast cancer cell lines [45].

It would suggest that telomerase inhibitors might be most effective in combinations with other conventional or experimental cancer treatments [2].

There are alternative mechanisms for telomerase maintenance (ALT) and some rare telomerase negative human cancers. Unfortunately, telomerase inhibitors might result in the emergence of drug resistant telomerase-independent cancer cells [2].

Telomerase inhibitors can be useful for the treatment of some other diseases. Blackburn proposed that telomerase might be target for drugs against eukaryotic pathogenic or parasitic microorganisms, such as parasitic protozoans or pathogenic yeast [4]. Actually, some studies about telomerase activities of eucaryotic pathogenic microorganisms were achieved. Cano and colleagues identified telomerase activity in extracts of *Trypanosoma brucei, Leishmania major*, and *Leishmania tarentolae* and they proposed as a target the inhibition of telomerase activity [46].

The catalytic subunit of telomerase (TERT) shows a striking similarity to retroviral reverse transcriptases (retroviral RTs) and viral RNA polymerase [8]. Rubomycin and some of its analogs were demonstrated to be potent inhibitors of retroviral RTs and also inhibitors of telomerase [20].

Telomerase inhibition is a good and specific target. Because the telomerase is not detected in most normal tissues [2, 17], differences in telomerase expression, telomere length, and cell kinetics between normal and cancer tissues suggest that targeting telomerase for cancer therapy may be relatively safe [19].

3. Telomerase activators and pharmaceutical importance

Proliferation of telomerase negative cells results in progressive telomere shortening. Cellular senescence is thought to serve as a protecting mechanism against cancer, but subsequent telomere dysfunction will be involved in tumorigenesis late in life [20]. Telomere shortening may cause aging and death. Some evidence suggests that the progressive loss of telomeric repeats of chromosomes may function as a molecular clock that triggers senescence [47–49]. Bodnar et al. analyzed two telomerase-negative normal human cell types. The cells were transfected with vectors encoding to human telomerase catalytic subunit. Telomerase expressing clones had elongated telomeres and showed reduced senescence signs [47]. Numerous epidemiological studies show that shorter telomeres in humans are associated with many age-related diseases [49, 50]. Humans with shorter than average telomere length are at risk of dying from heart disease, stroke, or infection. Individuals with chronic stress or infections have accelerated telomere shortening compared to age-matched counterparts [51].

Telomerase-related gene mutations result in some diseases. The first disease-associated with mutations in human telomerase is dyskeratosis congenita (DKC) [20]. The X-linked form of the DKC is caused by mutations in the gene encoding dyskerin (DKC1). It has been suggested that DKC may be caused by a defect in rRNA processing. Dyskerin is associated also with human telomerase RNA [52]. Autosomal dominant form of DKC is closely associated with the mutations in the TER and defective telomere maintenance [53]. Mitchell et al. find that primary fibroblasts and lymphoblasts from DKC affected males have a lower level of telomerase RNA, produce lower levels of telomerase activity, and have shorter telomeres [52].

More recently, telomerase mutations have been detected in the context of aplastic anemia, Hoyeraal-Hreidarsson syndrome, idiopaty pulmonary fibrosis, ataxia telangiectasia, Werner syndrome, Bloom syndrome, Nijmegen breakage syndrome, and ataxia telangiectasia-like disorder [20, 54]. The unifying molecular characteristics of these diseases are that patients harbor telomeres that are significantly shorter than the age-matched control subjects [54]. Not only the genetic modulation but also the epigenetic mechanisms may be responsible for the diverse expression status of telomerase in a tissue and cell-type dependent manner [55].

Also, telomere erosion occurred by excessive T-cell proliferation in AIDS or X-linked Lymphoproliferative syndrome. However, cardiovascular diseases have been recently linked with telomere-dependent senescence [20].

Because of that, telomerase activators are important for antiaging and telomerase-dependent disease treatments. Telomerase gene therapy in adult and old mice delays aging and increases longevity [56, 57]. TERT exhibits neuroprotective effects in experimental models of neurodegenerative disorders suggesting that inducing the telomerase activity in neurons may protect against age-related neurodegeneration and Alzheimer's disease [58]. Geron Corp. and TA Therapeutics developed a single molecule telomerase activator, TAT2 (cycloastragenol) [29]. Cycloastragenol is an aglycone of astragaloside IV (**Figure 3**). It was first defined when screening Astragalus membranaceus extracts for antiaging properties and a potent telomerase activator in neuronal cells [59]. The extract of *Astragalus membranaceus* was licensed as a nutritional supplement called TA 65 (TA sciences, Geron Corp.). This extract could elongate short telomeres and increase health span of adult mice without increasing cancer incidence [60]. Also, this natural-based product can elongate short telomeres in human leukocytes [61].

Furthermore, certain phytochemicals like resveratrol and genistein have been shown to activate telomerase (Figure 3). Genistein is a natural isoflavone found in soybean products. Genistein inhibits the transcription of hTERT in breast MCF10AT benign cells and MCF7 cancer cells [62]. Genistein also decreases telomerase activity in prostate cancer cells [63, 64]. Ouchi et al. showed that the expression of hTERT and c-myc mRNA was downregulated by genistein in prostate cancer cells [63]. But, physiologically achievable concentrations of genistein enhance telomerase activity in prostate cancer cells [65]. Genistein may activate telomerase activity at low concentrations and inhibit telomerase activity at higher treatment concentrations [29].

The *trans*-izomer of resveratrol is a natural phytoalexin present in a limited number of Spermatophyta, especially in grapes, fruits, and root extracts. It is synthesized in response to stress conditions. Resveratrol has a direct inhibitory effect on cell proliferation. Studies showed that resveratrol treatment downregulates the telomerase activity and levels of hTERT in MCF7 breast cancer cells [66]. Besides, relatively high concentrations of resveratrol were found to be able to downregulate telomerase activity in human colon cancer cells [67]. Several compounds like resveratrol have been shown to act as both inhibitors and activators

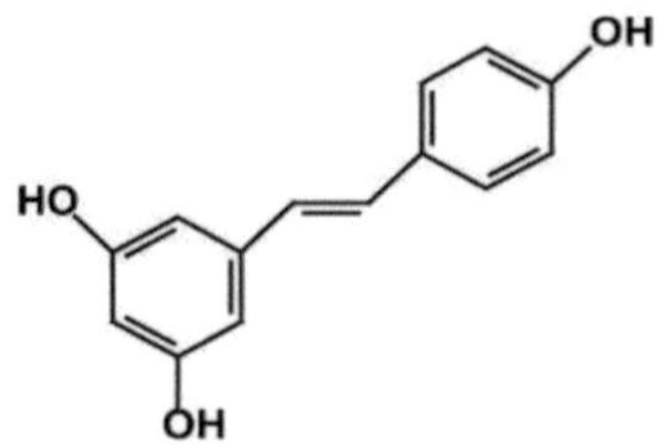

Resveratrol

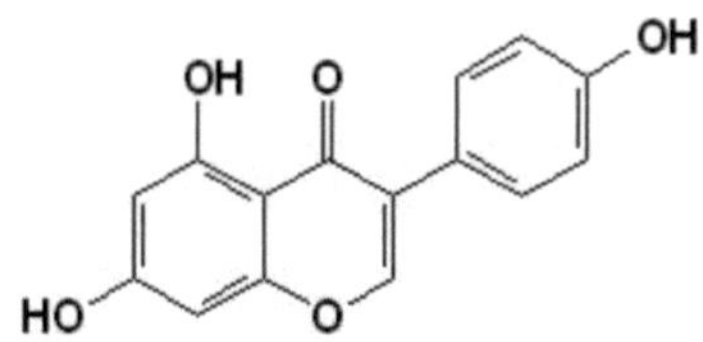

Genistein

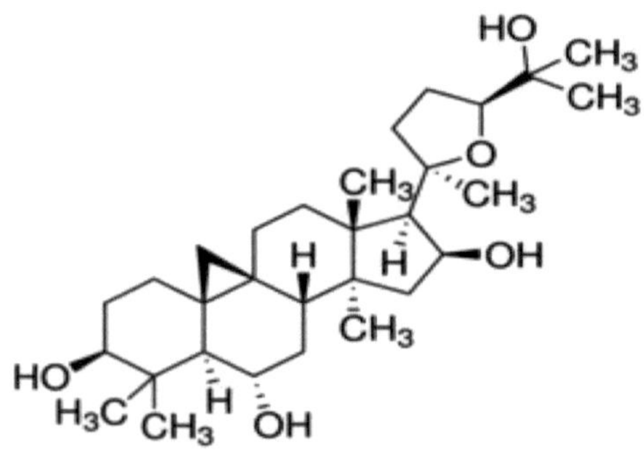

Cycloastragenol

Figure 3. Chemical formula of some natural telomerase activators.

of telomerase though this may be due to treatment concentration or cell type differences. Resveratrol has been shown to inhibit telomerase activity in cancer cells but activate telomerase in epithelial and endothelial progenitor cells [29, 68].

There are few studies about the effects of Gingko biloba on telomerase activity. Dong et al. showed that Gingko biloba extract increases telomerase activity in endothelial progenitor cell [69].

Also, HMG-CoA reductase inhibitor therapy and statin treatment are associated with delay of senescence and reduced cardiovascular diseases [29]. Moreover, in our study (unpublished) it was observed that dimetylsulfooxide (DMSO), which is used for solving chemical substances increases telomerase activity. In the study of Alfonso-De Matte et al., DMSO increased telomerase activity in some cell lines that is known to have no/low telomerase activity [70].

http://dx.doi.org/10.5772/65933

Also, in a study which was carried out about differentiation of embryonic stem cell on rats, TERT gene upregulated as a result of dimetylsulfooxide (DMSO) application on each individual and telomerase activity increased [71].

Telomere shortening correlates with cellular aging [6]. Telomerase-related gene mutations also result in some diseases [20]. Because of that, telomerase activators are important for anti-aging and telomerase-dependent disease treatments.

Acknowledgement

I thank Dr. Ismail Kayagil for photographs.

Author details

Ayse Gul Mutlu

Address all correspondence to: agmutlu@mehmetakif.edu.tr

Molecular Biology and Genetics Department, Mehmet Akif Ersoy University, Burdur, Turkey

References

[1] Blackburn EH. Switching and signaling at the telomere. Cell. 2001; 106: 661–673.

[2] Shay JW, Wright WE. Telomerase: a target for cancer therapeutics. Cancer Cell. 2002; 2: 257–262.

[3] Moyzis RK, Buckingam JM, Cram LS, Dani M, Deaven LL, Jones MD, Meyne J, Ratliff RL, Wu J. A highly conserved repetitive DNA sequence, (TTAGGG)n, present at the telomeres of human chromosomes. Proceeding of the National Academy of Sciences of USA. 1988; 85: 6622–6626.

[4] Blackburn EH. Structure and function of telomeres. Nature. 1991; 350: 569–573.

[5] Griffith JD, Comeau L, Rosenfield S, Stansel RM, Bianchi A, Moss H, de Lange T. Mammalian telomeres end in a large duplex loop. Cell. 1999; 97: 503–514.

[6] Sandin S, Rhodes D. Telomerase structure. Current Opinion in Structural Biology. 2014; 25: 104–110.

[7] Greider CW, Blackburn EH. Identification of a specific telomere terminal transferase activity in Tetrahymena extracts. Cell. 1985; 43: 405–413.

[8] Mitchell M, Gillis A, Futahashi M, Fujiwara H, Skordalakes E. Structural basis for telomerase catalytic subunit TERT binding to RNA template and telomeric DNA. Nature Structural and Molecular Biology. 2010; 17: 513–518.

[9] Banik SSR, Guo C, Smith AC, Margolis SS, Richardson DA, Tirado CA, Counter CM. C terminal regions of the human telomerase catalytic subunit essential for in vivo enzyme activity. Molecular and Cellular Biology. 2002; 22: 6234–6246.

[10] Jiang J, Miracco EJ, Hong K, Eckert B, Chan H, Cash DD, Min B, Zhou ZH, Collins K, Feigon J. The architecture of tetrahymena telomerase holoenzyme. Nature. 2013; 496: 187–192.

[11] Artandi SE, Depinho RA. Telomeres and telomerase in cancer. Carcinogenesis. 2010; 31: 9–18.

[12] Hiyama E, Hiyama K. Genetic and epigenetic modulation of telomerase activity in development and disease. British Journal of Cancer. 2007; 96: 1020–1024.

[13] Shay JW, Reddel RR,Wright WE. Cancer and telomeres an alternative to telomerase. Science. 2012; 336: 1388–1390.

[14] Kim NW, Piatyszek MA, Prowse KR, Harley CB, West MD, Ho PLC, Coviello GM, Wright WE, Weinrich SL, Shay JW. Specific association of human telomerase activity with immortal cells and cancer. Science. 1994; 266: 2011–2015.

[15] Tahara H, Nakanishi T, Kitamoto M, Nakashio R, Shay JW, Tahara E, Kajiyama G, Ide T. Telomerase activity in human liver tissues: comparison between chronic liver disease and hepatocellular carcinomas. Cancer Research. 1995; 55: 2734–2736.

[16] Hiyama E, Gollahon L, Kataoka T, Kuroi K, Yokoyama T, Gazdar AF, Hiyama K, Piatyszek MA, Shay JW. Telomerase activity in human breast tumors. Journal of the National Cancer Institute. 1996; 88: 116–122.

[17] Buseman CM, Wright WE, Shay JW. Is telomerase viable target in cancer. Mutation Research. 2012; 730: 90–97.

[18] Ferber MJ, Montoya DP, Yu C, Aderca I, McGee A, Thorland EC, Nagorney DM, Gostout BS, Burgart LJ et al. Integrations of the hepatitis B virus (HBV) and human papillomavirus (HPV) into the human telomerase reverse transcriptase (hTERT) gene in liver and cervical cancers. Oncogene. 2003; 22: 3813–3820.

[19] Harley CB, Telomerase and cancer therapeutics. Nature Reviews Cancer. 2008; 8: 167–179.

[20] Tarkanyi I, Aradi J. Pharmacological intervention strategies for affecting telomerase activity: Future prospects to treat cancer and degenerative disease. Biochimie. 2008; 90: 156–172.

[21] Saretzki G. Telomerase inhibition as cancer therapy. Cancer Letters. 2003; 194: 209–219.

[22] Damm K, Hemmann U, Garin-Chesa P, Hauel N, Kauffmann I, Priepke H, Niestroj C, Daiber C et al. A highly selective telomerase inhibitör limiting human cancer cell proliferation. The EMBO Journal. 2001; 20: 6958–6968.

[23] Leng Y, Lu T, Yuan HL, Liu HC, Lu S, Zhang WW, Jiang YL, Chen YD. QSAR studies on imidazopyrazine derivatives as Aurora A kinase inhibitors. SAR and QSAR in Environmental Research. 2012; 23: 705–730.

[24] Mitchell SA, Danca MD, Blomgren PA, Darrow JW, Currie KS, Kropf JE, Lee SH, Gallion SL, Xiong JM, Pippin DA, DeSimone RW, Brittelli DR, Eustice DC, Bourret A, Hill-Drzewi M, Maciejewski PM, Elkin LL. Imidazo[1,2-a]pyrazine diaryl ureas: Inhibitors of the receptor tyrosine kinase EphB4. Bioorganic and Medicinal Chemistry Letters. 2009; 19: 6991–6995.

[25] Matthews TP, McHardy T, Klair S, Boxall K, Fisher M, Cherry M, Allen CE, Addison GJ, Ellard J, Aherne GW, Westwood IM, Montfort R, Garrett MD, Reader JC, Collins I. Design and evaluation of 3,6-di(hetero)aryl imidazo[1,2-a]pyrazines as inhibitors of checkpoint and other kinases. Bioorganic and Medicinal Chemistry Letters, 2010; 20: 4045–4049.

[26] Rosse G. Imidazopyrazine derivatives as inhibitors of mTOR. ACS Medical Chemistry Letters. 2013; 4: 498–499.

[27] Chen L, Monti S, Juszczynski P, Ouyang J, Chapuy B, Neuberg D, Doench JG, Bogusz AM, Habermann TM, Dogan A, Witzig TE, Kutok JL, Rodig SJ, Golub T, Shipp MA. SYK inhibition modulates distinct PI3K/AKT dependent survival pathways and cholesterol biosynthesis in diffuse large B cell lymphomas. Cancer Cell. 2013; 23: 826–838.

[28] Baviskar AT, Madaan C, Preet R, Mohapatra P, Jain V, Agarwal A, Guchhait SK, Kundu CN, Banerjee UC, Bharatam PV. N-fused imidazoles as novel anticancer agents that inhibit catalytic activity of topoisomerase IIα and induce apoptosis in G1/S phase. Journal of Medical Chemistry. 2011; 54: 5013–5030.

[29] Sprouse AA, Steding CE, Herbert B. Pharmaceutical regulation of telomerase and its clinical potential. Journal of Cellular and Molecular Medicine. 2012; 16: 1–7.

[30] Sun L, Wang X. Effects of allicin on both telomerase activity and apoptosis in gastric cancer SGC-7901 cells. World Journal of Gastroenterology. 2003; 9: 1930–1934.

[31] Faezizadeh Z, Mesbah-Namin SAR, Allameh A. The effect of silymarin on telomerase activity in the human leukemia cell line K562. Planta Medica. 2012; 78: 899–902.

[32] Yurtcu E, Darcansoy Iseri O, Sahin FI. Effects of silymarin-doxorubicin applications on telomerase activity of human hepatocellular carcinoma cell line HepG2. Journal of Balkan Union Oncology. 2015; 20: 555–561.

[33] Parzonko A, Naruszewicz M. Silymarin inhibits endothelial progenitor cells' senescence and protects against the antiproliferative activity of rapamycin: preliminary study. Journal of Cardiavascular Pharmacology. 2010; 56: 610–618.

[34] Thelen P, Wuttke W, Jarry H, Grzmil M, Ringert RH. Inhibition of telomerase activity and secretion of prostate specific antigen by silibinin in prostate cancer cells. The Journal of Urology. 2004; 171: 1934–1938.

[35] Nasiri M, Zarghami N, Koshki KN, Mollazadeh M, Moghaddam MP, Yamchi MR, Esfahlan RJ, Barkhordari A, Alibakhsi A. Curcumin and silibinin inhibit telomerase expression in T47D human breast cancer cells. Asian Pacific Journal of Cancer Prevention. 2013; 14: 3449–3453.

[36] Chakraborty S, Ghosh U, Bhattacharyya NP, Bhattacharyya RK, Roy M. Inhibition of telomerase activity and induction of apoptosis by curcumin in K-562 cells. Molecular Mechanisms of Mutagenesis. 2006; 596: 81–90.

[37] Ramachandran C, Fonseca HB, Jhabvala P, Escalon EA, Melnick SJ. Curcumin inhibits telomerase activity through human telomerase reverse transcriptase in MCF-7 breast cancer cell line. Cancer Letters. 2002; 184: 1–6.

[38] Cosan DT, Soyocak A. Inhibiting telomerase activity and inducing apoptosis in cancer cells by several natural food compounds. In: Bibo Li, editör. Reviews on Selected Topics of Telomere Biology. InTech, Rijeka, Croatia; 2012. P. 123-148.

[39] Naasani I, Seimiya H, Tsuruo T. Telomerase inhibition, telomere shortening and senescence of cancer cells by tea catechines. Biochemical and Biophysical Research Communications. 1998; 249: 391–396.

[40] Naasani I, Oh-Hashi F, Oh-Tara T, Feng WY, Johnston J, Chan K, Tsuruo T. Blocking telomerase by dietary polyphenols is a major mechanism for limiting the growth of human cancer cells in vitro and in vivo. Cancer Research. 2003; 63: 824–830.

[41] Sadava D, Whitlock E, Kane SE. The green tea polyphenol, epigallocatechin-3-gallate inhibits telomerase and induces apoptosis in drug resistant lung cancer. Biochemical and Biophysical Communications. 2007; 360: 233–237.

[42] Yokoyama M, Noguchi M, Nakao Y, Pater A, Iwasaka T. The tea polyphenol, (−)-epigallocatechin gallate effects on growth, apoptosis, and telomerase activity in cervical cell lines. Gynecologic Oncology. 2004; 92: 197–204.

[43] Chen H, Landen CN, Li Y, Alvarez RD, Tollefsbol TO. Epigallocatechin gallate and sulforaphane combination treatment induce apoptosis in paclitaxel-resistant ovarian cancer cells through Htert AND Bcl-2 down-regulation. Experimental Cell Research. 2013; 319: 697–706.

[44] Moon D, Kang S, Kim K, Kim M, Choi YH, Kim G. Sulforaphane decreases viability and telomerase activity in hepatocellular carcinoma Hep3B cells through the reactive oxygen species-dependent pathway. Cancer Letters. 2010; 295: 260–266.

[45] Meeran SM, Patel SN, Tollefsbol TO. Sulforaphane causes epigenetic repression of Htert expression in human breast cancer cell lines. PLoS One. 2010; 5: e11457.

[46] Cano MIN, Dungan JM, Agabian N, Blackburn EH. Telomerase in kinetoplastid parasitic protozoa.Proceeding of the National Academy of Sciences of USA. 1999; 96: 3616–3621.

[47] Bodnar AG, Ouelette M, Frolkis M, Holt SE, Chiu C, Morin GB, Harley CB, Shay JW, Lichtsteiner S, Wright WE. Extension of life-span by introduction of telomerase into normal human cells. Science. 1998; 279: 349–352.

[48] Harley CB, Futcher AB, Greider CW. Telomerase shorten during ageing of human fibroblasts. Nature. 1991; 345: 458–460.

[49] Boccardi V, Paolisso G. Telomerase activation: a potential key modulator for human healthspan and longevity. Ageing Research Review. 2014; 15: 1–5.

[50] Blasco MA. Telomeres and human disease: ageing, cancer and beyond. Nature Reviews Genetics. 2005; 6: 611–622.

[51] Harley CB. Telomerase therapeutics for degenerative diseases. Current Molecular Medicine. 2005; 5: 29–38.

[52] Mitchell JR, Wood E, Collins K. A telomerase component is defective in the human disease dyskeratosis congenita. Nature. 1999; 402: 551–555.

[53] Marrone A, Walne A, Dokal I. Dyskeratosis congenita: telomerase, telomeres and anticipation. Current Opinion in Genetics and Development. 2005; 15: 249–257.

[54] Gomez DE, Armando RG, Farina HG, Menna PL, Cerrudo CS, Ghiringhelli PD, Alonso DF. Telomere structure and telomerase in health and disease. International Journal of Oncology. 2012; 41: 1561–1569.

[55] Liu L, Lai S, Andrews LG, Tollefsbol TO. Genetic and epigenetic modulation of telomerase activity in development and disease. Gene. 2004; 340: 1–10.

[56] De Jesus BB, Vera E, Schneeberger K, Tejera AM, Ayuso E, Bosch F, Blasco MA. Telomerase gene therapy in adult and old mice delays aging and increases longevity without increasing cancer. EMBO Molecular Medicine. 2012; 4: 691–704.

[57] De Jesus BB, Blasco MA. Telomerase at the intersection of cancer and aging. Trends Genet. 2013; 29: 513–520.

[58] Mattson MP. Emerging neuroprotective strategies for Alzheimer's disease: dietary restriction, telomerase activation, and stem cell therapy. Experimental Gerontology. 2000; 35: 489–502.

[59] Ip FCF, Ng YP, An HJ, Dai Y, Pang HH, Hu YQ, Chin AC, Harley CB, Wong YH, Ip NY. Cycloastragenol is a potent telomerase activator in neuronal cells: implications for depression management. Neurosignals. 2014; 22: 52–63.

[60] de Jesus BB, Schneeberger K, Vera E, Tejera A, Harley CB, Blasco MA. The telomerase activator TA-65 elongates short telomeres and increases health span of adult/old mice without increasing cancer incidence. Aging Cell. 2011; 10: 604–621.

[61] Harley CB, Liu W, Blasco M, Vera E, Andrews WH, Briggs LA, Raffaele JM. A natural product telomerase activator as part of a health maintenance program. Rejuvenation Research. 2011; 14: 45–56.

[62] Li Y, Liu L, Andrews LG, Tollefsbol TO. Genistein depletes telomerase activity through cross-talk between genetic and epigenetic mechanisms. International Journal of Cancer. 2009; 125: 286–296.

[63] Ouchi H, Ishiguro H, Ikeda N, Hori M, Kubota Y, Uemura H. Genistein induces cell growth inhibition in prostate cancer through the supression of telomerase activity. International Journal of Urology. 2005; 12: 73–80.

[64] Jagadeesh S, Kyo S, Banarjee PP. Genistein represses telomerase activity via both transcriptional and posttranslational mechanisms in human prostate cancer cells. Cancer Research. 2006; 66: 2107.

[65] Chau MN, El Touny LH, Jagadeesh S, Banerjee PP. Physiologically achievable concentrations of genistein enhance telomerase activity in prostate cancer cells via the activation of STAT3. Carcinogenesis. 2007; 28: 2282–2290.

[66] Lanzilli G, Fuggetta MP, Tricarico M, Cottarelli A, Serafino A, Falchetti R, Ravagnan G, et al. Resveratrol down-regulates the growth and telomerase activity of breast cancer cells in vitro. International Journal of Oncology. 2006; 28: 641–648.

[67] Fuggetta MP, Lanzilli G, Tricarico M, Cottarelli A, Falchetti R, Ravagnan G, Bonmassar E. Effect of resveratrol on proliferation and telomerase activity of human colon cancer cells in vitro. Journal of Experimental Clinical Cancer Research. 2006; 25: 189.

[68] Xia L, Wang XX, Hu XS, Guo XG, Shang YP, Chen HJ, Zeng CL, Zhang FR, Chen JZ. Resveratrol reduces endothelial progenitor cells senescence through augmentation of telomerase activity by Akt-dependent mechanisms. British Journal of Pharmacology. 2008; 155: 387–394.

[69] Dong XX, Hui ZJ, Xiang WX, Rong ZF, Jian S, Zhu CJ. Gingko biloba extract reduces endothelial progenitor-cell senescence through augmentation of telomerase activity. Journal of Cardiovascular Pharmacology. 2007; 49: 111.

[70] Alfonso-De Matte AMY, Cheng JQ, Kruk PA. Ultraviolet irradiation- and dimethyl sulfoxide-induced telomerase activity in ovarian epithelial cell lines. Experimental Cell Research. 2001; 267: 13–27.

[71] Armstrong L, Lako M, Lincoln J, Cairns MP, Hole N. mTERT expression correlates with telomerase activity during the differentiation of murine embryonic stem cells. Mechanisms of Development, 2000; 97: 109–116.

Chapter 6

Effect of Metal Ions, Chemical Agents and Organic Compounds on Lignocellulolytic Enzymes Activities

Josiani de Cassia Pereira, Ellen Cristine Giese,
Marcia Maria de Souza Moretti,
Ana Carolina dos Santos Gomes,
Olavo Micali Perrone, Maurício Boscolo,
Roberto da Silva, Eleni Gomes and
Daniela Alonso Bocchini Martins

Additional information is available at the end of the chapter

http://dx.doi.org/10.5772/65934

Abstract

Lignocellulolytic enzymes have been extensively studied due to their potential for industrial applications such as food, textile, pharmaceutical, paper, and, more recently, energy. The influence of metal ions, chemical agents, and organic compounds on these enzyme activities are addressed in this chapter, based on data available in the scientific literature.

Keywords: cellulases, hemicellulases, ligninases, enzymatic activities, metal ions, chemical agents, organic compounds

1. Introduction

Lignocellulolytic enzymes comprise cellulases, hemicellulases, and ligninases, which respectively degrade cellulose, hemicellulose, and lignin, the main constituents of plant cell wall, which collectively are named lignocellulose. Cellulases are employed in many industrial sectors, such as textile [1], detergents [2], animal feed, and vinification [2–5]. In the last years, the potential of these enzymes to saccharify cellulose from lignocellulosic residues has been extensively studied aiming the use of glucose for cellulosic ethanol production [6]. Hemicellulases are used in biobleaching of Kraft pulp for paper production [7, 8], bioclarification of fruit juices [9], and obtainment of C5 and C6 sugars from lignocellulosic residues, in the context of second-generation ethanol production [10]. Finally, ligninases are used in paper, textile, cosmetic, and pharmaceutical industries, in bioremediation and wastewater treatment [11, 12], in organic synthesis, and in biological pretreatment of lignocellulosic residues [13] to be used for cellulosic ethanol production.

Many studies have elucidated how cellulases bind to their substrates, as well as their catalytic mechanisms [14–17]. The modes of action of hemicellulases and ligninases have also been explored [18, 19]. The knowledge about these enzymes activators and inhibitors is also relevant, mainly in the context of industrial applications. Metal ions, for example, influence

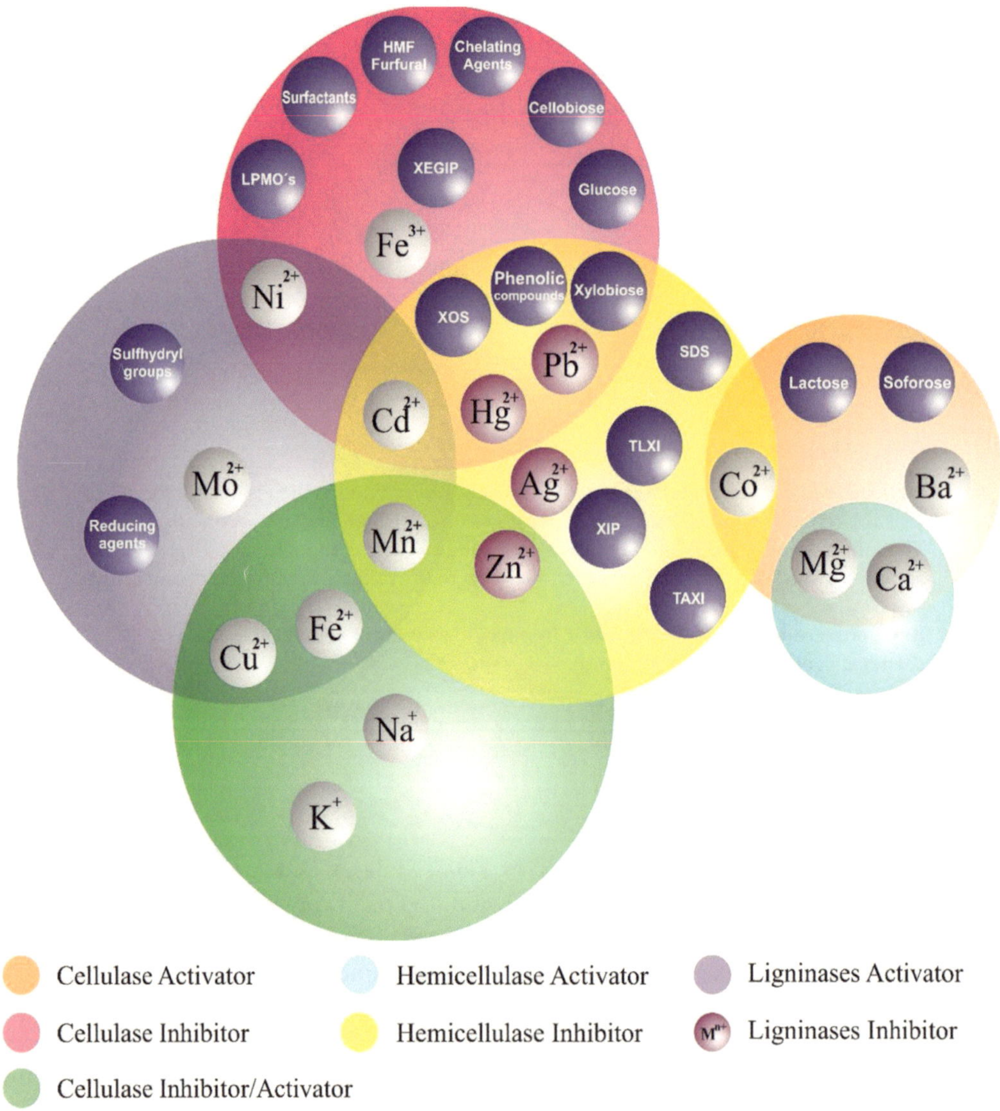

Figure 1. General distribution of activators and inhibitors of lignocellulases. HMF furfural: hydroxymethyl furfural; LPMOs: lytic polysaccharide monooxygenases; XEGIP: xyloglucan-endo-β-glucanase inhibitor proteins; XOS: xylooligosaccharides; SDS: sodium dodecyl sulfate; TAXI: *T.* xylanase inhibitor; XIP: xylanase inhibitor protein; TLXI: thaumatin-like xylanase inhibitor.

these enzymes activities and may be present in water and/or other reagents employed in industrial processes or may result from equipment corrosion [20]. However, the interference mechanisms are not well understood. There is also a lack of data to corroborate if the inhibition or activation occurs via allosteric or nonallosteric mechanism. So, this chapter presents a brief review of the main activators and inhibitors of lignocellulolytic enzymes, which are summarized in **Figure 1**.

2. Cellulases

Cellulases are glycoside hydrolases produced mainly by microorganisms, especially filamentous fungi. Microbial cellulases include endoglucanases, exoglucanases, and β-glucosidases, which synergistically degrade cellulose.

The glycosidic bonds in cellulose molecule are not easily accessible to the active site of cellulases. So, many of these enzymes are modular, consisting of one or more noncatalytic carbohydrate binding modules (CBMs). CBMs associate the enzyme with the insoluble substrate and are connected to the catalytic module by linker peptides varying in length and structure [21, 22].

Endoglucanases (EG, endo-1,4-β-endoglucanases, E.C. 3.2.1.4) hydrolyze the amorphous fraction of cellulose, releasing cellodextrins and cello-oligosaccharides [22] decreasing the substrate polymerization degree. They are classified into 11 families of glycosil-hydrolases: GH 5, 6, 7, 8, 9, 12, 44, 45, 48, 51, and 74 [23]. Some endoglucanases have affinity with others substrates, besides cellulose, such as xyloglucan, xylan, and mannan [24].

Exoglucanases or cellobiohydrolases (CBH, exo-1,4-β-exoglucanases, E.C. 3.2.1.91) degrade the crystalline fraction of cellulose, releasing cellobiose, and are named Type I or II (action in nonreducing or reducing ends, respectively). Exoglucanases are clustered in two families of glycosil-hydrolases: GH 7 (CBH I) and GH 6 (CBH II) [22].

β-Glucosidases or cellobiases (beta-D-glucosideglucohydrolase, BG, E.C. 3.2.1.21) hydrolyze cellobiose to glucose and also remove the nonreducing terminal β-D-glucosyl residue from glycoconjugates [25].

2.1. Metal ions associate to cellulases activities

Metal ions can be associated to proteins and can also form complexes with other molecules linked to enzymes acting as electron donors or acceptors as Lewis's acids, or as structural regulators [26]. These ions can either activate or inhibit the enzymatic activity by interacting with amine or carboxylic acid group of the amino acids [27].

Several studies have reported the activation or inactivation of microbial cellulases by metal ions (**Table 1**).

Mono-, di-, and trivalent metal ions such as Na^+, K^+, Ca^{2+}, Mg^{2+}, Mn^{2+}, Fe^{2+}, Co^{2+}, Cu^{2+}, Ni^{2+}, Zn^{2+}, Hg^{2+}, and Fe^{3+} are commonly studied in the characterization assays of cellulases [46]. Besides

Enzyme	Microorganism	Activator metals	Inactivating metals	Reference
Endoglucanase	*Aspergillus fumigatus*	Co^{2+} and Mg^{2+}	K^{+}, Mn^{2+}, Na^{+}, Cu^{2+}, Fe^{2+}, Fe^{3+}, Pb^{2+}, Ni^{2+}, Cd^{2+}, Hg^{2+}	[28]
Endoglucanase	*Penicillium simplicissimum* H-11	Mg^{2+} and Sn^{2+}	Cu^{2+}, Co^{2+}, Li^{2+}, Fe^{2+}, Mn^{2+}	[29]
Endoglucanase	*Aspergillus niger*	Ca^{2+} and Mn^{2+}	Co^{2+}, Fe^{2+}, Cu^{2+}	[30]
Endoglucanase	*Aspergillusniger*ANL301	Mn^{2+}, Fe^{2+},	Mg^{2+}, Ca^{2+}, Cu^{2+}, Zn^{2+}, Hg^{2+}	[31]
Endoglucanase exoglucanase	*Aspergillus niger NRRL 567*	Zn^{2+}, Ca^{2+}, Mn^{2+}, Co^{2+}	Mg^{2+}, Fe^{2+}, Hg^{2+}	[32]
Endoglucanase		–	Cu^{2+}	
Exoglucanase		Cu^{2+}		
Endoglucanase	*Daldiniaeschscholzii* (Ehrenb.:Fr.)	Ca^{2+}, Co^{2+}	Hg^{2+}, Cu^{2+}, Fe^{2+}	[33]
β-Glucosidase	*Melanocarpus* sp.	Na^{+}, K^{+}, Ca^{2+}, Mg^{2+}, Zn^{2+}	Cu^{2+}	[34]
Cellobiohydrolase	*Trichoderma reesei*	Mn^{2+}, Ba^{2+}, Ca^{2+}	Hg^{2+}	[35]
β-Glucosidase	*Aspergillus niger*322	–	Pb^{2+}, Hg^{2+}, Mn^{2+}, Fe^{2+}	[36]
Endoglucanase	*Penicillium pinophilim* MS20	Co^{2+}, Zn^{2+}, Mg^{2+}	Na^{+}, Cu^{2+}, Hg^{2+}, Fe^{2+}, Pb^{2+}, Ni^{2+}, Mn^{2+}, Cd^{2+}	[37]
Endoglucanase	*Mucor circinelloides*	Ca^{2+}, Mg^{2+}, Co^{2+}, Cu^{2+}	Mn^{2+}	[38]
β-Glucosidase	*Penicillium citrinum* YS40-5	Na^{+}	Zn^{2+}, Cu^{2+}	[39]
β-Glucosidase	*Fusarium oxysporum*	Mn^{2+}, Fe^{2+}, Ca^{2+}, Mg^{2+}, Cu^{2+}	Hg^{2+}	[40]
β-Glucosidase	*Monascus sanguineus*	–	Ca^{2+}, K^{+}	[41]
Exoglucanase	*Aspergillus fumigatus*	Ca^{2+}, Mg^{2+}, Zn^{2+}	–	[42]
Cellobiohydrolase	*Penicillium purpurogenum*KJS506	–	Fe^{2+}, Hg^{2+}	[43]
Cellobiohydrolase	*Agaricus arvencis*	Ca^{2+}, Cu^{2+}, Mg^{2+}	Zn^{2+}	[44]
Endoglucanase	*Aspergillus terreus*	Cu^{2+}, Mg^{2+}, Ca^{2+}, Na^{+}	Fe^{2+}, Mn^{2+}, Zn^{2+}, K^{+}	[45]

Table 1. Effect of metal ions on microbial cellulases activities.

ionic charge, ion radius size has a great influence on the activity and stability of the enzyme. It was demonstrated that larger radius has less influence on catalytic amino acids, while the smaller radius can more intensely attract charged amino acids changing the enzyme's overall conformation with damage on the catalytic site [47, 48].

The studies reported inhibitory effects of Fe^{2+} and Cu^{2+} on endoglucanases, exoglucanases, and β-glucosidases activities. However, the effect of other divalent ions on cellulases activities seems to be variable among the enzymes secreted by different microorganisms (e.g., **Table 1**). The effect of divalent ions on cellulases is not well elucidated, and possibly occurs by redox effects on the amino acids, increasing or decreasing their activities [49].

Inhibition of cellulases by Hg^{2+} is related to the interaction with catalytic amino acid residues containing sulfur, leading to oxidation and irregular formation of disulfide bonds [45, 46, 49]. Fe^{2+} can complex with D/L-lysine and L-methionine [50], Cu^{2+} with histidine [51], and Ba^{2+} with arginine, glutamine, proline, serine, and valine [52].

Sajadi [53] evaluated the interaction of amino acids, such as arginine and glutamine, with metal ions and established the following order of interaction degree: $Ca^{2+} < Mg^{2+} < Mn^{2+} < Co^{2+} < Cu^{2+} > Zn^{2+}$.

2.2. Chemical agents and organic compounds associate to cellulases activities

Cellulases activities may also be affected by drugs (2,3-dichloride-1,4-nafthoquinone, for example), fungicides (such as phenylmercury acetate and ethylen-bis-dithiocarbamate), antibiotics and disinfectants (Phenylmercury nitrate and 8-hydroxiquinoline, among others), sugars (final product inhibition), protein (such as those secreted by plant as defense mechanism), CBM-binding organic compounds, products from sugar and lignin degradation (such as phenolic compounds) [54], food additives (such as Octyl gallate), plant hormones (auxins, such as indoleacetic acid), and ionic solids (Sodium azide) [55–58].

Cellulose degradation products such as cello-oligosaccharides and cellobiose can inhibit endo- and exoglucanase activities, respectively. Endoglucanases that act on xyloglucan and xylan can be inhibited by the xylooligomers released [59]. The addition of xylanase to the reaction media is an alternative to remove these products [60]. The inhibition of β-glycosidases activities by glucose is frequently observed [6, 61]. Disaccharides such as cellobiose and xylobiose, and monosaccharides such as mannose and galactose can inhibit some exoglucanases activities [22, 59, 62].

Gluconolactone, resulting from cellulose oxidation by lytic polysaccharide monooxygenases (LPMOs) activities, can act as β-glycosidases inhibitor. Cellobiose and also other substrates of β-glycosidases compete with gluconolactone and other LPMO-degrading products [63–65]. On the other hand, β-glycosidases can be activated by soforose and lactose [66, 67].

It is relevant to consider that sugars released by enzymatic hydrolysis of lignocellulose can be degraded and converted into inhibitory compounds. Under acidic conditions, glucose, mannose, and galactose can be converted into furan aldehydes such as hydroxymethylfurfurals (HMF). HMF, in turn, can be converted into levulinic and formic acids [68].

Lignin degradation during the hydrolysis of some lignocellulosic materials such as alkali or acid pretreatment, or else during enzymatic hydrolysis (by laccases action) can release phenolic compounds [68] such as vanillin, syringaldehyde, trans-cinnamic acid, and hydroxybenzoic acid. These compounds are potential inhibitors of endo/exoglucanases and β-glycosidases activities due to the presence of hydroxyl, carbonyl, and methoxyl groups [69, 70].

As mentioned above, another class of cellulolytic inhibitors has a proteic origin. Specific xyloglucan endo-β-glucanase inhibitor proteins (XEGIPs) are presented in the cell walls of some vegetables such as tomatoes, tobacco, and wheat and inhibit endoglucanases that act on xyloglucan [71–73]. These proteins are part of the plant protecting mechanism against pathogens and act by forming high-affinity complexes with the enzyme [73].

Another factor that affects the catalysis by cellulases is the enzymes interaction with lignin, the phenomenon called "nonproductive adsorption" or "nonspecific binding." Cellulases can adsorb lignin through their CBMs [21, 74–77], more specifically by its alanine residues [76]. Some cellulases show higher catalytic activity when CBMs are removed by decreasing nonproductive adsorption on lignin [74].

Nonproductive adsorption of cellulases on lignin can also be decreased by adding surfactants to the reaction media, which increases the efficiency of enzymatic catalysis [78–81]. Tween 20, 40, 60, 80, and 100, Triton X-100, polyethylene glycol (PEG), among others surfactants, tend to decrease the surface tension of aqueous systems, which may alter the properties of liquids such as detergency, emulsification, greasing, and solubilization. Surfactant properties can decrease the nonproductive adsorption of cellulases on lignin, acting as "activators agents" of these enzymes [78].

Chelating agents such as EDTA (ethylene diamine tetra acetic acid), ethylene glycol (or β-mercaptoethanol), and DPPE (1,2-bis diphenylphosphino-ethylene) may activate some enzymes activities, especially cellulases, by sequestering inhibitors' metal ions from the aqueous system [82]. When chelating agents complex with metals in the reaction media, the active site of enzyme is available to react with the substrate, which represents the positive effect of these compounds on cellulases activities. In contrast, the negative effect of chelating agents on enzymatic activity suggests that enzyme activities depend on the inorganic ion that was sequestered [20, 33, 45].

3. Hemicellulases

Since hemicellulose is very heterogeneous, its complete degradation requires the synergic action of several enzymes, mainly endoxylanases and β-xylosidases as well as a variety of accessory enzymes that act in substituted xylans and include α-D-glucuronidases, acetyl xylan esterases, ferulic acid esterases, α-galactosidases, acetyl mannan esterases, and α-L-arabinofuranosidases [83].

α-L-Arabinofuranosidases (EC 3.2.1.55.; AFases) are exopolysaccharide hydrolases which remove side chains containing arabinose residues linked by α-1,2, α-1,3, and α-1,5 glycosidic bonds to the main chain of arabinananas or arabinoxylans [84]. AFases are grouped into six families of glycoside hydrolases: GH 3, 10, 43, 51, 54, and 62 [85]. A variety of AFases have been purified from fungi, bacteria, and plants [86–88]. These enzymes' activities can be affected by metal ions, ionic and nonionic detergents, and by chelating and reducing agents [85].

Xylans with acetyl and methyl glucuronic acid (MeGlcA) as substituents groups are named *O*-acetyl-4-*O*-methylglucuronoxylans. On the other hand, when α-4-0-methylglucuronic acid and α-arabinofuranose are the substituent groups, xylans are named as arabino 4-*O*-methylglucuronoxylan [89]. α-glucuronidases (EC 3.2.1.139.) hydrolyze α-1,2-glycosidic bond of MeGlcA in the side chain [90]. Among xylan-degrading enzymes, α-glucuronidases are the less studied and characterized ones. They are grouped into three families of glycosyl-hydrolases: GH 4, 67, and 115 [91].

Endoxylanases (E.C. 3.2.1.8; endo-β-1,4-xylanases) hydrolyze β-1,4 glycosidic linkages in the backbone of xylans that are composed of xylose residues [92]. According to the similarities of amino acid sequences, the majority of xylanases are grouped into glycoside hydrolases (GH) families 10 and 11 and are also classified into families GH 5, 7, 8, and 43 [93].

β-Xylosidases (E.C. 3.2.1.37; β-1,4-xylosidases) release β-D-xylopyranosyl residues from the nonreducing end of xylobiose and some small 4-β-D-xylooligosaccharides [92]. These enzymes have been classified into 10 families: GH 1, 3, 30, 39, 43, 51, 52, 54, 116, and 120, based on the predicted structural motifs of the enzyme's catalytic domain. β-Xylosidases play a crucial role in endoxylanases activities, since their substrates, such as xylobiose, can inhibit endoxylanases action [94, 95].

3.1. Metal ions associate to hemicellulases activities

The inhibitory effect of Hg^{2+} on AFases activities has been reported [96–99]. Besides Hg^{2+}, Ag^{2+}, and Pb^{2+} are mixed inhibitors, which do not bind to the active site, but to another region of the enzyme, and thus do not interfere with substrate binding to the catalytic site. In addition, Hg^{2+} is known to react with histidine and tryptophan residues, reducing the enzyme availability to metabolic function [100]. Zn^{2+}, Cd^{2+}, and Co^{2+} have also been described as potential inhibitors of AFases [88, 99, 101].

Most scientific works about α-glucuronidases purification and characterization report that these enzymes do not require metal ions for their activities [102–106]. On the other hand, various metal ions exert inhibitory effects on α-glucuronidases activities, such as Ag^{2+}, Zn^{2+}, Cd^{2+}, Hg^{2+}, Mn^{2+}, Fe^{2+}, and Fe^{3+} (e.g., **Table 2**).

Some GH 10 family enzymes require metal ions for their stability and activities. For example, *Pseudomonas fluorescens* sub sp. produces a xylanase that is one of the first GH 10 enzymes found to contain a calcium-binding site [93]. On the other hand, there are many GH 43 enzymes with crystal structures that showed tightly bound metal ions such as Ca^{2+}, with structural roles [107]. Besides, many studies have reported the apparent activation of fungal β-xylosidases by Mn^{2+} and Ca^{2+}, suggesting that these ions activate and protect the active site [95].

The negative effect of heavy metals, such as Hg^{2+}, Fe^{2+}, Co^{2+}, Mn^{2+}, Ag^{2+}, Cu^{2+}, and Pb^{2+} on xylanases activities have been reported [108]. Inhibition by heavy metal ions (such as Zn^{2+}, Pb^{2+}, and Hg^{2+}) may occur due to the formation of a complex with the reactive groups of the enzyme. Metals from group IIb exhibit high affinity for SH, $CONH_2$, NH_2, COOH, and PO_4 [109]. Furthermore, inhibition of xylanase by Hg^{2+} has been reported as related to the presence of tryptophan residues, which oxidize indole ring, thereby inhibiting the enzyme activity [110]. Xylanase from *Bacillus halodurans* TSEV1 was strongly inhibited by Hg^{2+}, Cu^{2+}, and Pb^{2+}, probably due to the catalysis of the cysteine thiol group autooxidation, which leads to the formation of intra- and intermolecular disulfide bonds or to the formation of sulfenic acid [111].

3.2. Chemical agents and organic compounds associate to hemicellulases activities

Some authors have reported that the addition of chelating agents such as EDTA and reducing agents such as β-mercaptoethanol and DTT (dithiothreitol) does not affect AFases activity [85,

Metal ions	Microorganism	Referees
Ag^{2+}	*Bacillus stearothermophilus*	[105]
	Saccharophagus degradans 2-40	[106]
Zn^{2+}	*Bacillus stearothermophilus*	[105]
Cd^{2+}	*Thermotoga maritime*	[104]
Hg^{2+}	*Thermotoga maritime*	[104]
	Bacillus stearothermophilus	[105]
	Aspergillus niger	[102]
	Helix pomatia	[103]
	Saccharophagus degradans 2-40	[106]
Mn^{2+}	*Thermotoga maritime*	[104]
	Bacillus stearothermophilus	[105]
	Aspergillus niger	[102]
Fe^{2+} and Fe^{3+}	*Aspergillus niger*	[102]
Ni^{2+}	*Bacillus stearothermophilus*	[105]
	Saccharophagus degradans 2-40	[106]
Cu^{2+}	*Thermotoga maritima*	[104]
	Bacillus stearothermophilus	[105]
K^{+}	*Geobacillus stearothermophilus*	[105]

Table 2. Metal ions that exert inhibitory effects on α-glucuronidases activities.

99, 112]. Such agents are well known as inhibitors of thiol groups, and these data suggest that sulfhydryl groups are not related to the active site of AFases.

There are few studies reporting the action of ionic detergents in AFases activities. At low concentrations (1–2 mM), ionic detergents such as SDS can stimulate the enzyme activity, whereas in higher concentrations (20 mM) they can cause an inhibitory effect [113]. Since SDS interferes in hydrophobic regions of the enzyme, it alters its three-dimensional structure [114], indicating that these concentrations may be critical and cause enzyme denaturation.

Among the compounds that significantly activate the enzyme activity there are 2-mercaptoethanol, DTT (dithiothreitol), L-cysteine, and NAD^+ indicating that these reducing agents are required for maximal activities of α-glucuronidases [115]. Some of the family 4 enzymes are known to be NAD^+ dependent. The role of NAD^+ for the activity of the hydrolytic GHF4 is not well known. The pyridine nucleotide cofactor could have structural and/or catalytic function and, in addition, could also be important for the regulation of enzyme activity [116].

Xylanases have received great attention in recent years, mainly due to their potential for the application in the processes of xylooligosaccharides (XOs) production, pulp bleaching,

http://dx.doi.org/10.5772/65934

removal of antinutritional factors of animal feeds, bread making (improving the separation of wheat or other cereal gluten from starch), juice extraction from fruits or vegetables, clarification of fruit juices and wines, and extraction of more fermentable sugar from barley to produce beer [111, 117].

Xylanase proteic inhibitors might hamper their efficacy when used in industrial application. Two distinct types of xylanase inhibitors have been identified in barley, wheat, and rye: XIP (xylanase inhibitor protein), a monomeric and glycosylated protein (XIP-I most widely studied in the XIP class), that can inhibit all GH 10 and GH 11 fungal xylanases, except that from *Aspergillus aculeatus*. The other type of xylanase inhibitor, TAXI (*Triticumaestivum* xylanase inhibitor) is a mixture of two proteins, TAXI I and TAXI II, which differ according to xylanase specificities and p*I*. TAXI inhibitors seem to be specific for GH 11 bacterial and fungal xylanases. More recently, a third class of inhibitor called TLXI (thaumatin-like xylanase inhibitor) also purified from wheat, showed variable activities against most of GH 11 xylanases and does not inhibit GH 10 microbial xylanases [117, 118].

Many other substances, such as EDTA (a chelating reagent), β-mercaptoethanol, and DTT (both disulfide bonds reducing agents) have been extensively investigated regarding their influence on xylanases activities. Xylanase from *Talaromyces thermophile* is inhibited by EDTA and DTT, suggesting that disulfide bonds are essential to maintain the enzyme conformation [119]. On the other hand, the activation of xylanases in the presence of β-mercaptoethanol and DTT was reported and indicates the presence of a reduced thiol group of cysteine in these enzymes [120].

The effect of different modulators on the activity of xylanase from *B. halodurans* TSEV1 has been investigated. These modulators include *N*-bromosuccinimide (N-BS), ethyl-3-(3-dimethyl aminopropyl) carbodiimide (EDAC), iodoacetate (IAA), and Woodward's reagent K (WRK). The inhibition of xylanase activity in the presence of NBS suggests the presence of tryptophan residues in their active site. EDAC and WRK inhibited the enzyme activity, which indicates the importance of carboxylic groups in enzyme catalysis [111].

Treatments for deconstruction of the lignocellulosic structure are frequently employed in the use of biomass as sugar's source for ethanol production and can generate besides soluble sugars, other sources such as furan derivatives, organic acids, and phenolic compounds that can act as xylanases inhibitors, as described for cellulase [121].

Significant inhibition of xylanase activity by vanillic acid, syringic acid, acetosyringone, and syringaldehyde has been observed [121]. Boukari et al. [122] reported that endoxylanase from *Thermobacillus xylanilyticus* was inhibited by phenolic compounds, including cinnamic acid, p-coumaricacid, caffeic acid, ferulic acid, and 3, 4, 5-trimethoxy-cinnamic acid by the noncompetitive multisite inhibition mechanism.

Studies on the inhibitory effect of sugars on xylanases (mainly β-xylosidases) are essential for a better understanding about the decrease in the enzyme activity during biomass conversion. This kind of inhibition was subject of research for a long time, bringing up many different opinions about its mechanism. Jordan et al. [123] studied the active site of the GH 43 β-xylosidase from *Selenomonas ruminantium* and reported that it comprises of two subsites and a single access route for ligands. The authors classified the inhibitors into two groups: I, single binding inhibitors including cellobiose (4-O-β-D-glucopyranosyl D-glucose), D-glucose, maltose (4-O-a-D-glucopyranosyl-D-glucose), D-xylose, and L-xylose; II, double binding inhibitors including D-arabinose, L-arabinose, D-erythrose, and D-ribose. Both groups have presented competitive or noncompetitive inhibition.

4. Ligninolytic enzymes

Microorganisms that colonize on living and decaying wood are capable of producing oxidative extracellular enzymes which together play a fundamental role in lignin biodegradation. The ligninases, or lignin-degrading enzymes, can oxidize lignin and several related compounds, e.g., environmental pollutants containing polycyclic aromatic hydrocarbons, dyes, and chlorophenols [124].

Lignin-peroxidase (LiP, E.C. 1.11.1.14), manganese-peroxidase (MnP, E.C. 1.11.1.13), and laccase (E.C. 1.10.3.2) are the major lignin-modifying enzyme systems of white-rot fungi and have also been described in actinomycetes and bacteria. These enzymes oxidize phenolic compounds and reduce molecular oxygen to water, generating intermediary radicals as illustrated in **Figure 2** [125, 126].

Accessory enzymes involved in the main reactions of degradation of lignin have also been described and comprise the following: cellobiose-quinoneoxireductase (E.C. 1.1.5.1), aryl alcohol oxidase (E.C. 1.1.3.7), glyoxal oxidase (GO, E.C. 1.2.3.5), manganese-independent peroxidase (E.C. 1.11.1.7), versatile peroxidase (VP, E.C. 1.11.1.16), and cellobiose dehydrogenase (E.C. 1.1.99.18) [127, 128].

Besides ligninolytic enzymes have been used to reduce the lignin content in several feedstock and to degrade recalcitrant aromatic compounds, due to the high chemical similarity of these compounds with lignin [13, 129, 130], the lignin-degrading enzymes have been applied in various industries such as textile dye bleaching, pulp and paper delignification, food, brewery, animal feed, laundry detergents, and xenobiotic compound degradation. Phenol oxidases such as laccases, particularly, have been applied in immunoassay, biosensors, biocatalysts, and oxygen cathode manufacturing [127, 131].

The performance of these enzymes is easily affected by environmental factors including metal ions and other chemical compounds usually found in the aforementioned industries. Ligninases with stronger tolerance to metal ions and organic solvents exhibit high potential for the application in the recalcitrant xenobiotics biodegradation and also improve the effectiveness of biotechnological and industrial enzymatic process [132, 133].

4.1. Laccases (E.C. 1.10.3.2)

Laccases are multicopper blue oxidases that catalyze the one-electron oxidation of a wide range of substrates with a concomitant four-electron reduction of molecular oxygen to water [126]. The active site of laccase comprises four copper atoms in three groups: T1 (mononuclear

http://dx.doi.org/10.5772/65934

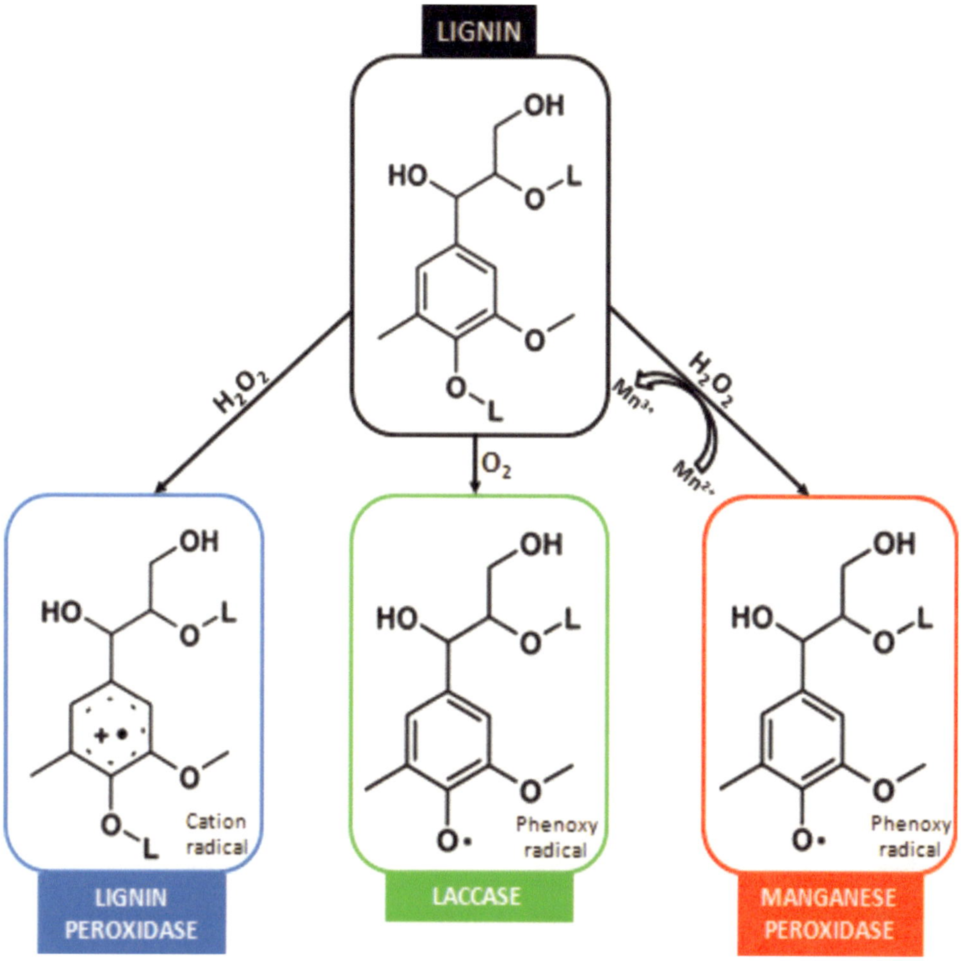

Figure 2. Simplified reactions of lignin peroxidase, manganese peroxidase, and laccase.

copper), T2 (normal copper), and T3 (coupled binuclear copper). The T1 and T2 Cu^{2+}-sites contribute as the primary electron acceptors while T3 is reduced by an intramolecular two-electron transfer from T1 and T2 Cu^{2+} sites [126, 134].

4.1.1. Metal ions associate to laccase activity

Although laccases are efficient on a wide range of substrates without cofactors, in most cases, the addition of Cu^{2+}, Cd^{2+}, Ni^{2+}, Mo^{2+}, and Mn^{2+} ions increases the activity of laccases, whereas Ag^{2+}, Hg^{2+}, Pb^{2+}, Zn^{2+}, NaN_3, NaCl, and H_2O_2 inhibit their activity [126].

Apart from the inhibition problem, the influence of metal ions on the performance of enzyme-catalyzed reaction is also important, in addition to the study of effects of single metal ions on the enzyme activity. Lu et al. [135] observed that monovalent and trivalent metal ions inhibited the 4-nitrophenol degradation by laccase-Cu^{2+}, as well as the addition of low concentrations of divalent ions. The suppressive effects of cations on laccase activity comprised $Mg^{2+} > Na^{+} > Al^{3+} > K^{+} > Mn^{2+} > Hg^{2+} > Co^{2+}$.

4.1.2. Chemical agents and organic compounds associate to laccase activity

The Michaelis-Menten equation has been suitably used to describe the laccase kinetics and apparent binding constant (K_m) and maximal reaction rate (V_{max}) values. In water-miscible solvents, these kinetic parameters can be affected by the changes in water thermodynamic activity. In the case of laccase from the white-rot fungus *Phlebiaradiata*, e. g., pK_I values show the linear dependence on solvent hydrophobicity (log*P*) in a system of 2,6 dimethoxyphenol as substrate in the presence of methanol, ethanol, *n*-propanol, acetonitrile, acetone, and DMSO [136].

Previously, the changes in V_{max} by the addition of solvents have been compared to free and immobilized laccases. The activity of laccase from *P. radiata* was rather similar to both forms of the enzyme in the presence of 10% of ethanol, methanol, acetone, DMSO, and dioxane. The immobilized laccase was less vulnerable to Cu-chelatorthioglycolic acid, 2,6-dimethoxy-1,4-benzoquinone [128, 137].

In the conditions of low water content, which is the case of water/organic mixtures, the values of the apparent K_m tend to grow exponentially with water concentration. The apparent V_{max} of immobilized laccase from *Coriolusversicolor* increased two orders of magnitude values with a linear increase in water content [138].

4.2. LiP (E.C. 1.11.1.14)

Lignin-peroxidases are heme-containing glycoproteins that contain Fe^{3+} in their active site. LiP catalyzes the H_2O_2-dependent oxidative depolymerization of nonphenolic lignin and lignin-model compounds as well as a variety of phenolic compounds [139].

4.2.1. Metal ions, chemical agents, and organic compounds associate to LiP activity

The decrease in LiP activity is described as inhibition or denaturation according to the concentration of inhibitor compounds in an aqueous reaction system. The hydrogen bonding and anion stabilization are important characteristics to describe the effect of compounds on the active sites of enzymes, as well as water activity (a_w), log*P*, and solvation [140].

The addition of Cu^{2+}, Mn^{2+}, and Fe^{2+} ions increases the activity of LiP, whereas Ag^{2+} inhibit their activity [141]. On the other hand, different solvents and organic compounds have been described as LiP potential inhibitors: alcohols, aldehydes, ketones, esters, ethers, amines, acids, amides, acetonitrile, cysteine, DMSO, EDTA, DMF, TEMED, CTAB, sodium azide, and H_2O_2 [140–144].

Vazquez-Duhalt et al. [145] chemically modified a LiP from the white-rot fungus *Phanerochaete chrysosporium* by reductive alkylation with benzyl, naphthyl, and anthracyl moieties, thereby increasing its superficial hydrophobicity. These modifications altered the kinetics and increased the yield of oxidation of pyrroles, pyridines, and aromatic amines in 10% acetonitrile.

4.3. MnP (E.C. 1.11.1.13)

Manganese-peroxidases catalyze the H_2O_2-dependent oxidation of Mn^{2+} into Mn^{3+}, which is stabilized by fungal chelators such as oxalic acid or different organic acids. Then, the oxidation of various phenolic substrates (e.g., amines, dyes, lignin related compounds) occurs under the action of chelated Mn^{3+} ions that comprise a diffusible charge-transfer mediator in these reactions [141, 146].

4.3.1. Metal ions associate to MnP activity

MnP activity is completely inhibited by Hg^{2+}, Pb^{2+}, Ag^{+}, lactate, NaN_3, $CaCl_2$, TEMED, ascorbic acid, β-mercaptoethanol, and dithreitol [147, 148]. Partial inhibition of MnP activity was observed with EDTA, a metal chelating compound that complexes with inorganic cofactors and prosthetic groups of enzymes. High concentrations of Cu^{2+} and Fe^{2+} (~4 mM) could enhance MnP activities [148]. Youngs et al. [149] related that Cd^{2+} is a reversible competitive inhibitor of Mn^{2+} to MnP activity. The inhibition was not observed in reaction systems containing 2,6-dimethoxyphenol or guaiacol in the absence of Mn^{2+}.

Acknowledgements

This work was supported by "Programa de Internacionalização da Pesquisa da UNESP" (PROINTER, Edital 06/2016), Pró-reitoria de Pesquisa (PROPe) of the Universidade Estadual Paulista "Júlio de Mesquita Filho" (UNESP), Brazil. Josiani de Cassia Pereira and Ana Carolina dos Santos Gomes thank to "Comissão de Aperfeiçoamento de Pessoal do Nível Superior (CAPES)", Brazil, for their PhD scholarships. Olavo Micali Perrone and Marcia Maria de Souza Moretti thank to ''Fundação de Amparo à Pesquisa do Estado de São Paulo – FAPESP", Brazil for their PhD and Post-doc Scholarships (Processes 2015/11588-4 and 2014/20521-8).

Author details

Josiani de Cassia Pereira[1]*, Ellen Cristine Giese[2], Marcia Maria de Souza Moretti[1], Ana Carolina dos Santos Gomes[1], Olavo Micali Perrone[1], Maurício Boscolo[1], Roberto da Silva[1], Eleni Gomes[1] and Daniela Alonso Bocchini Martins[3]

*Address all correspondence to: josianipereira@gmail.com

1 São Paulo State University, UNESP, Cristovão Colombo, CEP, São José do Rio Preto, SP, Brazil

2 Coordination for Metallurgical and Environmental Process, Centre for Mineral Technology, CETEM, Pedro Calmon, CEP, Rio de Janeiro, RJ, Brazil

3 São Paulo State University, UNESP, Prof. Francisco Degni, CEP, Araraquara, SP, Brazil

References

[1] Simic, K., Soljacic, I., Pusic, T., 2015. Application of cellulases in the process of fiishing [Uporaba celulaz v procesu plemenitenja]. Tekstilec, 58, 47–56. doi:10.14502/Tekstilec2015.58.47

[2] Podrepsek, G.H., Primozic, M., Knez, Z., Habulin, M., 2012. Immobilization of cellulase for industrial production. Chem. Eng. Transact. 27, 235–240. doi:10.3303/CET1227040

[3] Karmakar, M., Ray, R.R., 2011. Current trends in research and applications of microbial cellulases. Res. J. Microbiol. 6, 41–53. doi:10.3023/jm.2011.41.53

[4] Leite, R.S.R., Alves-Prado, H.F., Cabral, H., Pagnocca, F.C., Gomes, E., Da-Silva, R., 2008. Production and characteristics comparison of crude β-glucosidases produced by microorganisms Thermoascus aurantiacus e Aureobasidium pullulans in agricultural wastes. Enzyme Microb. Technol. 43, 391–395. doi:10.1016/j.enzmictec.2008.07.006

[5] Graminha, E.B.N., Gonçalves, A.Z.L., Pirota, R.D.P.B., Balsalobre, M.A.A., Da Silva, R., Gomes, E., 2008. Enzyme production by solid-state fermentation: Application to animal nutrition. Anim. Feed Sci. Technol. 144, 1–22. doi:10.1016/j.anifeedsci.2007.09.029

[6] de Cassia Pereira, J., Travaini, R., Paganini Marques, N., Bolado-Rodríguez, S., Bocchini Martins, D.A., 2016. Saccharification of ozonated sugarcane bagasse using enzymes from Myceliophthora thermophila JCP 1-4 for sugars release and ethanol production. Bioresour. Technol. 204, 122–129. doi:10.1016/j.biortech.2015.12.064

[7] Bajpai, P., 2004. Biological bleaching of chemical pulps. Crit. Rev. Biotechnol, 24, 1-58. doi:10.1080/07388550490465817

[8] Tavares, V.B., Gomes, E., Silva, R., 1997. Characterization of a cellulase-free xylanase producing *Bacillus* sp for biobleaching of kraft pulp. Rev. Microbiol. 28, 179–182.

[9] Rakesh Sharma (2012). Enzyme Inhibition: Mechanisms and Scope, Enzyme Inhibition and Bioapplications, Prof. Rakesh Sharma (Ed.), InTech, DOI: 10.5772/39273. Available from: http://www.intechopen.com/books/enzyme-inhibition-and-bioapplications/enzyme-inhibition-mechanisms-and-scope

[10] Sues, A., Millati, R., Edebo, L., Taherzadeh, M.J., 2005. Ethanol production from hexoses, pentoses, and dilute-acid hydrolyzate by Mucor indicus. FEMS Yeast Res. 5, 669–676. doi:10.1016/j.femsyr.2004.10.013

[11] Gomes, E., Aguiar, A.P., Boscolo, M., Carvalho, C.C., da Silva, R., Bonfá, M.R.B., 2007. Ligninases production by basidiomicetes strains on lignocellulosic agricultural residues and decolorization of synthetic dyes. J. Biotechnol. 131, S228. doi:10.1016/j.jbiotec.2007.07.414

[12] Sahadevan, L., Misra, C., Thankamani, V., 2013. Ligninolytic enzymes for application in treatment of effluent from pulp and paper industries. Univ. J. Environ. Res. Technol. 3, 14–26.

[13] Plácido, J., Capareda, S., 2015. Ligninolytic enzymes: A biotechnological alternative for bioethanol production. Bioresour. Bioprocess. 2, 23. doi:10.1186/s40643-015-0049-5

[14] Fong, M., Berrin, J.-G., Paës, G., 2016. Investigation of the binding properties of a multimodular GH45 cellulase using bioinspired model assemblies. Biotechnol. Biofuels 9, 12. doi:10.1186/s13068-016-0428-y

[15] Gao, D., Chundawat, S.P.S., Sethi, A., Balan, V., Gnanakaran, S., Dale, B.E., 2013. Increased enzyme binding to substrate is not necessary for more efficient cellulose hydrolysis. Proc. Natl. Acad. Sci. 110, 10922–10927. doi:10.1073/pnas.1213426110

[16] Jervis, E.J., Haynes, C.A., Kilburn, D.G., 1997. Surface diffusion of cellulases and their isolated binding domains on cellulose. J. Biol. Chem. 272, 24016–24023. doi:10.1074/jbc.272.38.24016

[17] Wahlström, R., Rahikainen, J., Kruus, K., Suurnäkki, A., 2014. Cellulose hydrolysis and binding with Trichoderma reesei Cel5A and Cel7A and their core domains in ionic liquid solutions. Biotechnol. Bioeng. 111, 726–733. doi:10.1002/bit.25144

[18] Nonaka, H., Kobayashi, A., Funaoka, M., 2013. Behavior of lignin-binding cellulase in the presence of fresh cellulosic substrate. Bioresour. Technol. 135, 53–57. doi:10.1016/j.biortech.2012.10.065

[19] Tenkanen, M., Buchert, J., Viikari, L., 1995. Binding of hemicellulases on isolated polysaccharide substrates. Enzyme Microb. Technol. 17, 499–505. doi:10.1016/0141-0229(94)00050-2

[20] Oviedo, C., Rodríguez, J., 2003. EDTA: The chelating agent under environmental scrutiny. Qum. Nova 26, 901–905.

[21] Boraston, A.B., Bolam, D.N., Gilbert, H.J., Davies, G.J., 2004. Carbohydrate-binding modules: Fine-tuning polysaccharide recognition. Biochem. J. 382, 769–81. doi:10.1042/BJ20040892

[22] Payne, C.M., Knott B.C., Mayes, H.B., Hansson, H., Himmel, M.E., Sandgren, M., Ståhlberg, J., Beckham, G.T., 2015. Fungal cellulases. Chem. Rev., 115, 1308-1448. doi:10.1021/cr500351c

[23] Vlasenko, E., Schülein, M., Cherry, J., Xu, F., 2010. Substrate specificity of family 5, 6, 7, 9, 12, and 45 endoglucanases. Bioresour. Technol. 101, 2405–2411. doi:10.1016/j.biortech.2009.11.057

[24] Nakazawa, H., Okada, K., Kobayashi, R., Kubota, T., Onodera, T., Ochiai, N., Omata, N., Ogasawara, W., Okada, H., Morikawa, Y., 2008. Characterization of the catalytic domains of Trichoderma reesei endoglucanase I, II, and III, expressed in Escherichia coli. Appl. Microbiol. Biotechnol. 81, 681–689. doi:10.1007/s00253-008-1667-z

[25] Sørensen, A., Lübeck, M., Lübeck, P.S., Ahring, B.K., 2013. Fungal beta-glucosidases: A bottleneck in industrial use of lignocellulosic materials. Biomolecules. 3, 612–631. doi:10.3390/biom3030612

[26] Riordan, J.F., 1977. The role of metals in enzyme activity. Ann. Clin. Lab. Sci. 7, 119–129.

[27] Ishida, N., Okubo, A., Kawai, H., Yamazaki, S., Toda, S., 1980. Interaction of amino acids with transition metal ions in solution (I) solution structure of L-lysine with Co(II) and Cu(II) ions as studied by nuclear magnetic resonance spectroscopy. Agric. Biol. Chem. 44, 263–270. doi:10.1080/00021369.1980.10863934

[28] Bagewadi, Z.K., Ninnekar, H.Z., 2015. Production, purifiation and characterization of endoglucanase from Aspergillus fumigatus and enzymatic hydrolysis of lignocellulosic waste. International Journal of Biotechnology and Biomedical Sciences, 1, 25–32.

[29] Bai, H., Wang, H., Sun, J., Irfan, M., Han, M., Huang, Y., Han, X., Yang, Q., 2013. Purification and characterization of beta 1,4-glucanases from Penicillium simplicissimum H-11. BioResources 8, 3657–3671.

[30] Bhavsar, N.H., Raol, B.V., Amin, S.S., Raol, G.G., 2015. Original research article production, optimization and characterization of fungal cellulase for enzymatic saccharifiation of lignocellosic agro-waste. Internation Journal of Current Microbiology and Applied Sciences, 4, 30–46.

[31] Nwodo, S., Obinna, C., Uzoma, C.N., Veronica, A.O., 2011. Kinetic study and characterization of 1,4-E-endoglucanase of Aspergillus niger ANL301. Dyn. Biochem. Process Biotechnol. Mol. Biol. 5, 2–7.

[32] Ghori, M.I., Ahmed, S., Malana, M.A., Jamil, A., 2012. Kinetics of exoglucanase and endoglucanase produced by Aspergillus niger NRRL 567. Afr. J. Biotechnol. 11, 7227–7231. doi:10.5897/AJB12.329

[33] Karnchanatat, A., Petsom, A., Sangvanich, P., Piapukiew, J., Whalley, A.J.S., Reynolds, C.D., Gadd, G.M., Sihanonth, P., 2008. A novel thermostable endoglucanase from the wood-decaying fungus Daldinia eschscholzii (Ehrenb.:Fr.) Rehm. Enzyme Microb. Technol. 42, 404–413. doi:10.1016/j.enzmictec.2007.11.009

[34] Kaur, J., Chadha, B.S., Kumar, B.A., Kaur, G.S., Saini, H.S., 2007. Purification and characterization of β-glucosidase from Melanocarpus sp. MTCC 3922. Electr. J. Biotechnol. 10, 260–270. doi:10.2225/vol10-issue2-fulltext-4

[35] Kim, D.W., Jang, Y.H., Kim, C.S., Lee, N.S., 2001. Effect of metal ions on the degradation and adsorption of two cellobiohydrolases on microcrystalline cellulose. Bull. Korean Chem. Soc. 22, 716–720.

[36] Peshin, a., Mathur, J.M.S., 1999. Purification and characterization of β-glycosidase from Aspergillus niger strain 322. Lett. Appl. Microbiol. 28, 401–404. doi:10.1046/j.1365-2672.1999.00533.x

[37] Pol, D., Laxman, R.S., Rao, M., 2012. Purification and biochemical characterization of endoglucanase from penicillium pinophilum MS 20. Indian J. Biochem. Biophys. 49, 189–194.

[38] Saha, B.C., 2004. Production, purification and properties of endoglucanase from a newly isolated strain of *Mucor circinelloides*. Process Biochem. 39, 1871–1876. doi: 10.1016/j.procbio.2003.09.013

[39] Ng, I.S., Li, C.W., Chan, S.P., Chir, J.L., Chen, P.T., Tong, C.G., Yu, S.M., Ho, T.H.D., 2010. High-level production of a thermoacidophilic β-glucosidase from Penicillium citrinum YS40-5 by solid-state fermentation with rice bran. Bioresour. Technol. 101, 1310–1317. doi:10.1016/j.biortech.2009.08.049

[40] Olajuyigbe, F.M., Nlekerem, C.M., Ogunyewo, O.A., 2016. Production and characterization of highly thermostable β-glucosidase during the biodegradation of methyl cellulose by Fusarium oxysporum. Biochem. Res. Int., 2016, 1-8. doi:10.1155/2016/3978124

[41] Dikshit, R., Tallapragada, P., 2015. Partial purification and characterization of β-glucosidase from Monascus sanguineus. Brazil. Arch. Biol. Technol. 58, 185–191.

[42] Mahmood, R.T., Asad, M.J., Mehboob, N., Mushtaq, M., Gulfraz, M., Asgher, M., Minhas, N.M., Hadri, S.H., 2013. Production, purification, and characterization of exoglucanase by Aspergillus fumigatus. Appl. Biochem. Biotechnol. 170, 895–908. doi:10.1007/s12010-013-0227-x

[43] Lee, K.M., Joo, A.R., Jeya, M., Lee, K.M., Moon, H.J., Lee, J.K., 2011. Production and characterization of cellobiohydrolase from a novel strain of penicillium purpurogenum KJS506. Appl. Biochem. Biotechnol. 163, 25–39. doi:10.1007/s12010-010-9013-1

[44] Lee, K.M., Moon, H.J., Kalyani, D., Kim, H., Kim, I.W., Jeya, M., Lee, J.K., 2011. Characterization of cellobiohydrolase from a newly isolated strain of Agaricus arvencis. J. Microbiol. Biotechnol. 21, 711–718. doi:10.4014/jmb.1102.02001

[45] Nazir, A., Soni, R., Saini, H.S., Manhas, R.K., Chadha, B.S., 2009. Purification and characterization of an endoglucanase from Aspergillus terreus highly active against barley β-glucan and xyloglucan. World J. Microbiol. Biotechnol. 25, 1189–1197. doi:10.1007/s11274-009-0001-y

[46] Mandels, M., Reese, E.T., 1965. Inhibition of cellulases. Annu. Rev. Phytopathol. 3, 85–102. doi:10.1146/annurev.py.03.090165.000505

[47] Coolbear, T., Whittaker, J.M., Daniel, R.M., 1992. The effect of metal ions on the activity and thermostability of the extracellular proteinase from a thermophilic Bacillus, strain EA.1. Biochem. J. 287 (Pt 2, 367–374.

[48] Zeng, J., Gao, X., Dai, Z., Tang, B., Tang, X.F., 2014. Effects of metal ions on stability and activity of hyperthermophilic pyrolysin and further stabilization of this enzyme by modification of a Ca^{2+}-binding site. Appl. Environ. Microbiol. 80, 2763–2772. doi:10.1128/AEM.00006-14

[49] Tejirian, A., Xu, F., 2010. Inhibition of cellulase-catalyzed lignocellulosic hydrolysis by iron and oxidative metal ions and complexes. Appl. Environ. Microbiol. 76, 7673–7682. doi:10.1128/AEM.01376-10

[50] Vassilev, K., 2013. Catalytic activity of amino acids-metal complexes in oxidation reactions. J. Biomater. Nanobiotechnol. 04, 28–36. doi:10.4236/jbnb.2013.42A004

[51] Kryukova, N.P., Bolotin, S.N., Panyushkin, V.T., 2003. Increased information leakage from text. Ceur Workshop Proc. 52, 1–4. doi:10.1023/A

[52] Bush, M.F., Oomens, J., Saykally, R.J., Williams, E.R., 2008. Effects of alkaline earth metal ion complexation on amino acid zwitterion stability: Results from infrared action spectroscopy. J. Am. Chem. Soc. 130, 6463–6471. doi:10.1021/ja711343q

[53] Sajadi, S.A.A., 2010. Metal ion-binding properties of L-glutamic acid and L-aspartic acid, a comparative investigation. Nat.Sci. 2, 85–90.doi 10.4236/ns.2010.22013 Natural.

[54] Kim, Y., Ximenes, E., Mosier, N.S., Ladisch, M.R., 2011. Soluble inhibitors/deactivators of cellulase enzymes from lignocellulosic biomass. Enzyme Microb. Technol. 48, 408–415. doi:10.1016/j.enzmictec.2011.01.007

[55] Reese, E.T., Mandels, M., 1957. Chemical inhibition of cellulases and B-glucosidases. Res. Rept. Pioneering Res. Div., QMRE Center, Natick, Mass., Microbial. Ser. 17, 60.

[56] Jermyn, M. A., 1952. Fungal cellulases. I. General properties of unpurified enzyme preparations from Aspergillus oryzae. Aust. J. of Sci. Res. Ser. B., 5, 409–432.

[57] Sison, B., Schubert, W., Nord, F.F., 1958. On the mechanism of enzyme action. LXV. A cellulolytic enzyme from the mold Poria vaillantii. Arch. Biochem. Biophys., 75, 260–272.

[58] Li, P., Zhu, J., Kong, Q., Jiang, B., Wan, X., Yue, J., Li, M., Jiang, H., Li, J., Gao, Z., 2013. The ethylene bis-dithiocarbamate fungicide mancozeb activates voltage-gated KCNQ2 potassium channel. Toxicol. Lett. 219, 211–217. doi:10.1016/j.toxlet.2013.03.020

[59] Qing, Q., Yang, B., Wyman, C.E., 2010. Xylooligomers are strong inhibitors of cellulose hydrolysis by enzymes. Bioresour. Technol. 101, 9624–9630. doi:10.1016/j.biortech.2010.06.137

[60] Qing, Q., Wyman, C.E., 2011. Supplementation with xylanase and β-xylosidase to reduce xylo-oligomer and xylan inhibition of enzymatic hydrolysis of cellulose and pretreated corn stover. Biotechnol. Biofuels 4, 18. doi:10.1186/1754-6834-4-18

[61] Pereira, J.C., Alonso, D., Martins, B., Gomes, E., Silva, R., 2015. Production and characterization of β-glucosidase obtained by the solid-state cultivation of the thermophilic fungus thermomucor indicae-seudaticae, N31. Appl. Biochem. Biotechnol., 175, 723-732. Doi: 10.1007/s12010-014-1332-1

[62] Hsieh, C.W.C., Cannella, D., Jørgensen, H., Felby, C., Thygesen, L.G., 2014. Cellulase inhibition by high concentrations of monosaccharides. J. Agric. Food Chem. 62, 3800–3805. doi:10.1021/jf5012962

[63] Kou, Y., Xu, J., Cao, Y., Lv, X., Zhao, G., Chen, G., Zhang, W., Liu, W., 2014. Gluconolactone induces cellulase gene expression in cellulolytic filamentous fungus Trichoderma reesei. RSC Adv. 4, 36057. doi:10.1039/C4RA06731B

[64] Mishra, S.K., Sangwan, N.S., Sangwan, R.S., 2013. Purification and physicokinetic characterization of a gluconolactone inhibition-insensitive β-glucosidase from *Andrographis paniculata* NEES. Leaf. Prep. Biochem. Biotechnol. 43, 481–499. doi:10.1080/10826068.2012.759966

[65] Tiwari, P., Misra, B.N., Sangwan, N.S., 2013. β-Glucosidases from the fungus Trichoderma: An effient cellulase machinery in biotechnological applications-glucosidases from the fungus trichoderma: An effient cellulase machinery in biotechnological applicationsglucosidases from the fungus Trichode. Biomed Res. Int., 2013, 1-10. doi:10.1155/2013/203735

[66] Dhillon, G.S., Kaur, S., Brar, S.K., Verma, M., 2012. Potential of apple pomace as a solid substrate for fungal cellulase and hemicellulase bioproduction through solid-state fermentation. Ind. Crops Prod. 38, 6–13. doi:10.1016/j.indcrop.2011.12.036

[67] Nisizawa, T., Suzuki, H., Nakayama, M., Nisizawa, K., 1971. Inductive formation of cellulase by sophorose in Trichoderma viride. J. Biochem. 70, 375–385.

[68] Jönsson, L.J., Alriksson, B., Nilvebrant, N.-O., 2013. Bioconversion of lignocellulose: Inhibitors and detoxification. Biotechnol. Biofuels 6, 16. doi:10.1186/1754-6834-6-16

[69] Qin, L., Li, W.-C., Liu, L., Zhu, J.-Q., Li, X., Li, B.-Z., Yuan, Y.-J., 2016. Inhibition of lignin-derived phenolic compounds to cellulase. Biotechnol. Biofuels. 9, 70. doi:10.1186/s13068-016-0485-2

[70] Ximenes, E., Kim, Y., Mosier, N., Dien, B., Ladisch, M., 2011. Deactivation of cellulases by phenols. Enzyme Microb. Technol. 48, 54–60. doi:10.1016/j.enzmictec.2010.09.006

[71] Juge, N., 2006. Plant protein inhibitors of cell wall degrading enzymes. Trends Plant Sci. 11, 359–367. doi:10.1016/j.tplants.2006.05.006

[72] Misas-Villamil, J.C., van der Hoorn, R. AL, 2008. Enzyme-inhibitor interactions at the plant-pathogen interface. Curr. Opin. Plant Biol. 11, 380–388. doi:10.1016/j.pbi.2008.04.007

[73] York, W.S., Qin, Q., Rose, J.K.C., 2004. Proteinaceous inhibitors of endo-β-glucanases. Biochim. Biophys. Acta: Proteins Proteom. 1696, 223–233. doi:10.1016/j.bbapap.2003.07.003

[74] Le Costaouëc, T., Pakarinen, A., Várnai, A., Puranen, T., Viikari, L., 2013. The role of carbohydrate binding module (CBM) at high substrate consistency: Comparison of Trichoderma reesei and Thermoascus aurantiacus Cel7A (CBHI) and Cel5A (EGII). Bioresour. Technol. 143, 196–203. doi:10.1016/j.biortech.2013.05.079

[75] Rahikainen, J.L., Evans, J.D., Mikander, S., Kalliola, A., Puranen, T., Tamminen, T., Marjamaa, K., Kruus, K., 2013. Cellulase-lignin interactions: The role of carbohydrate-

binding module and pH in non-productive binding. Enzyme Microb. Technol. 53, 315–321. doi:10.1016/j.enzmictec.2013.07.003

[76] Strobel, K.L., Pfeiffer, K.A., Blanch, H.W., Clark, D.S., 2015. Structural insights into the affinity of Cel7A carbohydratebinding module for lignin. J. Biol. Chem. 290, 22818–22826. doi:10.1074/jbc.M115.673467

[77] Li, Y., Sun, Z., Ge, X., Zhang, J., 2016. Effects of lignin and surfactant on adsorption and hydrolysis of cellulases on cellulose. Biotechnol. Biofuels 9, 20. doi:10.1186/s13068-016-0434-0

[78] Hsieh, C.-W.C., Cannella, D., Jørgensen, H., Felby, C., Thygesen, L.G., 2015. Cellobiohydrolase and endoglucanase respond differently to surfactants during the hydrolysis of cellulose. Biotechnol. Biofuels 8, 52. doi:10.1186/s13068-015-0242-y

[79] Pardo, A.G., 1996. Effect of surfactants on cellulase production by Nectria catalinensis. Curr. Microbiol. 33, 275–278. doi:10.1007/s002849900113

[80] Reese, E.T., Maguire, A., 1969. Surfactants as stimulants of enzyme production by microorganisms. Appl. Microbiol. 17, 242–245.

[81] Trinh, D.K., Quyen, D.T., Do, T.T., Nghiem, N.M., 2013. Purification and characterization of a novel detergent- and organic solvent-resistant endo-beta-1,4-glucanase from a newly isolated basidiomycete Peniophora sp. NDVN01. Turk. J. Biol. 37, 377–384. doi:10.3906/biy-1207-37

[82] Miyano, H., Toyo'oka, T., Imai, K., Nakajima, T., 1985. Influences of metal ions on the reaction of amino and imino acids with fluorogenic reagents. Anal. Biochem. 150, 125–130. doi:10.1016/0003-2697(85)90450-6

[83] Shallom, D., Shoham, Y., 2003. Microbial hemicellulases. Curr. Opin. Microbiol. 6, 219–228. doi:10.1016/S1369-5274(03)00056-0

[84] Koseki, T., Okuda, M., Sudoh, S., Kizaki, Y., Iwano, K., Aramaki, I., Matsuzawa, H., 2003. Role of two α-L-arabinofuranosidases in arabinoxylan degradation and characteristics of the encoding genes from shochu koji molds, Aspergillus kawachii and Aspergillus awamori. J. Biosci. Bioeng. 96, 232–241. doi:10.1016/S1389-1723(03)80187-1

[85] Numan, M.T., Bhosle, N.B., 2006. α-L-Arabinofuranosidases: The potential applications in biotechnology. J. Ind. Microbiol. Biotechnol. 33, 247–260. doi:10.1007/s10295-005-00721

[86] Bezalel, L., Shoham, Y., Rosenberg, E., Aviv, R., 1993. Applied microbiology biotechnology characterization and delignifiation activity of a thermostable α-L-arabinofuranosidase from Bacillus stearothermophilus. Appl. Microbiol. Biotechnol., 57–62.

[87] Hong, M.R., Park, C.S., Oh, D.K., 2009. Characterization of a thermostable endo-1,5-α-L-arabinanase from Caldicellulorsiruptor saccharolyticus. Biotechnol. Lett. 31, 1439–1443. doi:10.1007/s10529-009-0019-0

[88] Sakamoto, T., Kawasaki, H., 2003. Purification and properties of two type-B α-L-arabinofuranosidases produced by Penicillium chrysogenum. Biochim. Biophys. Acta: Gen. Subj. 1621, 204–210. doi:10.1016/S0304-4165(03)00058-8

[89] Sunna, a, Antranikian, G., 1997. Xylanolytic enzymes from fungi and bacteria. Crit. Rev. Biotechnol. 17, 39–67. doi:10.3109/07388559709146606

[90] Roy, N., Timell, T.E., 1964. The acid hydrolysis of glycosides. Can. J. Chem. 42, 1456–1472. doi:10.1016/S0008-6215(00)82576-8

[91] Henrissat, B., Vegetales, M., Grenoble, F., 1991. A classification of glycosyl hydrolases based sequence similarities amino acid. Biochem. J. 280, 309–316. doi:10.1007/s007920050009

[92] Daniela Alonso Bocchini Martins, Heloiza Ferreira Alves do Prado, Rodrigo Simões Ribeiro Leite, Henrique Ferreira, Márcia Maria de Souza Moretti, Roberto da Silva and Eleni Gomes (2011). Agroindustrial Wastes as Substrates for Microbial Enzymes Production and Source of Sugar for Bioethanol Production, Integrated Waste Management - Volume II, Mr. Sunil Kumar (Ed.), InTech, DOI: 10.5772/23377. Available from: http://www.intechopen.com/books/integrated-waste-management-volume-ii/agroindustrial-wastes-as-substrates-for-microbial-enzymes-production-and-source-of-sugar-for-bioetha

[93] Bhradwaj, A., Mahanta, P., Ramakumar, S., Ghosh, A., Leelavathi, S., Reddy, V.S., 2012. Emerging role of N-and C-terminal interactions in stabilizing $(\beta/\alpha)_8$ fold with special emphasis on family 10 xylanases. Comput. Struct. Biotechnol. J. 2, 1–10. doi:10.5936/csbj.201209014

[94] Lee, C.C., Braker, J.D., Grigorescu, A.A., Wagschal, K., Jordan, D.B., 2013. Divalent metal activation of a GH43 β-xylosidase. Enzyme Microb. Technol. 52, 84–90. doi:10.1016/j.enzmictec.2012.10.010

[95] Yang, X., Shi, P., Huang, H., Luo, H., Wang, Y., Zhang, W., Yao, B., 2014. Two xylose-tolerant GH43 bifunctional β-xylosidase/α- arabinosidases and one GH11 xylanase from Humicola insolens and their synergy in the degradation of xylan. Food Chem. 148, 381–387. doi:10.1016/j.foodchem.2013.10.062

[96] Gilead, S., Shoham, Y., 1995. Purification and characterization of alpha-L-arabinofuranosidase from Bacillus stearothermophilus T-6. Appl. Environ. Microbiol. 61, 170–174.

[97] Raweesri, P., Riangrungrojana, P., Pinphanichakarn, P., 2008. α-L-Arabinofuranosidase from Streptomyces sp. PC22: Purification, characterization and its synergistic action with xylanolytic enzymes in the degradation of xylan and agricultural residues. Bioresour. Technol. 99, 8981–8986. doi:10.1016/j.biortech.2008.05.016

[98] Khandeparker, R., Numan, M.T.H., Mukherjee, B., Satwekar, A., Bhosle, N.B., 2008. Purification and characterization of α-L-arabinofuranosidase from Arthrobacter sp. MTCC 5214 in solid-state fermentation. Process Biochem. 43, 707–712. doi:10.1016/j.procbio.2008.02.014

[99] Margolles, A., de los Reyes-Gavilán, C.G., 2003. Purification and functional characterization of a novel alpha-L-arabinofuranosidase from Bifidobacterium longum B667. Appl. Environ. Microbiol. 69, 5096–103. doi:10.1128/AEM.69.9.5096

[100] Volkin, D.B., Klibanov A.M.,. 1989. Minimizing protein inactivation, in: T. E. Creighton (ed.), Protein Function, A Practical Approach. IRL Press, Oxford, England. pp. 1–24.

[101] Tsujibo, H., Takada, C., Wakamatsu, Y., Kosaka, M., Tsuji, A., Miyamoto, K., Inamori, Y., 2002. Cloning and expression of an alpha-L-arabinofuranosidase gene (stxIV) from Streptomyces thermoviolaceus OPC-520, and characterization of the enzyme. Biosci. Biotechnol. Biochem. 66, 434–438. doi:10.1271/bbb.66.434

[102] Uchida, H., Nanri, T., Kawabata, Y., Kusakabe, I., Murakami, K., 1992. Purification and characterization of intracellular α-glucuronidase from *Aspergillus niger*. Biosci. Biotechnol. Biochem. 56, 1608–1615.

[103] Kawabata, Y., Ono, K., Gama, Y., Yoshida, S., Kobayashi, H., Kusakabe, I., 1995. Purification and characterization of alpha-glucuronidase from snail acetone powder. Biosci. Biotechnol. Biochem. 59, 1086–1090. doi:10.1271/bbb.59.1086

[104] Ruile, P., Winterhalter, C., Liebl, W., 1997. Isolation and analysis of a gene encoding alpha-glucuronidase, an enzyme with a novel primary structure involved in the breakdown of xylan. Mol. Microbiol. 23, 267–279.

[105] Zaide, G., Shallom, D., Shulami, S., Zolotnitsky, G., Golan, G., Baasov, T., Shoham, G., Shoham, Y., 2001. Biochemical characterization and identification of catalytic residues in alpha-glucuronidase from Bacillus stearothermophilus T-6. Eur. J. Biochem. 268, 3006–3016. doi:ejb2193 [pii]

[106] Wang, W., Yan, R., Nocek, B.P., Voung, T.V., Leo, R. Di, Xu, X., Cui, H., Gatenholm, P., Toriz, G., Tenkanen, M., Savchenko, A., Master, E.R., 2016. Biochemical and structural characterization of a fie-domain GH115 alpha-glucuronidase from the marine bacterium Saccharophagus degradans 2-40T. J. Biol. Chem., 291, 14120-14133. doi:10.1074/jbc.M115.702944

[107] Jordan, D.B., Lee, C.C., Wagschal, K., Braker, J.D., 2013. Activation of a GH43 β-xylosidase by divalent metal cations: Slow binding of divalent metal and high substrate specificity. Arch. Biochem. Biophys. 533, 79–87. doi.org/10.1016/j.abb.2013.02.020

[108] Juturu, V., Wu, J.C., 2012. Microbial xylanases: Engineering, production and industrial applications. Biotechnol. Adv. 30, 1219–1227. doi:10.1016/j.biotechadv.2011.11.006

[109] Heinen, P.R., Henn, C., Peralta, R.M., Bracht, A., De Cássia, R., Simão, G., Luís, J., Silva, C., De Lourdes, M., Polizeli, T.M., Kadowaki, M.K., 2014. African journal of biotechnology xylanase from Fusarium heterosporum: Properties and influence of thiol compounds on xylanase activity. 13, 1047–1055. doi:10.5897/AJB2013.13282

[110] Verma, D., Kawarabayasi, Y., Miyazaki, K., Satyanarayana, T., 2013. Cloning, expression and characteristics of a novel alkalistable and thermostable xylanase encoding gene (Mxyl) retrieved from compost-soil metagenome. PLoS One 8, 1-8. doi:10.1371/journal.pone.0052459

[111] Kumar, V., Satyanarayana, T., 2013. Biochemical and thermodynamic characteristics of thermo-alkali-stable xylanase from a novel polyextremophilic Bacillus halodurans TSEV1. Extremophiles 17, 797–808. doi:10.1007/s00792-013-0565-1

[112] Tuncer, M., 2000. Characterization of β-xylosidase and α-L-arabinofuranosidase activities from Thermomonospora fusca BD25. Turk. J. Biol. 24, 753–767.

[113] Debeche, T., Cummings, N., Connerton, I., Debeire, P., O'Donohue, M.J., 2000. Genetic and biochemical characterization of a highly thermostable alpha-L-arabinofuranosidase from Thermobacillus xylanilyticus. Appl. Environ. Microbiol. 66, 1734–1736. doi:10.1128/AEM.66.4.1734-1736.2000

[114] Nelson, D.L., Cox, M.M. Princípios de bioquímica de Lehninger. 5. ed. Artmed, Porto Alegre, 2011.

[115] Suresh, C., Kitaoka, M., Hayashi, K., 2003. A thermostable non-xylanolytic α-glucuronidase of *Thermotoga maritima* MSB8. Biosci. Biotechnol. Biochem. 67, 2359–2364. doi:10.1271/bbb.67.2359

[116] Raasch, C., Armbrecht, M., Streit, W., Höcker, B., Sträter, N., Liebl, W., 2002. Identification of residues important for NAD+ binding by the Thermotoga maritima α-glucosidase AglA, a member of glycoside hydrolase family 4. FEBS Lett. 517, 267–271. doi:10.1016/S0014-5793(02)02641-8

[117] Brutus, A., Villard, C., Durand, A., Tahir, T., Furniss, C., Puigserver, A., Juge, N., Giardina, T., 2004. The inhibition specificity of recombinant Penicillium funiculosum xylanase B towards wheat proteinaceous inhibitors. Biochim. Biophys. Acta: Proteins Proteom. 1701, 121–128. doi:10.1016/j.bbapap.2004.06.010

[118] Paës, G., Berrin, J.G., Beaugrand, J., 2012. GH11 xylanases: Structure/function/properties relationships and applications. Biotechnol. Adv. 30, 564–592. doi:10.1016/j.biotechadv.2011.10.003

[119] Maalej, I., Belhaj, I., Masmoudi, N.F., Belghith, H., 2009. Highly thermostable xylanase of the thermophilic fungus talaromyces thermophilus: Purification and characterization. Appl. Biochem. Biotechnol. 158, 200–212. doi:10.1007/s12010-008-8317-x

[120] Fialho, M.B., Carmona, E.C., 2004. Purification and characterization of xylanases from Aspergillus giganteus. Folia Microbiol. (Praha). 49, 13–18.

[121] Duarte, C.L., Ribeiro, M.a., Oikawa, H., Mori, M.N., 2013. Study of thermal treatment combined with radiation on the decomposition of polysaccharides in sugarcane bagasse. Radiat. Phys. Chem. 84, 191–195. doi:10.1016/j.radphyschem.2012.06.019

[122] Boukari, I., O'Donohue, M., Remond, C., Chabbert, B., 2011. Probing a family GH11 endo-β-1,4-xylanase inhibition mechanism by phenolic compounds: Role of functional phenolic groups. J. Mol. Catal. B: Enzym. 72, 130–138. doi:10.1016/j.molcatb.2011.05.010

[123] Jordan, D.B., Braker, J.D., 2007. Inhibition of the two-subsite β-D-xylosidase from Selenomonas ruminantium by sugars: Competitive, noncompetitive, double binding, and slow binding modes. Arch. Biochem. Biophys. 465, 231–246. doi:10.1016/j.abb.2007.05.016

[124] Eriksson, K-E.L., Bermek, H.L., Lignocellulose, Ligninase. Schaechter, M. Encyclopedia of Microbiology. 3. ed. Elsevier, San Diego, Volume 1, p. 373, 2009.

[125] Mester, T., Ming, T., 2000. Oxidation mechanism of ligninolytic enzymes involved in the degradation of environmental pollutants. Int. Biodeterior. Biodegrad. 46, 51–59.

[126] Manavalan, T., Manavalan, A., Heese, K., 2015. Characterization of lignocellulolytic enzymes from white-rot. Fungi Curr. Microbiol. 70, 485–498.

[127] Maciel, M.J.M., Castro e Silva, A., Ribeiro, H.C.T., 2010. Industrial and biotechnological applications of ligninolytic enzymes of the basidiomycota: A review. Electr. J. Biotechnol. 13, 1–13. doi: 10.2225/vol13-issue6-fulltext-2

[128] Pollegioni, L., Tonin, F., Rosini, E., 2015. Lignin-degrading enzymes. FEBS J. 282, 1190–1213. doi: 10.1111/febs.13224

[129] Ruiz-Dueñas, F.J., Martínez, A.T., 2009. Microbial degradation of lignin: How a bulky recalcitrant polymer is efficiently recycled in nature and how we can take advantage of this. Microbiol. Biotechnol. 2, 164–177. doi: 10.1111/j.1751-7915.2008.00078.x

[130] Coelho-Moreira, J.S., Maciel, G.M., Castoldi, R., Mariano, S.S., Inácio, F.D., Bracht, A., Peralta, R.M. Involvement of lignin-modifying enzymes in the degradation of herbicides, in: Price, A.J., Kelton, J.A. Herbicides—Advances in Research. Intech, 2013.

[131] Abdel-Hamid, A.M., Solbiati, J.O., Cann, I.K., 2013. Insights into lignin degradation and its potential industrial applications. Adv. Appl. Microbiol. 82, 1–28. doi:10.1016/B978-0-12-407679-2.00001-6

[132] Karigar, C.S., Rao, S.S., 2011. Role of microbial enzymes in the bioremediation of pollutants: A review. Enzyme Res. 2011, 1–11. doi:10.4061/2011/805187

[133] Viswanath, B., Rajesh, B., Janardhan, A., Kumar, A.P., Narasimha, G., 2014. Fungal laccases and their applications in bioremediation. Enzyme Res. 2014, 1–21. doi:10.1155/2014/163242

[134] Yaropolov, A.I., Skorobogat'ko, O.V., Vartanov, S.S., Varfolomeyev, S.D., 1994. Laccase: Properties, catalytic mechanism, and applicability. Appl. Biochem. Biotechnol. 49, 257–280. doi:10.1007/BF02783061

[135] Lu, C., Cao, L., Liu, R., Lei, Y., Ding, G., 2012. Effect of common metal ions on the rate of degradation of 4-nitrophenol by a laccase-Cu^{2+} synergistic system. J. Environ. Manage. 113, 1–6. doi: 10.1016/j.jenvman.2012.08.023

[136] Rodakiewicz-Nowak, J., Kasture, S.M, Dudek, B., Haber, J., 2000. Effect of various water-miscible solvents on enzymatic activity of fungal laccases. J. Mol. Catal. B: Enzym. 11, 1–11.

[137] Rogalski, J., Jozwik, E., Hatakka, A., Leonowicz, A., 1995. Immobilization of laccase from *Phlebiaradiata* on controlled porosity glass. J. Mol. Catal. A: Chem. 95, 99–108.

[138] vanErp, S.H., Kamenskaya, E.O., Khmelnitsky, Y.L., 1991. The effect of water content and nature of organic solvent on enzyme activity in low-water media. A quantitative description. Eur. J. Biochem. 2, 379–384.

[139] Dasthban, M., Schraft, H., Syed, T.A., Qin, W., 2010. Fungal biodegradation and enzymatic modification of lignin. Int. J. Biochem. Mol. Biol. 1, 36–50.

[140] Gorjup, B., Lampic, N., Pencaa, R., Perdiha, A., Perdih, M., 1999. Solvent effects on ligninases. Enzymc Microb. Technol. 25, 15–22.

[141] Asgher, M., Iqbal, H.M.N., Irshad, M., 2012. Characterization of purified and xerogel immobilized novel lignin peroxidase produced from trametes versicolor IBL-04 using solid state medium of corncobs. BMC Biotechnol. 12, 46. doi:10.1186/1472-6750-12-46

[142] Tuisel, H., Grover, T.A., Lancaster Jr., J.R., Bumpus, J.A., Aust, S.D., 1991. Inhibition of lignin peroxidase H_2 by sodium azide. Arch. Biochem. Biophys. 288, 456–462.

[143] Chang, H.C., Bumpus, J.A., 2001. Inhibition of lignin peroxidase mediated oxidation activity by ethylenediamine tetra acetic acid and N-N-N′-N′-tetramethylenediamine. Proc. Natl. Sci. Counc. Repub. China B 25, 26–33.

[144] Liu, A., Huang, X., Song, S., Wang, D., Lu, X., Qu, Y., Gao, P., 2003. Kinetics of the H_2O_2-dependent ligninase-catalyzed oxidation of veratryl alcohol in the presence of cationic surfactant studied by spectrophotometric technique.Spectrochim. Acta A: Mol. Biomol. Spectrosc. 59, 2547–2551.

[145] Vazquez-Duhalt, R., Westlake, D.W.S., Fedorak, P.M., 1994. Lignin peroxidase oxidation of aromatic compounds in systems containing organic solvents. Appl. Environ. Microbiol. 60, 459–466

[146] Wesenberg, D., Kyriakides, I., Agatho, S.N., 2003. White-rot fungi and their enzymes for the treatment of industrial dye effluents. Biotechnol. Adv. 22, 161–187.

[147] Kanayama, N., Tohru, S., Keiichi, K., 2002. Purification and characterization of an alkaline manganese peroxidase from Aspergillus terreus LD-1. J. Biosci. Bioeng. 93, 405–410.

[148] Asgher, M., Iqbal, H.M.N., 2011. Characterization of a novel manganese peroxidase purified from solid state culture of Trametes versicolor IBL-04. BioResources 6, 4302–4315.

[149] Youngs, H.L., Sundaramoorthy, M., Gold, M.H., 2000.Effects of cadmium on manganese peroxidase competitive inhibition of MnII oxidation and thermal stabilization of the enzyme. Eur. J. Biochem. 267, 1761–1769.

Chapter 7

Natural Products as a Potential Enzyme Inhibitors from Medicinal Plants

Abdur Rauf and Noor Jehan

Additional information is available at the end of the chapter

http://dx.doi.org/10.5772/67376

Abstract

Enzyme inhibitory agents are attractive because of their application in treating different ailments. The absence of enzymes produce a number of diseases. Medicinal plants are a rich source of producing secondary metabolites which showed broad-spectrum enzyme inhibitory potential. The position of enzyme inhibitors as new drugs is vast since these compounds have been used for the treatment of various physiological disorders. Bioactive secondary metabolites can deliver excellent pharmacophore patterns for drugs related to numerous illnesses. This book chapter is planned to document the enzyme inhibitory potential of natural compounds, medicinal plant extract, and its isolated compounds.

Keywords: natural products, enzyme inhibitors, medicinal plants

1. Introduction

Medicinal plants are a rich source of producing bioactive natural products in most precise and selective way. Since the mid-nineteenth century, many natural products have been purified from plants, and most of them exist to be used as active elements of the modern medication. The search of excellence, real, inexpensive, and simply accessible natural compound enzyme inhibitor is one of the drug finding and strategy investigation works in the study organizations through the world [1].

Medicinal plants, extracts, and its fractions are used by 80% of the world population for their simple health necessities. The association among human, medicinal plants, and derived drugs from medicinal plants defines the past of men. Medicinal plants are the significant basis of natural drug molecules. The medicinal plants are expected to comprise secondary metabolites which have properties to use in modern medication for the cure of various diseases which are not treatable. Throughout the historical period, old-style systems of medicine have developed a topic of global significance. Present approximations recommended that in numerous emerging republics a huge population trusts seriously on traditional specialist and medicinal plants to chance the main health care wants, though modern drug may be obtainable in these countries. Herbal medicines must frequently preserved approval for important and national reasons. Presently, several people in the advanced countries have initiated to go to another or complementary treatments, containing therapeutic herbs [2]. Ayurvedic medication for drug adjustment switches to medicinal plants. Ayurvedic medicine is a combination of numerous elements which is ready from medicinal plants, but the active compounds when purified from that medicinal natural plant source fail to provide the wanted activity. In the nonappearance of pharmacological data on several medicinal plants and isolated compounds which is not likely to regulate the vigorous compounds consuming wanted biological potency. Earlier trainings presented the poisonous properties of chemotherapy and radiation in handling of cancer by decrees by Ayurvedic medication, and wound curing might be complete by using Ayurvedic medicine. Modern discipline production is an important part in this procedure, to grow natively establishing materials for wanted quality [3].

Enzyme inhibitors are mainly bioactive secondary metabolites that bind with an enzyme and decrease its bioactivity. Subsequently, blocking enzyme activity can kill a pathogen or correct a metabolic imbalance; many drug molecules are enzyme inhibitors, and mainly enzyme activators connect to various enzymes, increase their enzymatic actions, and subtract link and subsequently distort to products in the catalytic cycle of the enzymes. The linking of inhibitors can finish a substrate from the enzyme-active site and stays the enzyme in catalyzing in chemical reaction. Enzyme inhibition is both an irreversible and reversible process. The irreversible inhibitors react with enzyme and adjust it chemically by a covalent likening formation. Then, these inhibitors adjust important amino acid remnants wanted from an enzymatic reversible inhibitors which are non-covalently bonded; different types of inhibition are shaped depending on whether inhibitors link non-covalently, and dissimilar types of inhibition are shaped depending on whether these inhibitors bind to the enzyme and produced enzyme substrate complex or both [4].

Many natural products are enzyme inhibitors; the finding and development are dynamic areas of pharmacology and biochemistry. Medicinal enzyme inhibitors are frequently mediated by its specificity and its effectiveness that designated the absorption desirable to inhibit the enzyme. Great specificity and potency confirm that a medicinal drug will have few side effects and possess low toxicity. Natural enzyme inhibitors are involved in the guideline of much metabolic procedure. Actually, enzyme is a metabolic pathway which can be inhibited by many downstream yields. These types of bad response slow the manufacture line when product activates to shape up and a significant way to reservation homeostasis in cell. An additional cellular enzyme inhibitor is protein which specially binds and inhibits an enzyme objective. These help regulator enzymes which may be harmful to cell alike proteases. The well-categorized example of this is the ribonuclease inhibitor that link ribonucleases in the tightest recognized protein contact. Many natural enzyme inhibitors may also be poisonous and are used as defenses besides predators as habits of killing several preys [4].

http://dx.doi.org/10.5772/67376

2. Discovery and design of new enzyme inhibitors

Discovery of new drugs is actually the product of a very long drug growth procedure; the first step among which is the discovery of new enzyme inhibitors. In the past time, the only way to discover new drugs was a trial-and-error method, which proceeds to screen enormous libraries of chemical constituents against a marked enzyme and expect that maybe some valuable lead drugs will arise. This physical force method is still fruitful and has been lengthy by combinatorial chemistry methods that rapidly yield huge statistics of new, known, and novel molecules and high-throughput screening expertise to quickly screen these enormous chemical libraries for valuable new inhibitors [5].

Recently, it is reported that an alternative approach has been documented: rational new drug uses the three-dimensional chemical structure of an enzyme-active position to expect which compound potency to be the new inhibitors [6]. These predictions are then screening, and some of these screenings of compounds may be proven as novel inhibitors. These new inhibitors are then used to attempt to get a chemical structure of enzyme in an inhibitor/enzyme complex to show how the chemical constituents are connecting to the active position, presenting alteration to be complete to the inhibitor to try to optimize binding. This test and recovered cycle are then repeated until a suitably strong inhibitor is formed [7]. The computer-based methods of expecting the attraction of an inhibitor for an enzyme are also existence advanced; these are molecular docking [8] and molecular mechanics.

3. Main uses of enzyme inhibitors

The enzyme inhibitors are abundantly original in nature and considered as well as produced as a main part of pharmacology and biochemistry. Natural poisons are frequently enzyme inhibitors which have grown to defend a plant or animal against predators. These natural toxins comprised certain known poisonous compounds. Artificial inhibitors are mostly used as new drugs but also be used as insecticides such as malathion, herbicides, or glyphosate and may be used as disinfectants like triclosan. Some other artificial enzyme inhibitors block acetylcholinesterase, an enzyme which disrupts dejected acetylcholine, and are used as nerve agents in chemical warfare [9]. Pistagremic acid and di-naphthodiospyrol are isolated compounds having enzyme inhibitory activity (**Figure 1**). Coenzyme folic acid is also linked to the anticancer drug methotrexate (**Figure 2**).

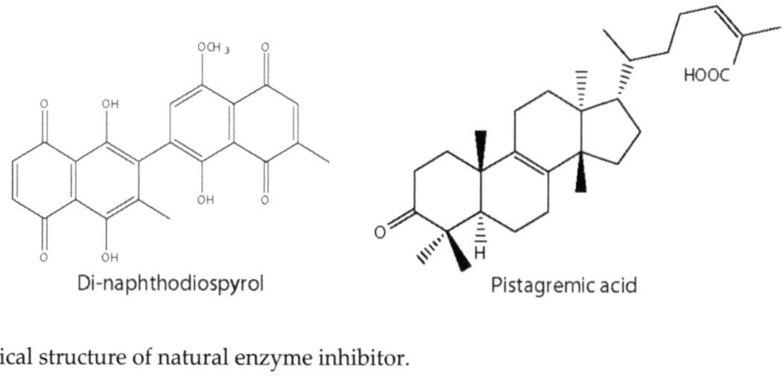

Figure 1. The chemical structure of natural enzyme inhibitor.

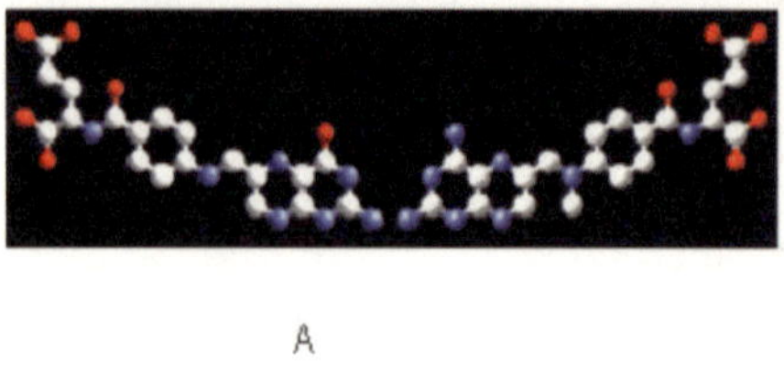

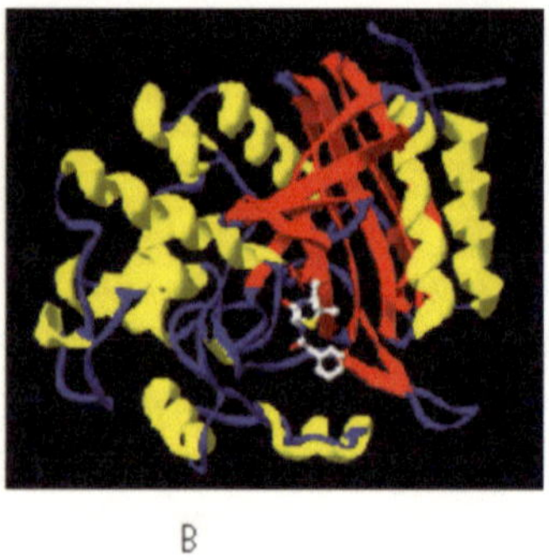

Figure 2. (A) Chemical structure of coenzyme folic acid (left side) associated to the anticancer drug methotrexate (right side). (B) The chemical structure of a complex between penicillin G and *Streptomyces* transpeptidase (produced from DB).

4. Results and discussion

Compounds isolated from medical plants and its extract and frication have potential enzyme inhibitor. Natural products derived from plants have excellent enzymatic action.

4.1. Natural products as urease inhibitors

For global nitrogen cycle, which can occur in medicinal plants, fungi and various bacteria urease (EC 3.5.1.5) are leading enzymes. Such type of hydrolase speeds up to 100 folds of the rate of urea hydrolysis and converts to ammonia and carbon dioxide [10–13]. Meanwhile, this finding in medicinal plants including *Canavalia ensiformis* which belongs to Fabaceae, the urease has been fully explored and converted the innovative in the field of biochemistry learning as the principal enzyme to be crystallized [14, 15] and likewise confirmed to be firmly needy on nickel ions [6]. The requirement on Ni^{2+} ions for catalytic action is a sole piece of urease between hydrolytic enzymes [10, 12].

In 1995, Jabri and coworkers succeeded to fully report the three-dimensional structure of urease enzyme from crystallography studies done with urease enzyme derived from *Klebsiella aerogenes* [16]. Later on, additional structures were revealed for ureases identified in *Bacillus pasteurii, Helicobacter pylori,* and maximum newly *C. ensiformis* [17–19]. Certainly, the characterization of the urease enzyme structure from a legume was central to improve and comprehend the supplies for ureolytic properties of this type of enzymes in diverse animals [20].

The countless resemblance of amino acid order among ureases since multiple origins recommends a mutual family for this enzyme [21]. Urease enzyme part an elementary trimeric collection with one, two, or three subunits that can combine creating dodecameric or hexameric. Each active position comprises two Ni^{2+} ions separately after all additional between 3.5 and 3.7 Å, linked by oxygen particles of a lysine carbamate rest and a hydroxide ion [22].

Medicinal plants and fungus ureases showed a solitary polypeptide chain though bacteria which must be 2/3 of dissimilar subunits (A, B, and C) [23]. The combination of Ni^{2+} ions in protein structure is supported by additional proteins, supposed to be specific urease chaperones [23].

Urease enzyme in the background of *H. pylori*, which raise the medium pH by the accretion of NH_3, is a urease trait of great medical position [13]. Gastrointestinal infections or urine by ureolytic bacteria can be a basis of health problems in humans and many other animals including pyelonephritis, kidney stone formation, ultimately hepatic coma, and hepatic encephalopathy [24]. Consequently, the main public health subjects are connected by *H. pylori*, Gram-negative bacteria that are bright to live in an environment as acidic as that of the stomach (pH 2–4). By way of significance, *H. pylori* poison can bring gastric irritation and raise the risk for the growth of duodenal, gastric adenocarcinoma, gastric ulcers, and gastric lymphoma [24].

Fifty percent of the universal population is dedicated by *H. pylori*. *H. pylori* can persevere in the stomach for the entire life of diseased persons without producing illness signs. The high occurrence of *H. pylori* in human population designates that such microorganism has established mechanisms for confrontation against host fortifications [25]. The urease enzyme present in cytoplasm or bound to *H. pylori* superficial is the chief virulence factor of such human pathogen [25]. It is suggested that the lyses of some pathogen cells tip to the issue of cytosolic ureases which connect to the superficial of intact bacterial cells and basis the hydrolysis of urea existing in human guts at an absorption of 3 mM. The NH_3 fashioned raises the medium pH, which produces an outgoing location for *H. pylori* survival [26]. Throughout the past 20 years, the endorsed first-line therapy for *H. pylori* abolition contained the mixture of the antibiotics amoxicillin and clarithromycin with omeprazole, a proton pump cell inhibitor. The upsurge of *H. pylori* resistance to these antibiotics (chiefly to clarithromycin) completes this therapy which is a non-attractive choice in new ages [27]. Additional action plans have arose to competition *H. pylori* infections, which comprise the usage of bismuth salts joint with a proton pump cell inhibitor [28]. Furthermore, urease inhibitors might be active therapies for the cure of diseases produced by urease-dependent pathogenic microorganisms. However, the commercially accessible urease inhibitors, including hydroxamic acid derivatives, phosphorodiamidates, and imidazoles, are toxic and have low stability, feature that stop their clinical usage [29]. The main search for new, known, novel, and bioactive urease inhibitors which enhanced stability and low toxicity is necessary to improve life excellence of human beings and animals.

4.2. Xanthine oxidase

Gout is a public illness with a universal spreading. Hyperuricemia, related with gout, is current in 5–30% of the overall people [30]. It appears to be growing universally and is measured as an important risk issue in thoughtful complaints similar to tophaceous gout, gouty nephropathy, and nephrolithiasis [31, 32]. Hyperuricemia consequences from the overproduction or under-excretion of uric acid and is importantly unfair by the high dietary consumption of foods ironic in nucleic acids, such as meats, leguminous seeds, and certain kinds of sea food. Throughout the previous step of purine metabolism, xanthine oxidase catalyzes the oxidation of xanthine and hypoxanthine into uric acid uricosuric drugs which development the urinary removal of uric acid, and xanthine oxidase inhibitors which block the mortal step in uric acid bio-synthesis, can minor the plasma uric acid concentration, and are usually working for the conduct of gout [33]. Furthermore, xanthine oxidase helps as a significant organic source of oxygen resulting to free radicals that pay to oxidative damage of existing materials producing several extreme positions like inflammation, carcinogenesis, hepatitis, ischemia reperfusion, as well as elderly [34, 35]. Allopurinol is the individual clinically used xanthine oxidase inhibitor in the cure of gout [36]. This drug bases countless side effects including nephropathy, hepatitis, and allergic responses [37]. Thus, the exploration for new xanthine oxidase inhibitors with advanced therapeutic potential and less side effects wanted not only to treat gout but also fight numerous additional diseases connected with xanthine oxidase action.

4.3. Angiotensin-converting enzyme

Angiotensin I-converting enzyme action (ACE, peptidyldipeptide hydrolase, kininase II, EC 3.4.15.1) plays a significant part in ruling of blood pressure [38]. Angiotensin I-converting enzyme is a significant blood pressure controller that catalyzes the release of His-Leu from the carboxyl irredeemable angiotensin I, which, in go, produces a strong vasopressor octapeptide, angiotensin II. Angiotensin I-converting enzyme is also complicated in the poverty of vasodilator bradykinin [39]. The greatest if not all commercialized angiotensin I-converting enzyme inhibitors have developed peptides from the venom of the Brazilian viper *Bothrops jararaca* as classical materials [40]. Also, this animal basis, microorganisms, and plants deliver chemical substances with angiotensin I-converting enzyme inhibitory activity which might help as perfect materials in the growth of new angiotensin I-converting enzyme inhibitors. Angiotensin I-converting enzyme inhibitors prevent the formation of angiotensin II by ACE and thus decrease outlying vascular confrontation and blood pressure. These synthetic drugs are supposed to have sure side effects such as cough, taste conflicts, and skin rashes [41]. Consequently, for safe and cost-effective use, curiosity in finding food sources as angiotensin I-converting enzyme inhibitor has improved. Further compelling angiotensin I-converting enzyme inhibitors have been intended and synthesized to indulgence of hypertension excellently. Oral management of these drugs regularly results in unsolicited side effects; nutritional method strength be a healthier medium by which blood pressure in skillful. Besides, some trainings obligate been made on single plant species where several classes of angiotensin I-converting enzyme inhibitory molecules must be recognized, such as flavonoids, xanthones, proanthocyanidins, secoiridoids, and peptides for a complete evaluation of natural compounds [42–48].

4.4. α-Amylase

α-Amylase is a protein enzyme (EC 3.2.1.1) which hydrolyses α bonds of big, α-bonded polysaccharides, including starch and glycogen, elastic glucose, and maltose [49]. That is the main form of the amylase existing in humans and other mammals [50, 51]. It is also existing in seeds covering starch as a food reserve, which is secreted by many fungi.

Though it originates in numerous tissues, amylase is greatest projecting in saliva and pancreatic juice, and all of them have its individual isoform of human α-amylase. They act inversely on isoelectric focusing, and also be detached in testing by consuming precise monoclonal antibodies. Amylase is created in saliva and disruption starch into maltose and dextrin. This form of amylase is also called ptyalin. It will break down bulky, insoluble starch molecules into soluble starches making consecutively minor starches and finally maltose. Ptyalin performances on linear α(1–4) glycosidic linkages, but parts hydrolysis wants an enzyme that presentations on cleft products. Gastric acid deactivates the salivary amylase in the stomach. In gastric juice agreed to pH near to 3.3–4, where ptyalin was deactivated completely at 37°C in 20 min. In difference, 50% of amylase activity is sustained after 150 min of introduction to gastric juice at pH 4.3 [52, 53]. Together, starch and substrate for ptyalin and then product (glucose of short chains) are capable to partly defend it touching in-activation by gastric acid. Ptyalin additional to buffer at pH 3.0 underwent whole inactivation in 120 min; adding of the starch at a 0.1% level caused 10% of the activity residual, and similar addition of starch to a 1.0% level produced about 40% of the activity residual at 120 min [54].

4.5. α-Glucosidase enzymes

Alpha-glucosidase; α-glucopyranoside; glucoinvertase; including glucoamylase, maltase, glucosidosucrase, maltase, glucosidoinvertase, alpha-D-glucosidase, hydrolase, α-1,4-glucosidase, and α-D-glucoside glucohydrolase are glucosidases positioned in the brush end of the small intestine which acts upon α(1–4) bonds [55–61]. In dissimilarity to β-glucosidase, the α-glucosidase decomposes starch and disaccharides in to glucose. In the meanwhile maltase enzyme decompose maltose is nearly functionally equal.

The key role of α-glucosidase is to hydrolyze incurable nonreducing (1–4) attached α-glucose which remains to release a lonely α-glucose molecule [62]. α-Glucosidase is mainly a carbohydrate which hydrolyzes the releases of α-glucose which is different to β-glucose. β-Glucose remains unconfined through glucoamylase, a similar enzyme. The substrate molecule discrimination of α-glucosidase is outstanding to subsite attractions of the active site of enzymes [62]. The main proposed mechanism comprises a nucleophilic shift of intermediate (oxocarbenium ion) [62].

Blood-sucking insect (*Rhodnius prolixus*) produces hemozoin when it digests hemoglobin of the host. The synthesis of hemozoin is reliant on the substrate connected to the site of α-glucosidase [63].

It has been documented in literature that α-glucosidases were extracted and then characterized from trout liver which exhibited maximum activity of the enzyme with increase rate that is 80% throughout workout in contrast to a latent trout.

This alteration was shown in correlation to the effects rise for liver glycogen phosphorylase.

From this it was offered that α-glucosidases in the glucosidic track play a significant portion in adding the phosphorolytic path in the livers' metabolic action to energy pressures of exercise.

The slight intestine of rat and yeast has α-glucosidases which exhibited to be reserved by numerous groups of flavonoid moiety [63].

Our research group has also reported phosphodiesterase-1 inhibitory, urease inhibition, and β-secretase enzyme effect of many natural products [64–67]. Novel glycine and phenylalanine sulfonamide derivatives have been reported for carbonic anhydrase inhibition activity [68]. It has recently documented that bovine liver tissue on glutathione reductase (GR) enzyme resolves the effects of adrenaline, thiamine, tyrosine, and dopamine. The bovine liver GR also effects on some natural amine [69]. Furthermore, the effects of particular catecholamines of the properties of carbonic anhydrase enzyme purified from bovine kidney tissue are also documented. Some synthetic sulfonamides which contain molecules have also documented in literature for urease inhibition potential [70].

5. Conclusion

In conclusion medicinal plants and derived natural products from plants have potential enzyme inhibitor. This book chapter directed researcher to isolate bioactive compounds from plants which have excellent enzymatic action. Thus, the exploration for new, known, and novel enzyme inhibitors with progressive therapeutic potential and fewer side effects are required to treat diseases in mankind as well as in animals.

Author details

Abdur Rauf[1]* and Noor Jehan[2]

*Address all correspondence to: mashaljcs@yahoo.com

1 Department of Chemistry, University of Swabi, Ambar, Khyber Pakhtunkhwa, Pakistan

2 Department of Geology, University of Swabi, Ambar, Khyber Pakhtunkhwa, Pakistan

References

[1] Uddin G, Rauf A, Siddiqui BS, Shah S. Preliminary comparative phytochemical screening of *Diospyros lotus* Stewart. Mid. J. Sci. Res. 2011;10:78–81.

[2] Uddin G, Rauf A, Rehman T, Qaisar M. Phytochemical screening of *Pistacia chinensis* var. *integerrima*. Mid. J. Sci. Res. 2011;7:707–11.

[3] Rauf A, Jehan N. 2015. Book chapter: The Folkloric Use of Medicinal Plants in Public Health Care. SM Group Open Access eBooks, 160 Greentree Drive, Suite 101, Dover, DE 19904, USA. 1–13.

[4] Jump up, Shapiro R, Vallee BL. Interaction of human placental ribonuclease with placental ribonuclease inhibitor. Biochemistry. 1991;30(8):2246–55.

[5] Koppitz M, Eis K. Automated medicinal chemistry. Drug Discov. Today. 2006;11 (11–12):561–8.

[6] Jump up, Scapin G. Structural biology and drug discovery. Curr. Pharm. Des. 2006; 12(17):2087–97.

[7] Jump up, Gohlke H, Klebe G. Approaches to the description and prediction of the binding affinity of small-molecule ligands to macromolecular receptors. Angew. Chem. Int. Ed. Engl. 2002;41(15):2644–76.

[8] Jump up, Glen RC, Allen SC. Ligand-protein docking: cancer research at the interface between biology and chemistry. Curr. Med. Chem. 2003;10(9):763–7.

[9] Jump up, Maggi M, Filippi S, Ledda F, Magini A, Forti G. Erectile dysfunction: from biochemical pharmacology to advances in medical therapy. Eur. J. Endocrinol.. Bioscientifica. 2000;143(2):143–54.

[10] Krajewska B, Ureases I. Functional, catalytic and kinetic properties: a review. J. Mol. Catal. B Enzym. 2009;59(1):9–21.

[11] Follmer C. Ureases as a target for the treatment of gastric and urinary infections. J. Clin. Pathol. 2010;63:424–30.

[12] Maroney MJ, Ciurli S. Nonredox nickel enzymes. Chem. Rev. 2014;114:4206–28.

[13] Takeuchi T. On the occurrence of urease in higher plants. J. Coll. Agric. Tokyo Imp. Univ. 1909;1:1–14.

[14] Sumner JB. The isolation and crystallization of the enzyme urease. J. Biol. Chem. 1926;69:435–41.

[15] Dixon NE, Gazzola C, Watters JJ, Blakeley RL, Zerner B. Jack Bean Urease (EC 3.5.1.5) [letter]. A metalloenzyme. A simple biological role for nickel?. J. Am. Chem. Soc. 1975;97(14):4131–3.

[16] Jabri E, Carr MB, Hausinger RP, Karplus PA. The crystal structure of urease from *Klebsiella aerogenes*. Science. 1995;268:998–1004.

[17] Benini S, Rypniewski WR, Wilson KS, Miletti S, Ciurli S, Mangani S. A new proposal for urease mechanism based on the crystal structures of the native and inhibited enzyme from *Bacillus pasteurii*: why urea hydrolysis costs two nickels. Structure. 1999;7(2):205–16.

[18] Ha NC, Oh ST, Sung JY, Cha KA, Lee MH, Oh BH. Supramolecular assembly and acid resistance of *Helicobacter pylori* urease. Nat. Struct. Biol. 2001;8(6):505–9.

[19] Balasubramanian A, Ponnuraj K. Crystal structure of the first plant urease from jack bean: 83 years of journey from its first crystal to molecular structure. J. Mol. Biol. 2010;400:274–83.

[20] Sirko A, Brodzik R. Plant ureases: roles and regulation. Acta Biochim. Pol. 2000;47:1189–95.

[21] Boer JL, Mulrooney SB, Hausinger RP. Nickel-dependent metalloenzymes. Arch Biochem. Biophys. 2014;544:142–52.

[22] Yathindra N. Structure of an enzyme revealed 80 years after it was crystallized – differential functional behaviour of plant and microbial ureases uncovered. Curr. Sci. 2010;99:566–8.

[23] Algood HMS, Cover TL. *Helicobacter pylori* persistence. An overview of interactions between *H. pylori* and host immune defenses. Clin. Microbiol. Rev. 2006;19(4):597–613.

[24] Sachs G, Weeks DL, Wen Yi, Marcus EA, Scott DR. Acid acclimation by *Helicobacter pylori*. Physiology. 2005;20:429–38.

[25] Stingl K, Altendorf K, Bakker EP. Acid survival of *Helicobacter pylori*: how does urease activity trigger cytoplasmic pH homeostasis?. Trends Microbiol. 2002;10(2):70–4.

[26] Graham DY, Fischbach L. *Helicobacter pylori* treatment in the era of increasing antibiotic resistance. Gut. 2010;59:1143–53.

[27] Mégraud F. The challenge of *Helicobacter pylori* resistance to antibiotics: the comeback of bismuth-based quadruple therapy. Therap. Adv. Gastroenterol. 2012;5(2):103–9.

[28] Malfertheiner P, Megraud F, O'Morain C, Bazzoli F, El-Omar E, Graham D, et al. Current concepts in the management of *Helicobacter pylori* infection: the Maastricht III consensus report. Gut. 2007;56:772–81.

[29] Azizian H, Nabati F, Sharifi A, Siavoshi F, Mahdayi M, Amanlou M. Large-scale virtual screening for the identification of new *Helicobacter pylori* urease inhibitor scaffolds. J. Mol. Model. 2012;18:2917–27.

[30] Riches PL, Wright AF, Ralston SH. Recent insights into the pathogenesis of hyperuricaemia and gout. Hum. Mol. Genet. 2009;18:R177–84.

[31] Forbes JM, Coughlan MT, Cooper ME. Oxidative stress as a major culprit in kidney disease in diabetes. Diabetes. 2008;57:1446–54.

[32] Watanabe S, Kanellis J, Nakagawa T, Han L, Ohashi R, Lan H, Feng L, Johnson R J. Reducing uric acid as a means to prevent cardiovascular and renal disease. Expert Opin. Ther. Pat. 2002;12:193–9.

[33] Vazquez-Mellado J, Alvarez Hernandez E, Burgos-Vargas R. Primary prevention in rheumatology: the importance of hyperuricemia. Best Pract. Res. Clin. Rheumatol. 2004;18:111–24

[34] Schlesinger N, Dalbeth N, Perez-Ruiz F. Gout – what are the treatment options?. Expert Opin. Pharmacother. 2009;10:1319–28.

[35] Matata BM, Elahi MM. Sources of reactive oxidants species in biology and disease. Oxid. Stress 2007, 14: 23–38.

[36] Hoey BM, Butler J, Halliwell B. On the specificity of allopurinol and oxypurinol as inhibitors of xanthine oxidase. A pulse radiolysis determination of rate constants for reaction of allopurinol and oxypurinol with hydroxyl radicals. Free Radic. Res. Commun. 1988;4:259–63.

[37] Fagugli RM, Gentile G, Ferrara G, Brugnano R. Acute renal and hepatic failure associated with allopurinol treatment. Clin. Nephrol. 2008;70:523–6.

[38] Gohlke P, Linz W, Scholkens BA, Kuwer I, Bartenbach S, Schnell A, et al. Angiotensin-converting enzyme inhibition improves cardiac function. Role of bradykinin. Hypertension. 1994;23(4):411–8.

[39] Skeggs LT, Jr., Kahn JR, Shumway NP. The preparation and function of the hypertension converting enzyme. J. Exp. Med. 1956;103(3):295–9.

[40] Cushman DW, Ondetti MA. History of the design of captopril and related inhibitors of angiotensin converting enzyme. Hypertension. 1991;17(4):589–92.

[41] Atkinson AB, Robertson JI. Captopril in the treatment of clinical hypertension and cardiac failure. Lancet. 1979;2(8147):836–9.

[42] Inokuchi J, Okabe H, Yamauchi T, Nagamatsu A, Nonaka G, Nishioka I. Antihypertensive substance in seeds of *Areca catechu* L. Life Sci. 1986;38(15):1375–82.

[43] Wagner H, Elbl G, Lotter H, Guinea M. Evaluation of natural products inhibitors of the angiotensin converting enzyme (ACE). A review between 1980-2000. Pharm. Pharmacol. Lett.. 1991;1:15–8.

[44] Wagner H, Elbl G. ACE-inhibitory procyanidins from *Lespedeza capitata*. Planta Med. 1992;58(3):297.

[45] Hansen K, Nyman U, Smitt UW, Adseren A, Christensen SB, Schwantner, Wagner H. Angiotensin converting enzyme (ACE) inhibitory flavonoids from *Erythroxylum laurifolium*. Phytomedicine. 1996;2:313–7.

[46] Chen CH, Lin JY, Lin CN, Hsu SY. Inhibition of angiotensin-I-converting enzyme by tetrahydroxyxanthones isolated from *Tripterospermum lanceolatum*. J. Nat. Prod. 1992;55(5):691–5.

[47] Hansen K, Adsersen A, Christensen SB, Jensen SR, Nyman U, Smith UW. Isolation of an angiotensin converting enzyme (ACE) inhibitor from olea europaea and olea lancea. Phytomedicine. 1996;2:319-

[48] Hansen K. In vitro screening for the detection of angiotensin converting enzyme (ACE) inhibitors in selected medicinal plants and identification of the active principles [PhD]. Copenhagen: The Royal School of Pharmacy; 1995.

[49] Ramasubbu N, Paloth V, Luo Y, Brayer GD, Levine MJ. Structure of human salivary α-amylase at 1.6 Å resolution: implications for its role in the oral cavity. Acta Crystallogr. Sect. D. 1996;52(3):435–46.

[50] Jump up, Pugh MB, ed. (2000). Stedman's Medical Dictionary (27th ed.). Baltimore, Maryland, USA: Lippincott Williams & Wilkins. p. 65. ISBN 978-0-683-40007-6.

[51] Jump up, Voet D, Voet JG. (2005). Biochimie. (2e éd.). Bruxelles: De Boeck. 1583 p.

[52] Jump up which was named by Swedish chemist Jöns Jacob Berzelius. The name derives from the Greek word πτυω (I spit), because the substance was obtained from saliva. See: J. Berzelius (Ms. Esslinger, trans.), Traité de Chimie (Paris, France: Firmin Didot Frerès, 1833), Vol. 7, p. 156.

[53] Jump up, Fried M, Abramson S, Meyer JH. Passage of salivary amylase through the stomach in humans. Dig. Dis. Sci. 1987;32:1097–103.

[54] Jump up, Rosenblum JL, Irwin CI, Alpers DH. Starch and glucose oligosaccharides protect salivary-type amylase activity at acid pH. Am. J. Physiol. 1988;254(Gastrointestinal

and Liver Physiology 17):G775–80. alpha-Glucosidases at the US National Library of Medicine Medical Subject Headings (MeSH)

[55] Jump up, Bruni CB, Sica V, Auricchio F, Covelli I. Further kinetic and structural characterization of the lysosomal α-D-glucoside glucohydrolase from cattle liver. Biochim. Biophys. Acta. 1970;212(3):470–7.

[56] Jump up, Flanagan PR, Forstner GG. Purification of rat intestinal maltase/glucoamylase and its anomalous dissociation either by heat or by low pH. Biochem. J. 1978;173(2):553–63.

[57] Jump up, Larner J, Lardy H, Myrback K. (1960). Other glucosidases. In Boyer PD. The Enzymes. Vol. 4 (2nd ed.). New York: Academic Press. pp. 369–78.

[58] Jump up, Sivikami S, Radhakrishnan AN. Purification of rabbit intestinal glucoamylase by affinity chromatography on Sephadex G-200. Indian J. Biochem. Biophys. 1973;10(4):283–4.

[59] Jump up, Sørensen SH, Norén O, Sjöström H, Danielsen EM. Amphiphilic pig intestinal microvillus maltase glucoamylase. Structure and specificity. Eur. J. Biochem. 1982;126:559–68.

[60] Jump up, Chiba S. Molecular mechanism in alpha-glucosidase and glucoamylase. Biosci. Biotechnol. Biochem. 1997;61(8):1233–9.

[61] Jump up, Mury FB, da Silva JR, Ferreira LS, et al. Alpha-glucosidase promotes hemozoin formation in a blood-sucking bug: an evolutionary history. PLoS One. 2009;4(9):e6966.

[62] Jump up, Mehrani H, Storey KB. Characterization of alpha-glucosidases from rainbow trout liver. Arch. Biochem. Biophys. 1993;306(1):188–94.

[63] Jump up, Tadera K, Minami Y, Takamatsu K, Matsuoka T. Inhibition of alpha-glucosidase and alpha-amylase by flavonoids. J. Nutr. Sci. Vitaminol. 2006;52(2):149–53.

[64] Rauf A, Orhan IK, Ertas A, Temel H, Hadda TB, Saleem M, Raza M, Khan H. *In vitro* and *in silico* studies on phosphodiesterase-1 inhibitory effect of natural polyphenolics. Curr. Top. Med. Chem. 2016. [Epub ahead of print]

[65] Rauf A, Uddin G, Raza M, Patel S, Bawazeer S, Hadda TB, Jehan N, Mabkhot YN, Khan A, Mubarak M. Urease inhibition potential of Di-naphthodiospyrol from *Diospyros lotus* roots. Nat. Prod. Res. 2016:1–5. [Epub ahead of print].

[66] Rauf A, Patel S. Pistagremic acid as a broad spectrum natural inhibitor from *Pistacia integerrima* Stewart. Nat. Prod. Res. 2016;4:367–368.

[67] Rauf A, Uddin G, Khan A, Siddiqui BS, Arfan M, Dalvandi K, Hadda TB. Pistagremic acid, a novel β-secretase enzyme (BACE1) inhibitor from *Pistacia integerrima* Stewart. Nat. Prod. Res. 2015;29:1735–8.

[68] Fidan I, Salmas RE, Arslan M, Senturk M, Durdagi S, Ekinci D, Senturk E, Cosgun S, Supuran CT. Carbonic anhydrase inhibitors: design, synthesis, kinetic, docking and molecular dynamics analysis of novel glycine and phenylalanine sulfonamide derivatives. Bioorg. Med. Chem. 2015;23:7353–8.

[69] Sentuk E, Urcar H, Senturk M, Yildrim S, Gul M. Bovine liver tissue on glutathione reductase enzyme determination of effects of thiamine, tyrosine, dopamine and adrenaline. Acta Physiol. 2016;218:58.

[70] Urcar H, Senturk E, Senturk M, Gul M, Yildirim S. Investigation of effects of some catecholamines on the activity of carbonic anhydrase enzyme purified from bovine kidney tissue. 42st National Physiology Congress. Acta Physiol. 2016;218:57.

Section 2

Enzyme Activation

Chapter 8

Microbial Glycosidases for Nondigestible Oligosaccharides Production

Thais Bezerra, Rubens Monti, Egon B. Hansen and Jonas Contiero

Additional information is available at the end of the chapter

http://dx.doi.org/10.5772/65935

Abstract

There is much interest in the study and production of nondigestible oligosaccharides (NDOs), due to their bioactivities and beneficial effects to the human health. The main approach in the production of NDOs relies on the action of glycosidases performing hydrolysis or transglycosylation of polysaccharides and sugars. In this chapter, a description of the main microbial glycosidases used for NDOs production, their sources, their principal properties, and a description of the production processes with the better results obtained are discussed.

Keywords: glycosidases, transglysosylation, enzymatic hydrolysis, oligosaccharides

1. Introduction

The concept of nondigestible oligosaccharides (NDOs) came from the observation that the human body does not have the necessary enzymes to hydrolyze β-glycosidic linkages present in some sugars of the human diet. Thus, these carbohydrates can arrive intact in the intestine where they are fermented selectively stimulating the growth and/or activity of bacteria in the colon acting as prebiotics [1]. In this context, nondigestible oligosaccharides have received much attention since they have important biological properties promoting health beneficial effects. Stimulation of the intestinal microbiota growth associated with low cariogenic and caloric value are some of these properties. Also noteworthy is a stimulation of the immune system leading to a reduced risk of diarrhea and other infections. The benefits are obtained by a decrease in intestinal pH due to the fermentation of NDOs, decreasing the proliferation of pathogenic microorganisms, and an increase of the bifidobacteria population [2]. The bioactive properties of NDOs can be influenced by monosaccharide composition, type of glycosidic linkage, and degree of polymerization [2].

Nondigestible oligosaccharides can be produced using chemical or enzymatic processes. The synthesis using chemical methods are complicated, with numerous protection and deprotection steps required in order to achieve regioselectivity [3]. Other challenges of chemical synthesis are the low yields, expensive chemicals, and impossibility for scale-up. For those reasons with few exceptions, most of the NDOs are produced by enzymatic processes.

The enzymatic production of NDOs can be achieved by two different approaches, the use of glycosyltransferases or glycosidases. Glycosyltransferases catalyze the stereospecific and regiospecific transfer of a monosaccharide from a donor substrate (glycosyl nucleotide) to an acceptor substrate. Some of the difficulties associated with the application of glycosyltransferases are availability of enzymes and sugar nucleotide donors, product inhibition, and reagent costs. These factors decrease the applications of these enzymes in the production of NDOs [4]. The glycosidases offer a good alternative for enzymatic production of NDOs, where they can be synthetized from monosaccharides using transglycosylation reactions, or formed by controlled enzymatic hydrolysis of polysaccharides. Some advantages of the glycosidases in relation to glycosyltransferases are availability, good stability, and the fact that they act on easily found substrates and do not need cofactors [3].

The transglycosylation route can be performed by the use of a good glycosyl donor that can be a disaccharide, in high concentrations. This donor will form an intermediate glycosyl-enzyme that can be intercepted by an acceptor to give a new glycoside or oligosaccharide [3]. When the substrate is a monosaccharide, it will be acting as a donor and acceptor. Some glycosidases used to produce NDOs using this approach are α-galactosidases, β-fructofuranosidase, cyclomaltodextrin glucanotransferase, and α-glucosidase [4].

The production of NDOs by controlled hydrolysis of polysaccharides involves the break of glycosidic bonds, the reaction is acid base catalyzed by an oxocarbenium ion-like transition state and involves two carboxylic groups at the active site [5]. The glycosidases can be divided into inverting or retaining depending on the configuration of the glycosidic linkage after the hydrolysis. Inverting glycosidases operate through direct displacement of the leaving group by water. The two carboxylic groups are responsible for the reaction, one provides base catalytic assistance to the attack of water and the other provides acid catalytic assistance to cleavage of the glycosidic bond. Retaining glycosidases use a double displacement mechanism involving the formation of a covalent glycosyl enzyme intermediate, where one carboxylic group acts as acid catalyst for the glycosylation step and base catalyst for the deglycosylation step [3]. The second carboxylic group acts as a nucleophile and a leaving group. The enzymes inulinase, pullulanase, amylase, xylanase, endogalactanase, rhamnogalacturonase, endogalacturonase, and chitosanase are used for NDOs production using the controlled hydrolysis approach [4].

2. Production of NDOs through glycosyl transfer reaction

2.1. Galactosidases

β-Galactosidases (EC 3.2.1.23) hydrolyze the nonreducing terminal of β-D-galactose residues in β-D-galactosides. The enzyme can be used in the production of galacto-oligosaccarides (GOs) by transgalactosylation reaction in which a galactosyl is transferred into the hydroxyl group of the galactose residue of lactose [6]. Due to the strong prebiotic factor, GOs can modulate the grown of microorganisms of the gut flora, increasing the population of bifidobacteria, this enhancement is associated with beneficial effects, inhibition the grown of potentially pathogens, improvement, elimination, prevention, stimulation mineral adsorption, and decrement cholesterol and lipids [7].

When using concentrated solutions of lactose (40%), high yields of GOs can be achieved. The β-galactosidase of *Pseudozyma tsukubaensis* showed high transgalactosylation capability, yielding of 18.28% of GOs with concentration of 73.12 g/L from a 40% lactose solution [8]. The immobilization of chemically aminated β-galactosidase from *Aspergillus oryzae* onto Purolite® A-109 leads to an increase in the operational stability and transgalactosylation capacity of the enzyme, producing in the optimum conditions (400 g/L lactose, pH 4.5, 50°C) 100 g/L of GOs in a fluidized bed reactor [9]. The utilization of an ultrafiltration membrane bioreactor, allows the synthesis and separation in one system. Using high lactose concentrations (470 g/L) and β-galactosidase from *A. oryzae*, the system yielded 1.88 gGOS/mgE that is 2.44-fold higher than the conventional batch (0.77 gGOS/mgE) [10].

The milk whey, a by-product from the dairy industry, is a valuable substrate for GOs productions due to its lactose contend (45–60%). The whey is produced by the processing and manufacturing of raw milk into products such as yogurt, ice cream, butter, and cheese through processes such as pasteurization, coagulation, filtration, centrifugation, chilling, etc. [11]. Depending on the procedure used to precipitate the casein, two types of whey are formed, the acid whey (pH < 5) is obtained after fermentation or addition of organic or mineral acids, whereas the sweet whey (pH 6–7) is obtained by addition of proteolytic enzymes like chymosin [12]. The production of GOs from milk whey using a two-dimensional packed bed bioreactor yielded 97% [13], while a yield of 29.9% of GOs with a concentration of (119.8 mg/mL) was achieved using cheese whey as substrate in a 4 h process [14]. When whey permeate was used as substrate in a membrane reactor system, a mixture of GOs with 77–78% of purity was produced [15]. A high lactose conversion was achieved (70–80%), when using whey as a substrate in the production of GOs, yielding 10–20% of total sugars and producing oligomers with DP3, DP4, and DP5 [16]. The GOs production from whey permeate yielded 50% corresponding to 322 g prebiotics/kg whey permeate, presenting tagatose and lactulose in the oligosaccharides mixture [17]. Galacto-oligosaccharides were synthesized by enzymatic transgalactosylation in UF-skimmed milk permeate fortified with lactose (40% w/w). The GOs yields, expressed as a percentage of the initial lactose content, were 41, 21, 13, and 11% with β-galactosidase from *Bacillus circulans, A. oryzae, Aspergillus aculeatus,* and *Kluyveromyces lactis,* respectively, under optimal conditions [18].

2.2. β-fructofuranosidases

The β-D-fructofuranosidases catalyze the hydrolysis of β-D-fructofuranoside residues at the nonreducing end of β-D-fructofuranosides [19]. Fructooligosaccharides (FOs) can be produced by transfructosylation of sucrose by β-fructofuranosidases, which is carried out through the breaking of the β(2-1) glycosidic bond and the transfer of the fructosyl moiety onto any acceptor other than water, such as sucrose or a FO. The sucrose is used as substrate acting as the glycosyl donor and as the glycosyl acceptor in competition with water (hydrolysis) in a glycosyl transfer reaction [20]. Besides the strong prebiotic factor, many bioactivities have been associated with FOs as anti-inflammatory effect on Crohn's disease and ulcerative colitis, antimicrobial activity against gut flora pathogens, and prevention of colon cancer [21].

A β-fructofuranosidase from *Penicillium oxalicum* was able to produce neokestose from a 500 g/L sucrose solution, giving 94.2 and 224.7 g/L of neokestose and total FOs, respectively [22]. An invertase produced by *Aspergillus niger* using salt-deoiled cake as substrate was able to form kestose during enzymatic hydrolysis using glucose (50%) [23]. *Penicillium sizovae* and *Cladosporium cladosporioides* were used to produce FOs from a 600 g/L of sucrose solution with maximum yield of 184 and 339 g/L, respectively [24]. The filamentous fungus *Gliocladium virens* was able to produce 6-kestose with a yield of 3in media containing 150 g/L sucrose after 4–5 days of culture [25]. An extracellular β-fructofuranosidase from *Rhodotorula dairenensis* produced a varied type of FOs containing β(2→1)- and β(2→6)-linked fructose oligomers with a maximum concentration of 87.9 g/L (75% sucrose conversion) [26]. A fructosyltransferase from *Aureobasidium pullulans* presented maximum transfructosylation rate at 600 g/L [27].

2.3. Cyclomaltodextrin glucanotransferase

Cyclomaltodextrin glucanotranferase (CGTase, EC 2.4.1.19) catalyze the cyclization of oligosaccharides composed of D-glucose monomers joined by α(1-4) glycosidic linkages. This enzyme catalyzes mainly transglycosylation reactions leading to the formation of nonreducing cyclic oligosaccharides, named cyclodextrins. The main types are α-, β-, and γ-cyclodextrins consisting of six, seven, and eight glucose monomers in cycles, respectively. The majority of the CGTases usually produce a mixture of α-, β-, and γ-cyclodextrins, and the product ratio can vary depending on condition and reaction time [28].

The CGTase can produce cyclodextrins from starch, amylose, and other polysaccharides by catalyzing different transglycosylation steps: intermolecular coupling and disproportionation and modification of the length of noncyclic dextrins [29]. Between main microbial sources of CGTases, the *Bacillus, Geobacillus, and Paenibacillus* species are highlighted. The optimum temperature and pH for this enzyme range from 4 to 10.3°C and 10 to 85°C, respectively, whereas the molecular weight ranges from 33 to 200 kDa.

The products of the CGTases α, β, and γ-cyclodextrins are not completely digested in the gastrointestinal tract, rising to the colon where they are fermented by the intestinal microflora and for this reason are considered prebiotics. The microbial degradation results in linear malto-oligosaccharides, which are further hydrolyzed and fermented to absorbable and metabolize short-chain fatty acids. Several studies showed that CDs reduce the digestion of

http://dx.doi.org/10.5772/65935

starch and the glycemic index of food. Other bioactivities include hypocholesterolemic and antithrombotic activity [30].

The most frequently used raw material for CDs production is starch. The product inhibition effect of cyclodextrins on CGTases, make the complete conversion of starch a challenge. Strategies to decrease this effect involve the continual removal of CDs by filtration or the precipitation using agents that forms a specific insoluble complex with CDs. Filtration devices can be coupled to the production systems, hollow fiber and [31]. **Table 1** shows the yields or concentration of CDs obtained through the action of microbial CGTase on different substrates.

Enzyme source	Substrate	Conditions	Yield (%)	Concentration (g/L)	Reference
α-cyclodextrin					
B. circulans STB01	5% maltodextrin	9 h; 50°C	25	4.3	[32]
B. lehensis	Cassava starch	55°C; 35 h	–	0.32	[33]
P. macerans	5% soluble starch	10 h; 45°C; pH 5.5	–	10.3	[34]
T. thermosulfurigenes	10% paselli SA2	0.1 U/mL; pH5.9; 60°C;8 h	33	13.0*	[35]
β-cyclodextrin					
B. lehensis	Cassava starch	55°C; 35 h	–	6.33	[33]
Bacillus sp. C26	Starch		26.5	10.6	[36]
B. firmus strain 37	5% starch	24 h	–	15.3	[37]
B. firmus strain 37	10% maltodextrin	24 h	–	21.6	[37]
Bacillus sp. C26	4% starch	72 h; 50	–	8.2	[38]
B. circulans STB01	5% maltodextrin	9 h; 50°C	58	10.1	[32]
B. firmus strain 37	5% corn starch	3 days; 60°C	–	15.0	[39]
B. firmus strain 37	5% maltodextrin	3 days; 60°C	–	10.1	[39]
Bacillus sp.	10% dextrin	90 min; 50°C; pH 8	–	6.0	[40]
Thermoanaerobacter sp.	4% soluble starch	30 s; 60°C; pH 6	7.9	1.3	[41]
A. gottschalkii	10% starch	24 h; 40°C; pH 8	45		[42]
B. macerans	Soluble starch		24	4.7	[43]
P. macerans	5% soluble starch	10 h; 45°C; pH 5.5	–	4.1	[34]
T. thermosulfurigenes	10% paselli SA2	0.1 U/mL; pH 5.9; 60°C; 8 h	54	20.0*	[35]
γ-cyclodextrin					
B. lehensis	Cassava starch	55°C; 35 h	–	1.02	[33]
B. cereus	5% starch	1 h; 20% CGTase	81.9	1.6	[44]
B. circulans STB01	5% maltodextrin	9 h; 50°C	17	3.0	[32]
Bacillus sp.	10% dextrin	90 min; 50°C; pH 8	–	1.5	[40]

Enzyme source	Substrate	Conditions	Yield (%)	Concentration (g/L)	Reference
B. clarkii 7364	Potato starch	10 h; 50°C; pH 7	72.5		[45]
P. macerans	5% soluble starch	10 h; 45°C; pH 5.5	–	1.8	[34]
B. clarkii 7364	15% soluble starch	55°C; pH 12	47		[46]
T. thermosulfurigenes	10% paselli SA2	0.1 U/mL; pH 5.9; 60°C; 8 h	13	5.0*	[35]
Mixture (α, β, and γ)					
B. macerans	Glucans	24 h; 40°C	21.1	15.1	[47]
P. macerans	5% soluble starch	22 h	36.9		[34]
B. circulans DF 9R	5% cassava starch	4 h; 56°C	55.6	99.5[a]	[48]
Toruzyme 3.0 l	Tapioca starch	4 h; 60°C	85	23.0	[49]
T. fusca	15% potato starch	24 h; 30°C; pH 5.6	84		[50]
B. cereus	6% sago starch	8 h; 55°C	–	13.7	[51]
Toruzyme 3.0 l	8% tapioca starch	2 h; 70°C;pH 5	–	12.1	[52]
Toruzyme 3.0 l	8% tapioca starch	3 h; 60°C	25	40.0	[49]
B. megaterium	50 g/L corn starch	pH 7; 45°C; 12 h; 2 U/g CGTase	30	–	
B. macerans	30% potato starch	pH 5.5–8.5; 40–55°C; 120 h;1000 U/g CGTase	30–35	–	[53]
B. macerans	7.5% corn starch	48 U/g CGTase; pH 6; 60°C; 24 h	25	–	[54]
B. circulans 251	10% potato starch	pH 6; 50°C; 45–50 h	40	–	[55]
Bacillus sp. 277	10% potato starch	400 U/g CGTase; pH 8; 60°C; 12 h	34	–	[56]
B. clausii E16	1% soluble starch	10 U/g; pH 5.5; 55°C; 24 h	80	–	[28]
B. macerans	10% tapioca starch	0.4 mmol cyclodecanone; pH 7; 25°C; 5–10 days	91–93	–	[57]
Mutant CGTase H43T	1% tapioca starch	1% toluene; pH 6; 60°C; 18 h	15.2	–	[58]
K. pneumoneae	12.5% wheat starch	20 U/g CGTase; 2% butanol; pH 7.5; 40°C; 6 h	42.5	–	[59]
Thermoanaerobacter sp.	5% soluble starch	60°C; pH 6	29	74.0	[60]
B. stearothermophilus	5% soluble starch	500U/g; 65°C; pH 6; 24 h	22	–	[61]
E. coli NV601	5% soluble starch	60°C; pH 6	30	75.0	[60]

Table 1. Production of cyclodextrins by microbial CGTases.

Bacillus sp. species are the main microbial source of CGTase, in some cases thermophiles are used to obtain enzymes with unusual characteristics. Most of studies are focused on the β-cyclodextrin or mixture production and higher concentrations are usually obtained for β-and γ-cyclodextrins. The substrate is usually corn starch, although tapioca, cassava,

http://dx.doi.org/10.5772/65935

wheat, and potato starches are also observed. The conditions for cyclodextrin production are usually 40–60°C, pH 6–7, and aqueous media, however, depending on the microbial source of the CGTase some unusual condition may be observed, as 25°C or pH 12. In some cases, the organic media is used to decrease the inhibition of the CD. The highest productivity is reported to the production of a mixture by a recombinant CGTase of *Thermoanaerobacter* using soluble starch that yielded 75 g/L.

2.4. Alpha-glucan acting enzymes

Alpha-glucans are polysaccharides consisting of glucose units connected by α(1-4) or α(1-6) glycosidic linkages. Pullulan, a glucan produced by the fungus *A. pullulans* of α(1-4) linked maltotriose repeats connected by α(1-6) linkages, amylopectin, formed by shorter α(1-4) glucan chains connected by α(1-6) branch points, and dextran are some examples of alpha-glucans [62].

Enzymes that act as hydrolyzing or debranching alpha-glucans are suitable for nondigestible oligosaccharides production. Pullulanase, dextransucrase, and starch acting enzymes can be used in the preparation of maltooligosaccharides and isomalto-oligosaccharides. Maltooligosaccharides contain α-D-glucose residues linked by α(1-4) glycosidic linkages, while isomaltooligosaccharides (IMOs) contain two to five glucose units with one or more α(1-6) linkages. While MO may exhibit immunoregulatory activity [63], the intake of IMO decreases serum cholesterol concentrations and improve bowel movement, stool output, and microbial fermentation in the colon [64]. IMOs also upregulate the Th1 response that play a triggering role in allergic diseases, such as rhinitis, asthma, and eczema [65].

Dextransucrases (EC 2.4.1.5) catalyze the synthesis of high molecular weight D-glucose polymers from sucrose to form a glucan called dextran. The synthesis of dextran occurs by successive transfer of glucosyl units to the polymer, while the presence of acceptor molecules in the reaction medium, the transfer of glucosyl units is made onto these molecules, leading to oligosaccharide synthesis. They can also transfer glucosyl units onto water molecules and simply hydrolyze sucrose [66]. *Leuconostoc citreum* KACC 91035 produced panose (8.63 mM), isomaltosyl maltose (6.56 mM), and isomaltotriosyl maltose (1.74 mM) after 12 days (10°C), using glucose (29 mM) as donor and maltose (28 mM) as acceptor through the transglycosylation activity of the dextransucrase [67]. An endodextranase D8144 from *Penicillium* sp. immobilized on epoxy produced IMOs (DPs 8–10) from dextran T40 in an enzymatic reactor [68]. A productivity of 42.95 mmol/L.h was obtained using 100 mmol/L of sucrose and 200 mmol/L of maltose, using dextransucrase (1 U/mL) from *Leuconostoc mesenteroides* NRRL B-512F [69]. A productivity of 7.26 mmol/L.h of IMOs was obtained a using an immobilized mixture of dextransucrase and dextranase [70], while a purified dextransucrase yielded 35 mmol/L.h of panose [71]. A productivity of 55.6 mmol/L of oligosaccharides was obtained by fermentation with *L. mesenteroides* B-742 [72]. Higher yields (70–90%) of IMOs were obtained from maltose/sucrose solutions using dextransucrase of *L. mesenteroides* B-512F [73]. Isomalto-oligosaccharides of controlled molecular weight were produced using an *L. mesenteroides* NRRL B-512F dextransucrase with a yield of 58% by the acceptor reaction with glucose, and reached a degree of polymerization of at least 27 glucosyl units [74]. The use of dextransucrase associated with dextranase in the production of IMOs lead to oligosaccharide mixtures containing mainly sugars (up to 36%) with DP varying between 10 and 60 together lower and higher molecular weight sugars [75].

Alpha-amylase (EC 3.2.1.1) also can be used to obtain maltooligosaccharides. This enzyme hydrolyses the internal α(1,4) linkages in starch in a random fashion, leading to the formation of soluble maltooligosaccharides, maltose, and glucose. A protein engineering approach of the amylase from *Bacillus lehensis* G1-produced mutated proteins with an increase in the transglycosylation to hydrolysis ratio of up to 4.0-fold and reduction in the concentration of maltotriose required for use as a donor/acceptor for transglycosylation. A reduction of steric interference and hydrolysis suppression introduced a synergistic effect to produce MOs with a higher degree of polymerization [76]. Amylases from *Streptomyces sp.* were able to produce mainly maltotriose (55–75%) from soluble starch at 20–30°C pH 6.5 [77]. The *Bacillus subtilis* strain SDP1 amylase hydrolyses starch to produce maltotriose and maltotetraose along with maltose after prolonged reactions of 5 h [78]. A recombinant alpha-amylase (145 mg/mL) from *Streptomyces avermitilis* was able to yield maltose (4.49) and maltotriose (1.77 g/L) from 10.0 g/L of soluble starch [79]. An amylase from Bacillus megaterium produced a maltooligomer mixture with high proportion of maltopentaose (G5) and maltotriose (G3) during hydrolysis of starch, amylopectin, and amylose [80]. Malto-oligosaccharide production by commercial α-amylase (liquefying amylase 6 T) using freeze-thaw infusion resulted in a maximum production of 6.5 g/L after 60 min at 1.0% (w/v) enzyme [81]. A productivity of 8.9 g/L of maltopentaose was achieved using a *Bacillus* sp. AIR-5 amylase and a 40 g/L solution of soluble starch [82]. A *S. solfataricus* KM1 amylase was able to give an 80% yield of trealose from a 10% amylose solution [83].

Pullulanase (EC 3.2.1.41), a debranching enzyme, hydrolyses the α(1-6) linkage in pullulan and branched polysaccharides, producing maltotriose. An amylopullulanase from the hyperthermophilic archaeon *Caldivirga maquilingensis* was able to act on a wide range of substrates. Assays with the enzyme produced linear MOs (≤G8–G1) from cyclodextrins, amylodextrins (DP6-96) from amylose, and amylodextrins (DP1-76) from amylopectin and potato starch [84]. A one-step method using neopullulanase and α-amylase for the bioconversion of purified rice starch slurry (30% w/w) resulted in a syrup containing 59.2% of IMO (dry basis) after 72 h of bioconversion (Lin et al. 2011).

Alpha-glucosidase (EC 3.2.1.20), an exo-acting hydrolase, attacks the substrates from the non-reducing end producing α-D-glucose and presents some transglycosylation activity that can be used in the production of oligosaccharides [85]. Liquefied banana slurries were used for IMO synthesis by Transglucosidase L, producing after 12 h of transglucosylation, a yield of 76.6% with a concentration of 70.74 g/L. The IMOs mixture was composed of 53 isomaltotriose, 21 isomaltotetraose, and 26% maltooligoheptaose and larger oligomers [86]. A yield of 58.1% with a concentration of 93 g/L was obtained for IMOs production from a immobilized glucosidase using as substrate a maltose solution (160 mg/mL) in a membrane reactor system [87]. Partially purified a-glucosidase from *Aspergillus carbonarious*, immobilized on glutaraldehyde-activated chitosan beads in a packed bed reactor, produced isomaltooligosaccharides at a yield of 60% (w/w) using 30% (w/v) maltose solution. Using intact mycelia attached with polyethyleneimine-glutaraldehyde, a yield of 46% (w/w) was obtained using 30% (w/v) maltose solution [88]. A high yield of IMOs (67%) with concentration of 2 g/L was obtained when 30% (w/v) of soluble tapioca starch was incubated with amylomaltase (120 U) for 0.5 h (pH 7.0; 40°C). While a yield of 53% and concentration of 1.63 g/L was obtained using transglucosidase (6 U) in the same condition for 1 h [89]. When amylomaltase (1.5 U) and transglucosidase (8 U) of were incubated with 20% (w/v) maltotriose for 30 min at 40°C, 9.9 mg/mL of IMOs were produced to with DP 2–7 [90].

3. Productions of NDOs through polysaccharide hydrolysis

3.1. Inulinase

Fructooligosaccharides can be produced by the controlled hydrolysis of fructans. Fructans are fructose-based polysaccharides, representing the major reserve carbohydrates in about 15% of flowering plant species [91]. According to differences in glycosidic linkages they can be classified in many types, being linear inulin the most studied and best-characterized fructan. Inulin consists of β(2-1)-linked fructose units terminating at the reducing end with a glucose residue attached through a sucrose-type linkage [92]. Inulinases can hydrolyze the β(2-1) linkages in inulin and can present endo- or exo-activity. Exo-acting inulinases (EC 3.2.1.80) produce fructose as the main end product, whereas endoinulinases (EC 3.2.1.7) act randomly and hydrolyze internal linkages of inulin to yield FOs and minor amounts of monosaccharides [93].

The highest yield (92%) for the conversion of chicory inulin (50 g/L) in to FOs was reported by the application of a dual system of *Xanthomonas* sp. and *Pseudomonas* sp. endoinulinases [94]. On another approach, an endoinulinase from *Xanthomonas* sp. yielded 86% of FOs from dahlia tubers inulin (10 g/L) after 10 h [95]. A production of 78% and 79% of FOs was achieved from a solution (100 g/L) of chicory inulin and chicory juice, respectively [96]. An endoinulinase produced by *Streptomyces rochei* E87 yielded 70% of FOs after 3 days of incubation with inulin producing mainly inulotriose [97]. A maximum yield of 75.6% in total of FOs was obtained by hydrolysis of a solution containing 50 g/L of inulin by *Pseudomonas* sp. inulinase, producing a mixture of oligosaccharides with DP2-7 [98]. A commercial inulinase preparation yielded 96% of FOs from dahlia tubers inulin (pH 6.0; 100 g/L). The product presented FOs with DP ranging from 1 to 6 but the major products were DP3 (23%) and DP4 (24%) [99].

The production of FOs by a inulinase from *A. niger* immobilized in montmorillonite lead to a yield of 18.32% on aqueous media and 16.03% in organic media [100], while high yields of DP3 (70.3 mM), DP4 (38.8 mM), and DP5 (3.5 mM) FOs were obtained through the enzymatic hydrolysis of inulin (150 mg/ml; 60°C; pH 6.0; 48 h) by other *A. niger* inulinase (60 U/mL) [101]. When a commercial endoinulinase preparation (Novozym®960) from *A. niger* was used in the production of FOs from inulin (60°C; pH 6.0), a productivity of F3 (70.3mM), F4 (38.8mM), and F5 (12.43 mM) was achieved [102]. Inulinases from *K. marxianus* NRRL Y 7571 produced DP2 (11.89%) and DP3 (20.83%) oligomers using inulin (20%) as substrate at 24 h at 50°C [103]. A maximum FOs production of 11.9 g/L.h and specific productivity of 72 g/g.h were observed when a mutant X. *campestris* pv. *phaseoli* grown in a 5 L fermenter containing 3%inulin and 2.5% tryptone [104]. A continuous production of FOs from inulin was carried in a bioreactor packed with immobilized cells of *Escherichia. coli* expressing a *Pseudomonas* sp. endoinulinase. Under the optimal operation conditions, continuous production of FOs was achieved by 150 g/L.h (17 days; 50°C) [105]. Continuous production of FOs from chicory juice (100 g/L) was carried out using the polystyrene-bound endoinulinase in an enzymatic reactor achieving an oligosaccharide yield of 82% [106]. *Aspergillus ficuum* endoinulinase (10 U/g) yielded 50% of FOs from Jerusalem artichoke inulin (50 g/L; 45°C; pH 6.0) after 72 h. With Jerusalem artichoke the yield reached 89% and the maximum IOS production was up to 80% after 72 h [107].

3.2. Xylanases

Xylan is also a heteropolysaccharide with a backbone formed by xylose homopolymer subunits linked through β(1-4) linkages. This polymer can be found in the hemicellulose fraction of lignocellulosic materials associated with lignin and cellulose. Through the hydrolysis of xylan with xylanases, xylooligosaccharides (XOs) can be produced. The intake of XOs is associated with many health benefits as improvement of bowel function, immunomodulatory, and anti-inflammatory activities, preventive effects on cancer and inhibitory effects on carcinogenesis, antimicrobial, antiallergic, and antioxidant activities [108].

The xylanase (β-1,4-d xylan xylanohydrolase, EC 3.2.1.8) is the main enzyme applied for xylan hydrolysis and XOs production, due its action on the main chain of xylan and release of oligosaccharides. Before the enzymatic hydrolysis of xylan, the hemicellulosic materials can be submitted to a pretreatment to enhance the xylan availability. Many types of pretreatments that can be performed, one approach uses NaOH or H_2SO_4 solutions associate with high temperatures to disrupt the hemicellulose structure. Between the substrates used for XOs production agroresidues and food by-products are highlighted due to their high contends of hemicellulose [109].

Hydrolysis of alkali pretreated corncob powder using a commercial endoxylanase produced 81 ± 1.5% of XOs in the hydrolysate equivalent to 5.8 ± 0.14 mg/mL of XOs. Reaction parameters for the production of XOs from corncob using endoxylanase from *A. oryzae* MTCC 5154 were optimized and an XOs yield of 10.2 ± 0.14 mg/mL corresponding to 81 ± 3.9% with 73.5% xylobiose [110]. The optimization of the XOs production from corncob using the thermostable endoxylanase from *Streptomyces thermovulgaris* TISTR1948, showed that for an enzyme concentration of 129.43 U/g of substrate, 53.80°C, and pH 6.17, the yield of XOs reached 162.97 mg/g of substrate or 752.15 mg/g of hemicellulose in KOH-pretreated corncob [111]. When corncob was hydrolyzed with a xylanase from *Aspergillus foetidus* MTCC 4898 a yield of 6.73 ± 0.23 mg/mL was obtained after 8 h of reaction time using 20 U of xylanase at 45°C [112]. A commercial xylanase produced 1.208 mg/mL of xylobiose and 0.715 mg/mL of xylotriose, using 5.83 U for 16.59 h of incubation (pH 5.91; 40.8°C) [113]. Steam-exploded liquor of corncobs was treated using a thermostable xylanase from *Paecilomyces themophila* J18 resulting in a XOs yield of 28.6 g/100 g xylan [114]. After a pretreatment with H_2SO_4 (60°C; 12 h), the corncob was hydrolyzed by xylanase, yielding 67.7% of XOs with

70% of purity [115]. Three commercial xylanase preparations (Rapidase Pomaliq, Clarex ML, and Validase) were evaluated for the enzymatic production of pentoses from the hemicellulose fraction of corn husks and corn cobs. Rapidase Pomaliq produced 104.1 g of XOs for each kg of corn husks or 133 g of XOs for each kg of corn cobs (480 min of reaction at pH 5.0 and 50°C) [116].

The application of agroresidues as a source of xylan for XOs production is a strategy that has been produced excellent results. The xylan obtained by alkali extraction from cotton stalk, was hydrolyzed using a commercial xylanase preparation produced XOs in the DP range of 2–7 (X6 ≈ X5 > X2 > X3) and also minor quantities of xylose, yielding 53% (40°C; 24 h) [117]. Tobacco stalks were hydrolyzed by xylanase producing a XOs yield of 8.2%after 8 h and 11.4% after 24 h reaction period [118]. Another process yielded 7.28 and 4.52 g/L of XOs from wheat straw and rice straw xylan, respectively, after hydrolysis with a from *A. foetidus* MTCC 4898 [119]. Using xylanases from two glycoside hydrolase families, yields of 60% and 40% were obtained for rye bran arabinoxylan hydrolysis by GH10 and GH11, respectively [120]. Wheat straw xylan was hydrolyzed using a variant of the alkali-tolerant *Bacillus halodurans* S7 endoxylanase A, resulting in 36% conversion of the xylan to predominantly xylobiose [121]. The XOs produced from garlic straw hemicelluloses hydrolyzed with xylanase secreted by *B. mojavensis* were composed of xylobiose, xylotriose, and xylotetrose, together with a small amount of xylopentaose and xylohexose yielding 29 ± 1.74% after 8 h [122]. Xylan extracted of *Mikania micrantha* was hydrolyzed by a recombinant *Paenibacillus* xylanase, yielding 68% of XOs [123]. Oil palm empty fruit bunch fiber was hydrolyzed by *Aspergillus terreus* xylanase with a maximum 262 mg of xylobiose was produced from 1.0 g of pretreated fiber [124]. Several crop by-products were subjected to an enzymatic treatment to obtain a XOs through the action of a Buzyme 2511 (R). The hydrolysis lead to a concentration of 5.3 (apple pomace), 1.3 (white poplar), 2.9 (giant cane), and 6.5 g/L (grape stalk) [125]. The enzymatic hydrolysis of hard shell almond yielded 34.0% of XOs with 70% of purity [126]. A process for producing XOs from *Sehima nervosum* grass through enzymatic hydrolysis yielded 11 g/100 g xylan of xylobiose [127]. The treatment of wheat bran with the commercial xylanase preparation enzymes, produced a yield of approximately 31.2% of XOs, with a purity of 95% (w/w) and degree of polymerization of 2–7 [128]. Viscose fiber mills were used as substrate in the production of XOs yielding 68.9% after enzymatic hydrolysis [129].

When sugarcane bagasse was hydrolyzed with a crude xylanase secreted by *Pichia stipites*, XOS accumulated with a maximum yield of 31.8% of the total xylan was achieved at 12 h, which contained 29.8% xylobiose, 47.1% xylotriose, and 18.4% xylotetraose [130]. The hydrolysis of sugarcane bagasse with a *B. subtilis* xylanase produced xylotriose (X3), xylotetraose (X4), and xylopentaose (X5) and also is less amounts xylooligomers (X11). The process yielded was 113 and 119 mg/g sugarcane bagasse for 7 and 8 h, respectively [131]. In another approach using sugarcane bagasse treated with hydrogen peroxide, the enzymatic hydrolysis by crude extracts from *Thermoascus aurantiacus* produced a maximum yield of 37.1 with 2.6% of substrate and xylanase load of 60 U/g [132]. A productivity of 2.36, 2.76, 2.03, and 2.17 mg/mL of X2, X3, X4, and X5, respectively, was obtained after hydrolysis of sugarcane bagasse by *Streptomyces rameus* L2001 xylanase [133]. A maximum yield of 5.96% was obtained for the conversion of sugarcane bagasse being xylobiose and xylotriose the main products [134]). The enzymatic hydrolysis of *Camellia oleifera* shell pretreated with NaOH produced 1.76 g/L of xylooligosaccharides (DP 2–6) [135].

3.3. Pectinases

Pectins are components of the cell walls of most higher plants, this heteropolysaccharide is characterized by a high content of galacturonic acid (GalA) monomers bonded together by α(1-4) linkages, showing acetylatilation or esterification with methyl groups. They are composed of homogalacturonans, xylogalacturonanes, rhamnogalacturonans, arabinans galactans, and arabinogalactans. Depending on how these polysaccharides are associated, pectin can be classified as homogalacturonan and rhamnogalacturonans I and II [136].

Studies using piglets showed that POs can modulate the grown of microbial communities in the ileum increasing, for example, the *Lactobacillus* counts [137, 138]. POs were also able to interfere with the toxicity of Shiga-like toxins from *E. coli* O157:H7, which play a key hole in diarrhea and hemorrhagic colitis, hemolytic uremic syndrome (HUS), and thrombotic thrombocytopenic purpura [139].

Enzymes that act on pectins with a hydrolyzing or debranching activity have the potential to produce nondigestible oligosaccharides. The pectinolytic enzymes can be divided into: pectinesterases, pectin-methylesterases, and depolymerases being this last one more suitable for POs production. Endopolygalacturonases are depolymerases produced by various microorganisms such as bacteria, yeasts, and molds. They are also found in some plants and especially in fruits. In general, they release mono-, di-, and tri-galacturonic acid by a multiple attack mechanism single chain. Rhamnogalacturonases produce linear oligomeric compounds of alternating rhamnose and galacturonic acid (4–6 residues) with galactose residues connected to some or all the rhamnose residues. Galactanases can be divided in to endo-β-1,4-galactanases and exo-β-1,3-galactanases. The difference between these enzymes lies in their ability to hydrolyze the β(1-3), β(1-4), or β(1-6) linkages between the galactose residues [136].

Because of its high pectin content, potato, sugar beet, and apple by-products are often used as substrate for POs production. The hydrolysis of sugar beet pectin by combining endopolygalacturonase and pectinmethylesterase produced POs with a DP 1–9, with a maximum yield of trigalacturonic acid of 3.7% [140]. POs were obtained by the action of commercial enzymes on the potato rhamnogalacturonan, with a yield of 93.9 and 66.2% using Depol 670L and endo-β-1,4-galactanase, respectively. The hydrolysates yielded up to 50.6% of oligomers with DP of 13–70. Major oligosaccharides obtained with Depol 670L were DP 5 (26.3%) and DP6 (24.9%), whereas the endo-β-1,4-galactanase were DP3 (19.0%), DP5 (10.6%), and DP8 (12.6%) [141]. A high yield (93.9%) of POs was achieved using multienzymatic preparation (Depol 670 L) to hydrolyze a potato pulp by-product rich in galactan-rich rhamnogalacturonan I. Main products were oligosaccharides with DP of 2–12 (79.8–100%), whereas the oligomers with DP of 13–70 comprised smaller proportion (0.0–20.2%) [142]. A pool of pectinases was used to produce POs with degree of polymerization from 2 to 8 and six different rhamnogalacturonide structures. Total recoveries were 200 (homogalacturonides) and 67 mg/g (rhamnogalacturonides) [143]. The use of commercial pectinase preparations (Endopolygalacturonase M2,

Pectinase, Viscozyme L, Pectinex Ultra SP-L, Pectinase 62 L, and Macer8 FJ) to produce POs from polygalacturonic acid. Best results were obtained with endopolygalacturonase M2 after 2 h of reaction, yielding 58, 18, and 13% of DP3 > DP2 > DP1, respectively [144].

In some cases, other food by-products were applied in the production of POs. A initial amount of 100 kg of orange peel can yield 7.5 kg of gluco-oligosaccharides, 4.5 kg of galacto-oligosaccharides, 6.3 kg of arabino-oligosaccharides, and 13 kg of oligogalacturonides [145]. Through the action commercial enzymes (EPG-M2, Viscozyme, and Pectinase) on onion skins a yield 5.6% of pectic oligosaccharides (POS) was obtained [146].

3.4. Chitosanase

Chitin is a polysaccharide formed by *N*-acetyl-glucosamine monomers, joined by β(1-4) linkages and chitosan is the N-deacetylated form of chitin. Chitosanases (EC 3.2.1.132) are glycosyl hydrolases that catalyze the hydrolysis of β(1-4) glycosidic bond in chitosan to produce glucosamine oligosaccharides [147]. Studies using pigs indicated a modulating effect of chito-oligosaccharide (COs) inhibiting growth of harmful bacteria in the gut [67]. Strong antibacterial activity was also reported with complete inhibition of *E. coli* growth with a 0.5% solution [148]. They can also inhibit the growth of tumor cells by exerting immunoenhancing effect [149] and stimulate the growth of *Lactobacillus* sp. and *B. bifidium* KCTC 3440 indicating considerable bifidogenic potential [150].

A chitosanase (EC 3.2.1.132) from *Aspergillus* sp. Y2K showed preference for higher deacetylated chitosan as substrate, producing chitotriose, chitotetraose, and chitopentaose as the major products after hydrolysis with a total yield of 115% [151]. The chitosanolytic enzymes of *Metarhizium anisopliae* produced dimers (0.2 g/L), trimers (0.19 g/L), tetramers (0.06 g/L), and pentamers (0.04 g/L) from chitosan hydrolysis [147]. The enzymatic hydrolysis of chitosan by a chitosanase from *Bacillus* sp. yielded 60% of COs with 95% of purity [152], whereas *Bacillus pumilus* BN-262 chitosanase yielded above 80% in a UF membrane reactor [148]. Through the action of a *B. pumilus* BN-262 chitosanase, a COs productivity of 20 and 15 g/L was obtained in a batch and membrane reactor, respectively [153]. The hydrolysis with *B. pumilus* chitosanase yielded 52% of COs, producing mainly pentameric and hexameric chitosan oligosaccharides was steadily produced at 2.3 g/L (46% yield) for a month [154].

4. Concluding remarks

Glycosidases are widely applied in the production of nondigestible oligosaccharides presenting easy-handed processes with high efficiency. The application of molecular biology tools to produce enzymes with new characteristics has increased the yield and productivity of NDOs. The immobilization of the enzymes and application of membrane and batch reactors are also highlighted for improvements in the production processes. Nowadays alternative substrates have been used frequently in co-products and by-products from food and agroindustry. This approach can lead to a decrease in the cost of the process and help in the correct management of these residues.

Author details

Thais Bezerra[1], Rubens Monti[2], Egon B. Hansen[3] and Jonas Contiero[1,*]

*Address all correspondence to: jconti@rc.unesp.br

1 Department of Biochemistry and Microbiology, Institute of Biological Sciences, UNESP-Universidade Estadual Paulista, Rio Claro, SP, Brazil

2 Faculty of Pharmaceutical Sciences, Department of Food and Nutrition, Sao Paulo State University, Araraquara, SP, Brazil

3 National Food Institute, Technical University of Denmark, Søborg, Denmark

References

[1] Manning TS, Gibson GR. Prebiotics. Best Practice & Research Clinical Gastroenterology. 2004;18(2):287–98.

[2] Tymczyszyn EE, Santos MI, Costa MdC, Illanes A, Gómez-Zavaglia A. History, synthesis, properties, applications and regulatory issues of prebiotic oligosaccharides. In: Carbohydrates Applications in Medicine, M. Helena Gil, ed., Kerala, India: Research Signpost, Chapter 5, 2014, 127–154.

[3] Shetty K, Paliyath G, Pometto A, Levin RE. Functional foods and biotechnology: London CRC Press; 2006.

[4] Barreteau H, Delattre C, Michaud P. Production of oligosaccharides as promising new food additive generation. Food Technology and Biotechnology. 2006;44(3):323.

[5] Mussatto SI, Mancilha IM. Non-digestible oligosaccharides: a review. Carbohydrate Polymers. 2007;68(3):587–97.

[6] Gosling A, Stevens GW, Barber AR, Kentish SE, Gras SL. Recent advances refining galactooligosaccharide production from lactose. Food Chemistry. 2010;121(2):307–18.

[7] Park A-R, Oh D-K. Galacto-oligosaccharide production using microbial beta-galactosidase: current state and perspectives. Applied Microbiology and Biotechnology. 2010;85(5):1279–86.

[8] Cavalcante Fai AE, Resende Simiqueli AP, de Andrade CJ, Ghiselli G, Pastore GM. Optimized production of biosurfactant from Pseudozyma tsukubaensis using cassava wastewater and consecutive production of galactooligosaccharides: an integrated process. Biocatalysis and Agricultural Biotechnology. 2015;4(4):535–42.

[9] Carevic M, Corovic M, Mihailovic M, Banjanac K, Milisavljevic A, Velickovic D, et al. Galacto-oligosaccharide synthesis using chemically modified beta-galactosidase from

Aspergillus oryzae immobilised onto macroporous amino resin. International Dairy Journal. 2016;54:50–7.

[10] Cordova A, Astudillo C, Vera C, Guerrero C, Illanes A. Performance of an ultrafiltration membrane bioreactor (UF-MBR) as a processing strategy for the synthesis of galacto-oligosaccharides at high substrate concentrations. Journal of Biotechnology. 2016;223:26–35.

[11] Benavente R, Pessela BC, Antonio Curiel J, de las Rivas B, Munoz R, Manuel Guisan J, et al. Improving properties of a novel beta-galactosidase from Lactobacillus plantarum by covalent immobilization. Molecules. 2015;20(5):7874–89.

[12] Panesar PS, Kennedy JF, Gandhi DN, Bunko K. Bioutilisation of whey for lactic acid production. Food Chemistry. 2007;105(1):1–14.

[13] Sen P, Bhattacharjee C, Bhattacharya P. Experimental studies and two-dimensional modelling of a packed bed bioreactor used for production of galacto-oligosaccharides from milk whey. Bioprocess and Biosystems Engineering. 2016;39(3):361–80.

[14] Lisboa CR, Martinez LD, Trindade RA, Costa FAD, Burkert JFD, Burkert CAV. Response surface methodology applied to the enzymatic synthesis of galacto-oligosaccharides from cheese whey. Food Science and Biotechnology. 2012;21(6):1519–24.

[15] Das R, Sen D, Sarkar A, Bhattacharyya S, Bhattacharjee C. A comparative study on the production of galacto-oligosaccharide from whey permeate in recycle membrane reactor and in enzymatic batch reactor. Industrial & Engineering Chemistry Research. 2011;50(2):806–16.

[16] Jovanovic-Malinovska R, Fernandes P, Winkelhausen E, Fonseca L. Galacto-oligosaccharides synthesis from lactose and whey by beta-galactosidase immobilized in PVA. Applied Biochemistry and Biotechnology. 2012;168(5):1197–211.

[17] Padilla B, Frau F, Isabel Ruiz-Matute A, Montilla A, Belloch C, Manzanares P, et al. Production of lactulose oligosaccharides by isomerisation of transgalactosylated cheese whey permeate obtained by beta-galactosidases from dairy Kluyveromyces. Journal of Dairy Research. 2015;82(3):356–64.

[18] Frenzel M, Zerge K, Clawin-Raedecker I, Lorenzen PC. Comparison of the galacto-oligosaccharide forming activity of different beta-galactosidases. Lwt-Food Science and Technology. 2015;60(2):1068–71.

[19] Alvaro-Benito M, de Abreu M, Fernandez-Arrojo L, Plou FJ, Jimenez-Barbero J, Ballesteros A, et al. Characterization of a beta-fructofuranosidase from Schwanniomyces occidentalis with transfructosylating activity yielding the prebiotic 6-kestose. Journal of Biotechnology. 2007;132(1):75–81.

[20] Yun JW. Fructooligosaccharides: occurrence, preparation, and application. Enzyme and Microbial Technology. 1996;19(2):107–17.

[21] Rastall RA. Functional oligosaccharides: application and manufacture. Annual Review of Food Science and Technology. 2010;1:305–39.

[22] Xu QS, Zheng XQ, Huang MP, Wu M, Yan YS, Pan JM, et al. Purification and biochemical characterization of a novel beta-fructofuranosidase from Penicillium oxalicum with trans-fructosylating activity producing neokestose. Process Biochemistry. 2015;50(8):1237–46.

[23] Singh S, Gupta N, Kaur J, Gupta A. Valorization of sal deoiled cake as media for acidic amylase and invertase co-production by aspergillus niger nj-1: optimization by response surface methodology and application in oligosaccharide synthesis. Journal of Food Processing and Preservation. 2015;39(6):2548–61.

[24] Zambelli P, Fernandez-Arrojo L, Romano D, Santos-Moriano P, Gimeno-Perez M, Poveda A, et al. Production of fructooligosaccharides by mycelium-bound transfructosylation activity present in Cladosporium cladosporioides and Penicillium sizovae. Process Biochemistry. 2014;49(12):2174–80.

[25] Fialho MB, Simoes K, Barros CD, Pessoni RAB, Braga MR, Figueiredo-Ribeiro RDL. Production of 6-kestose by the filamentous fungus Gliocladium virens as affected by sucrose concentration. Mycoscience. 2013;54(3):198–205.

[26] Gutierrez-Alonso P, Fernandez-Arrojo L, Plou FJ, Fernandez-Lobato M. Biochemical characterization of a beta-fructofuranosidase from Rhodotorula dairenensis with trans-fructosylating activity. Fems Yeast Research. 2009;9(5):768–73.

[27] Antosova M, Illeova V, Vandakova M, Druzkovska A, Polakovic M. Chromatographic separation and kinetic properties of fructosyltransferase from Aureobasidium pullulans. Journal of Biotechnology. 2008;135(1):58–63.

[28] Alves-Prado HF, Carneiro AAJ, Pavezzi FC, Gomes E, Boscolo M, Franco CML, et al. Production of cyclodextrins by CGTase from Bacillus clausii using different starches as substrates. Applied Biochemistry and Biotechnology. 2008;146(1–3):3–13.

[29] Biwer A, Antranikian G, Heinzle E. Enzymatic production of cyclodextrins. Applied Microbiology and Biotechnology. 2002;59(6):609–17.

[30] Fenyvesi E, Vikmon M, Szente L. Cyclodextrins in food technology and human nutrition: benefits and limitations. Critical Reviews in Food Science and Nutrition. 2016;56(12):1981–2004.

[31] Zhekova B, Dobrev G, Stanchev V, Pishtiyski I. Approaches for yield increase of beta-cyclodextrin formed by cyclodextrin glucanotransferase from Bacillus megaterium. World Journal of Microbiology & Biotechnology. 2009;25(6):1043–9.

[32] Li ZF, Ban XF, Gu ZB, Li CM, Huang M, Hong Y, et al. Mutations enhance beta-cyclodextrin specificity of cyclodextrin glycosyltransferase from Bacillus circulans. Carbohydrate Polymers. 2014;108:112–7.

[33] Blanco KC, Moraes FFd, Bernardi NS, Vettori B, Palmuti MH, Monti R, et al. Cyclodextrin production by Bacillus lehensis isolated from cassava starch: characterisation of a novel enzyme. Czech Journal of Food Sciences. 2014;32:48–53.

[34] Wang L, Duan XG, Wu J. Enhancing the alpha-cyclodextrin specificity of cyclodextrin glycosyltransferase from Paenibacillus macerans by mutagenesis masking subsite-7. Applied and Environmental Microbiology. 2016;82(8):2247–55.

[35] Wind RD, Liebl W, Buitelaar RM, Penninga D, Spreinat A, Dijkhuizen L, et al. Cyclodextrin formation by the thermostable alpha-amylase of thermoanaerobacterium thermosulfurigenes em1 and reclassification of the enzyme as a cyclodextrin glycosyltransferase. Applied and Environmental Microbiology. 1995;61(4):1257–65.

[36] Rakmai J, Cheirsilp B. Continuous production of beta-cyclodextrin by cyclodextrin glycosyltransferase immobilized in mixed gel beads: comparative study in continuous stirred tank reactor and packed bed reactor. Biochemical Engineering Journal. 2016;105:107–13.

[37] Fenelon VC, Aguiar MFA, Miyoshi JH, Martinez CO, Matioli G. Ultrafiltration system for cyclodextrin production in repetitive batches by CGTase from Bacillus firmus strain 37. Bioprocess and Biosystems Engineering. 2015;38(7):1291–301.

[38] Rakmai J, Cheirsilp B, Prasertsan P. Enhanced thermal stability of cyclodextrin glycosyltransferase in alginate-gelatin mixed gel beads and the application for beta-cyclodextrin production. Biocatalysis and Agricultural Biotechnology. 2015;4(4):717–26.

[39] Moriwaki C, Mangolim CS, Ruiz GB, de Morais GR, Baesso ML, Matioli G. Biosynthesis of CGTase by immobilized alkalophilic bacilli and crystallization of beta-cydodextrin: effective techniques to investigate cell immobilization and the production of cyclodextrins. Biochemical Engineering Journal. 2014;83:22–32.

[40] De Souza M, de Faria SHB, Zanin GM, Moraes FF. Kinetics of the simultaneous production of beta- and gamma-cyclodextrins catalyzed by CGTase from alkalophilic Bacillus sp. Acta Scientiarum Technology. 2013;35(4):687–93.

[41] Schoffer JD, Klein MP, Rodrigues RC, Hertz PF. Continuous production of beta-cyclodextrin from starch by highly stable cyclodextrin glycosyltransferase immobilized on chitosan. Carbohydrate Polymers. 2013;98(2):1311–6.

[42] Tesfai BT, Wu D, Chen S, Chen J, Wu J. Effect of organic solvents on the yield and specificity of cyclodextrins by recombinant cyclodextrin glucanotransferase (CGTase) from Anaerobranca gottschalkii. Journal of Inclusion Phenomena and Macrocyclic Chemistry. 2013;77(1–4):147–53.

[43] Son YJ, Rha CS, Park YC, Shin SY, Lee YS, Seo JH. Production of cyclodextrins in ultrafiltration membrane reactor containing cyclodextrin glycosyltransferase from Bacillus macerans. Journal of Microbiology and Biotechnology. 2008;18(4):725–9.

[44] Lin YK, Show PL, Yap YJ, Ariff AB, Annuar MSM, Lai OM, et al. Production of gamma-cyclodextrin by Bacillus cereus cyclodextrin glycosyltransferase using extractive bioconversion in polymer-salt aqueous two-phase system. Journal of Bioscience and Bioengineering. 2016;121(6):692–6.

[45] Wang L, Wu D, Chen J, Wu J. Enhanced production of gamma-cyclodextrin by optimization of reaction of gamma-cyclodextrin glycosyltransferase as well as synchronous use of isoamylase. Food Chemistry. 2013;141(3):3072–6.

[46] Wu D, Chen S, Wang N, Chen J, Wu J. Gamma-cyclodextrin production using cyclodextrin glycosyltransferase from Bacillus clarkii 7364. Applied Biochemistry and Biotechnology. 2012;167(7):1954–62.

[47] Koh DW, Park MO, Choi SW, Lee BH, Yoo SH. Efficient biocatalytic production of cyclodextrins by combined action of amylosucrase and cyclodextrin glucanotransferase. Journal of Agricultural and Food Chemistry. 2016;64(21):4371–5.

[48] Gaston JAR, Costa H, Ferrarotti SA. Continuous production of cyclodextrins in an ultrafiltration membrane reactor, catalyzed by cyclodextrin glycosyltransferase from Bacillus circulans DF 9R. Biotechnology Progress. 2015;31(3):695–9.

[49] Sakinah AMM, Ismail AF, Illias RM, Zularisam AW, Hassan O, Matsuura T. Effect of substrate and enzyme concentration on cyclodextrin production in a hollow fibre membrane reactor system. Separation and Purification Technology. 2014;124:61–7.

[50] Duan XG, Chen S, Chen J, Wu J. Enhancing the cyclodextrin production by synchronous utilization of isoamylase and alpha-CGTase. Applied Microbiology and Biotechnology. 2013;97(8):3467–74.

[51] Ng HS, Ooi CW, Mokhtar MN, Show PL, Ariff A, Tan JS, et al. Extractive bioconversion of cyclodextrins by Bacillus cereus cyclodextrin glycosyltransferase in aqueous two-phase system. Bioresource Technology. 2013;142:723–6.

[52] Sakinah AMM, Ismail AF, Hassan O, Zularisam AW, Illias RM. Influence of starch pretreatment on yield of cyclodextrins and performance of ultrafiltration membranes. Desalination. 2009;239(1–3):317–33.

[53] Yamamoto K, Zhang ZZ, Kobayashi S. Cycloamylose (cyclodextrin) glucanotransferase degrades intact granules of potato raw starch. Journal of Agricultural and Food Chemistry. 2000;48(3):962–6.

[54] Kim TJ, Kim BC, Lee HS. Production of cyclodextrins using moderately heat-treated cornstarch. Enzyme and Microbial Technology. 1995;17(12):1057–61.

[55] van der Veen BA, Uitdehaag JCM, Penninga D, van Alebeek G, Smith LM, Dijkstra BW, et al. Rational design of cyclodextrin glycosyltransferase from Bacillus circulans strain 251 to increase alpha-cyclodextrin production. Journal of Molecular Biology. 2000;296(4):1027–38.

[56] Cao XZ, Jin ZY, Wang X, Chen F. A novel cyclodextrin glycosyltransferase from an alkalophilic Bacillus species: purification and characterization. Food Research International. 2005;38(3):309–14.

[57] Rendleman JA. Enhanced production of gamma-cyclodextrin from corn syrup solids by means of cyclododecanone as selective complexant. Carbohydrate Research. 1993;247:223–37.

[58] Goh KM, Mahadi NM, Hassan O, Rahman R, Illias RM. The effects of reaction conditions on the production of gamma-cyclodextrin from tapioca starch by using a novel recombinant engineered CGTase. Journal of Molecular Catalysis B-Enzymatic. 2007;49(1–4):118–26.

[59] Gawande B, Patkar A. Alpha-cyclodextrin production using cyclodextrin glycosyltransferase from Klebsiella pneumoniae AS-22. Starch-Starke. 2001;53(2):75–83.

[60] Jorgensen ST, Tangney M, Starnes RL, Amemiya K, Jorgensen PL. Cloning and nucleotide sequence of a thermostable cyclodextrin glycosyltransferase gene from Thermoanaerobacter sp. ATCC 53627 and its expression in Escherichia coli. Biotechnology Letters. 1997;19(10):1027–31.

[61] Fujiwara S, Kakihara H, Sakaguchi K, Imanaka T. Analysis of mutations in cyclodextrin glucanotransferase from bacillus-stearothermophilus which affect cyclization characteristics and thermostability. Journal of Bacteriology. 1992;174(22):7478–81.

[62] Moller MS, Henriksen A, Svensson B. Structure and function of alpha-glucan debranching enzymes. Cellular and Molecular Life Sciences. 2016;73(14):2619–41.

[63] Zhu AP, Romero R, Huang JB, Clark A, Petty HR. Maltooligosaccharides from JEG-3 trophoblast-like cells exhibit immunoregulatory properties. American Journal of Reproductive Immunology. 2011;65(1):54–64.

[64] Chen HL, Lu YH, Lin JJ, Ko LY. Effects of isomalto-oligosaccharides on bowel functions and indicators of nutritional status in constipated elderly men. Journal of the American College of Nutrition. 2001;20(1):44–9.

[65] Mizubuchi H, Yajima T, Aoi N, Tomita T, Yoshikai Y. Isomalto-oligosaccharides polarize Th1-like responses in intestinal and systemic immunity in mice. Journal of Nutrition. 2005;135(12):2857–61.

[66] Plou FJ, Martin MT, de Segura AG, Alcalde M, Ballesteros A. Glucosyltransferases acting on starch or sucrose for the synthesis of oligosaccharides. Canadian Journal of Chemistry-Revue Canadienne De Chimie. 2002;80(6):743–52.

[67] Cho SK, Shin SY, Lee SJ, Li L, Moon JS, Kim DJ, et al. Simple synthesis of isomaltooligosaccharides during sauerkraut fermentation by addition of Leuconostoc starter and sugars. Food Science and Biotechnology. 2015;24(4):1443–6.

[68] Bertrand E, Pierre G, Delattre C, Gardarin C, Bridiau N, Maugard T, et al. Dextranase immobilization on epoxy CIM (R) disk for the production of isomaltooligosaccharides from dextran. Carbohydrate Polymers. 2014;111:707–13.

[69] Rabelo MC, Honorato TL, Goncalves LRB, Pinto GAS, Rodrigues S. Optimization of enzymatic synthesis of isomalto-oligosaccharides production. Journal of Food Biochemistry. 2009;33(3):342–54.

[70] Kubik C, Sikora B, Bielecki S. Immobilization of dextransucrase and its use with soluble dextranase for glucooligosaccharides synthesis. Enzyme and Microbial Technology. 2004;34(6):555–60.

[71] Heincke K, Demuth B, Jordening HJ, Buchholz K. Kinetics of the dextransucrase acceptor reaction with maltose—experimental results and modeling. Enzyme and Microbial Technology. 1999;24(8–9):523–34.

[72] Chung CH, Day DF. Efficacy of Leuconostoc mesenteroides (ATCC 13146) isomaltooligosaccharides as a poultry prebiotic. Poultry Science. 2004;83(8):1302–6.

[73] Lee MS, Cho SK, Eom HJ, Kim SY, Kim TJ, Han NS. Optimized substrate concentrations for production of long-chain isomaltooligosaccharides using dextransucrase of Leuconostoc mesenteroides B-512F. Journal of Microbiology and Biotechnology. 2008;18(6):1141–5.

[74] Moulis C, Medina GV, Suwannarangsee S, Monsan P, Remaud-Simeon M, Potocki-Veronese G. One-step synthesis of isomalto-oligosaccharide syrups and dextrans of controlled size using engineered dextransucrase. Biocatalysis and Biotransformation. 2008;26(1–2):141–51.

[75] Goulas AK, Fisher DA, Grimble GK, Grandison AS, Rastall RA. Synthesis of isomaltooligosaccharides and oligodextrans by the combined use of dextransucrase and dextranase. Enzyme and Microbial Technology. 2004;35(4):327–38.

[76] Manas NHA, Jonet MA, Murad AMA, Mahadi NM, Illias RM. Modulation of transglycosylation and improved malto-oligosaccharide synthesis by protein engineering of maltogenic amylase from Bacillus lehensis G1. Process Biochemistry. 2015;50(10):1572–80.

[77] Kashiwagi N, Miyake M, Hirose S, Sota M, Ogino C, Kondo A. Cloning and starch degradation profile of maltotriose-producing amylases from Streptomyces species. Biotechnology Letters. 2014;36(11):2311–7.

[78] Ozturk HU, Denizci AA, Ogan A, Kazan D. A Maltooligosaccharides producing alpha-amylase from Bacillus subtilis SDP1 Isolated from rhizosphere of Acacia cyanophylla Lindley. Food Biotechnology. 2014;28(4):309–32.

[79] Hwang SY, Nakashima K, Okai N, Okazaki F, Miyake M, Harazono K, et al. Thermal stability and starch degradation profile of alpha-amylase from Streptomyces avermitilis. Bioscience Biotechnology and Biochemistry. 2013;77(12):2449–53.

[80] Jana M, Maity C, Samanta S, Pati BR, Islam SS, Das Mohapatra PK, et al. Salt-independent thermophilic alpha-amylase from Bacillus megaterium VUMB109: an efficacy testing for preparation of maltooligosaccharides. Industrial Crops and Products. 2013;41:386–91.

[81] Shibata K, Sakamoto K, Nakatsu S, Kajihara R, Shimoda M. Enzymatic production of malto-oligosaccharide in potato by freeze-thaw infusion. Food Science and Technology Research. 2010;16(4):273–8.

[82] Kim YH, Kwon TK, Park S, Seo HS, Cheong JJ, Kim CH, et al. Trehalose synthesis by sequential reactions of recombinant maltooligosyltrehalose synthase and maltooligosyltrehalose trehalohydrolase from Brevibacterium helvolum. Applied and Environmental Microbiology. 2000;66(11):4620–4.

[83] Kato M. Trehalose production with a new enzymatic system from Sulfolobus solfataricus KM1. Journal of Molecular Catalysis B-Enzymatic. 1999;6(3):223–33.

[84] Li XL, Li D. Preparation of linear maltodextrins using a hyperthermophilic amylopullulanase with cyclodextrin- and starch-hydrolysing activities. Carbohydrate Polymers. 2015;119:134–41.

[85] Zhang L, Jiang YJ, Jiang ZY, Sun XH, Shi JF, Cheng W, et al. Immobilized transglucosidase in biomimetic polymer-inorganic hybrid capsules for efficient conversion of maltose to isomaltooligosaccharides. Biochemical Engineering Journal. 2009;46(2):186–92.

[86] Chockchaisawasdee S, Poosaran N. Production of isomaltooligosaccharides from banana flour. Journal of the Science of Food and Agriculture. 2013;93(1):180–6.

[87] Zhang L, Su YL, Zheng Y, Jiang ZY, Shi JF, Zhu YY, et al. Sandwich-structured enzyme membrane reactor for efficient conversion of maltose into isomaltooligosaccharides. Bioresource Technology. 2010;101(23):9144–9.

[88] Sheu DC, Huang CI, Duan KJ. Production of isomaltooligosaccharides by alpha-glucosidase immobilized in chitosan beads and by polyethyleneimine-glutataldehyde treated mycelia of Aspergillus carbonarious. Biotechnology Techniques. 1997;11(5):287–91.

[89] Kaulpiboon J, Rudeekulthamrong P, Watanasatitarpa S, Ito K, Pongsawasdi P. Synthesis of long-chain isomaltooligosaccharides from tapioca starch and an in vitro investigation of their prebiotic properties. Journal of Molecular Catalysis B-Enzymatic. 2015;120:127–35.

[90] Rudeekulthamrong P, Sawasdee K, Kaulpiboon J. Production of long-chain isomaltooligosaccharides from maltotriose using the thermostable amylomaltase and transglucosidase enzymes. Biotechnology and Bioprocess Engineering. 2013;18(4):778–86.

[91] Hendry GAF. Evolutionary origins and natural functions of fructans - a climatological, biogeographic and mechanistic appraisal. New Phytologist. 1993;123(1):3–14.

[92] Van den Ende W. Multifunctional fructans and raffinose family oligosaccharides. Frontiers in Plant Science. 2013;4:247–258.

[93] Singh RS, Singh RP. Production of fructooligosaccharides from inulin by endoinulinases and their prebiotic potential. Food Technology and Biotechnology. 2010;48(4):435–50.

[94] Cho YJ, Sinha J, Park JP, Yun JW. Production of inulooligosaccharides from inulin by a dual endoinulinase system. Enzyme and Microbial Technology. 2001;29(6–7):428–33.

[95] Park JP, Bae JT, You DJ, Kim BW, Yun JW. Production of inulooligosaccharides from inulin by a novel endoinulinase from Xanthomonas sp. Biotechnology Letters. 1999;21(12):1043–6.

[96] Park JP, Kim DH, Kim DS, Yun JW. Enzymatic production of inulo-oligosaccharides from chicory juice. Biotechnology Letters. 1998;20(4):385–8.

[97] Yokota A, Yamauchi O, Tomita F. Production of inulotriose from inulin by inulin-degrading enzyme from streptomyces-rochei-e87. Letters in Applied Microbiology. 1995;21(5):330–3.

[98] Kim DH, Choi YJ, Song SK, Yun JW. Production of inulo-oligosaccharides using endo-inulinase from a Pseudomonas sp. Biotechnology Letters. 1997;19(4):369–71.

[99] Yun JW, Kim DH, Uhm TB, Song SK. Production of high-content inulo-oligosaccharides from inulin by a purified endoinulinase. Biotechnology Letters. 1997;19(9):935–8.

[100] Kuhn GD, Dalla Rosa C, Silva MF, Treichel H, de Oliveira D, Oliveira JV. Synthesis of fructooligosaccharides from Aspergillus niger commercial inulinase immobilized in montmorillonite pretreated in pressurized propane and LPG. Applied Biochemistry and Biotechnology. 2013;169(3):750–60.

[101] Mutanda T, Wilhelmi BS, Whiteley CG. Response surface methodology: synthesis of inulooligosaccharides with an endoinulinase from Aspergillus niger. Enzyme and Microbial Technology. 2008;43(4–5):362–8.

[102] Mutanda T, Wilhelmi BS, Whiteley CG. Biocatalytic conversion of inulin and sucrose into short chain oligosaccharides for potential pharmaceutical applications. African Journal of Science Technology Innovation & Development. 2015;7(5):371–80.

[103] Silva MF, Rigo D, Mossi V, Golunski S, Kuhn GD, Di Luccio M, et al. Enzymatic synthesis of fructooligosaccharides by inulinases from Aspergillus niger and Kluyveromyces marxianus NRRL Y-7571 in aqueous-organic medium. Food Chemistry. 2013;138(1):148–53.

[104] Naidoo K, Ayyachamy M, Permaul K, Singh S. Enhanced fructooligosaccharides and inulinase production by a Xanthomonas campestris pv. phaseoli KM 24 mutant. Bioprocess and Biosystems Engineering. 2009;32(5):689–95.

[105] Yun JW, Choi YJ, Song CH, Song SK. Microbial production of inulo-oligosaccharides by an endoinulinase from Pseudomonas sp expressed in Escherichia coli. Journal of Bioscience and Bioengineering. 1999;87(3):291–5.

[106] Yun JW, Park JP, Song CH, Lee CY, Kim JH, Song SK. Continuous production of inulo-oligosaccharides from chicory juice by immobilized endoinulinase. Bioprocess Engineering. 2000;22(3):189–94.

[107] Zhengyu J, Jing W, Bo J, Xueming X. Production of inulooligosaccharides by endoinulinases from Aspergillus ficuum. Food Research International. 2005;38(3):301–8.

[108] Jain I, Kumar V, Satyanarayana T. Xylooligosaccharides: an economical prebiotic from agroresidues and their health benefits. Indian Journal of Experimental Biology. 2015;53(3):131–42.

[109] Carvalho AFA, Neto PD, Da Silva DF, Pastore GM. Xylo-oligosaccharides from lignocellulosic materials: chemical structure, health benefits and production by chemical and enzymatic hydrolysis. Food Research International. 2013;51(1):75–85.

[110] Aachary AA, Prapulla SG. Value addition to corncob: production and characterization of xylooligosaccharides from alkali pretreated lignin-saccharide complex using Aspergillus oryzae MTCC 5154. Bioresource Technology. 2009;100(2):991–5.

[111] Boonchuay P, Techapun C, Seesuriyachan P, Chaiyaso T. Production of xylooligosaccharides from corncob using a crude thermostable endo-xylanase from Streptomyces thermovulgaris TISTR1948 and prebiotic properties. Food Science and Biotechnology. 2014;23(5):1515–23.

[112] Chapla D, Pandit P, Shah A. Production of xylooligosaccharides from corncob xylan by fungal xylanase and their utilization by probiotics. Bioresource Technology. 2012;115:215–21.

[113] Samanta AK, Jayapal N, Kolte AP, Senani S, Sridhar M, Dhali A, et al. Process for enzymatic production of xylooligosaccharides from the xylan of corn cobs. Journal of Food Processing and Preservation. 2015;39(6):729–36.

[114] Teng C, Yan QJ, Jiang ZQ, Fan GS, Shi B. Production of xylooligosaccharides from the steam explosion liquor of corncobs coupled with enzymatic hydrolysis using a thermostable xylanase. Bioresource Technology. 2010;101(19):7679–82.

[115] Yang R, Xu S, Wang Z, Yang W. Aqueous extraction of corncob xylan and production of xylooligosaccharides. Lwt-Food Science and Technology. 2005;38(6):677–82.

[116] Yoon KY, Woodams EE, Hang YD. Enzymatic production of pentoses from the hemicellulose fraction of corn residues. Lwt-Food Science and Technology. 2006;39(4):388–92.

[117] Akpinar O, Ak O, Kavas A, Bakir U, Yilmaz L. Enzymatic production of xylooligosaccharides from cotton stalks. Journal of Agricultural and Food Chemistry. 2007;55(14):5544–51.

[118] Akpinar O, Erdogan K, Bakir U, Yilmaz L. Comparison of acid and enzymatic hydrolysis of tobacco stalk xylan for preparation of xylooligosaccharides. Lwt-Food Science and Technology. 2010;43(1):119–25.

[119] Chapla D, Dholakiya S, Madamwar D, Shah A. Characterization of purified fungal endoxylanase and its application for production of value added food ingredient from agroresidues. Food and Bioproducts Processing. 2013;91(C4):682–92.

[120] Falck P, Aronsson A, Grey C, Stalbrand H, Karlsson EN, Adlercreutz P. Production of arabinoxylan-oligosaccharide mixtures of varying composition from rye bran by a combination of process conditions and type of xylanase. Bioresource Technology. 2014;174:118–25.

[121] Faryar R, Linares-Pasten JA, Immerzeel P, Mamo G, Andersson M, Stalbrand H, et al. Production of prebiotic xylooligosaccharides from alkaline extracted wheat straw using the K80R-variant of a thermostable alkali-tolerant xylanase. Food and Bioproducts Processing. 2015;93:1–10.

[122] Kallel F, Driss D, Chaabouni SE, Ghorbel R. Biological activities of xylooligosaccharides generated from garlic straw xylan by purified xylanase from Bacillus mojavensis UEB-FK. Applied Biochemistry and Biotechnology. 2015;175(2):950–64.

[123] Ko CH, Shih TL, Jhan BT, Chang FC, Wang YN, Wang YC. Production of xylooligosaccharides from forest waste by membrane separation and Paenibacillus xylanase hydrolysis. Bioresources. 2013;8(1):612–27.

[124] Lakshmi GS, Rajeswari BU, Prakasham RS. Biosynthesis of xylobiose: a strategic way to enrich the value of oil palm empty fruit bunch fiber. Journal of Microbiology and Biotechnology. 2012;22(8):1084–91.

[125] Mazzaferro LS, Cuna MM, Breccia JD. Production of xylo-oligosaccharides by chemo-enzymatic treatment of agricultural by-products. Bioresources. 2011;6(4):5050–61.

[126] Rehman SU, Babu I, Zahoor T, Nawaz H, Bhatti IA, Latif F, et al. Extraction of xylooligosaccharides from hard-shell almond variety ('Wirin') and their utilisation in cookies. Journal of Food Science and Technology-Mysore. 2008;45(6):527–30.

[127] Samanta AK, Jayapal N, Kolte AP, Senani S, Sridhar M, Suresh KP, et al. Enzymatic production of xylooligosaccharides from alkali solubilized xylan of natural grass (Sehima nervosum). Bioresource Technology. 2012;112:199–205.

[128] Wang J, Sun BG, Cao YP, Tian Y. Enzymatic preparation of wheat bran xylooligosaccharides and their stability during pasteurization and autoclave sterilization at low pH. Carbohydrate Polymers. 2009;77(4):816–21.

[129] Zhang YD, Yu G, Li B, Mu XD, Peng H, Wang HS. Hemicellulose isolation, characterization, and the production of xylo-oligosaccharides from the wastewater of a viscose fiber mill. Carbohydrate Polymers. 2016;141:238–43.

[130] Bian J, Peng F, Peng XP, Peng P, Xu F, Sun RC. Enzymatic preparation of xylooligosaccharides by pichia stipitis from sugarcane bagasse. Proceeding of the 4th international conference on pulping, papermaking and biotechnology. Nanjing, China, v. I and II, p. 915–920, 2012.

[131] Bragatto J, Segato F, Squina FM. Production of xylooligosaccharides (XOS) from delignified sugarcane bagasse by peroxide-HAc process using recombinant xylanase from Bacillus subtilis. Industrial Crops and Products. 2013;51:123–9.

[132] Brienzo M, Carvalho W, Milagres AMF. Xylooligosaccharides production from alkali-pretreated sugarcane bagasse using xylanases from Thermoascus aurantiacus. Applied Biochemistry and Biotechnology. 2010;162(4):1195–205.

[133] Li XT, Li E, Zhu YP, Teng C, Sun BG, Song HL, et al. A typical endo-xylanase from Streptomyces rameus L2001 and its unique characteristics in xylooligosaccharide production. Carbohydrate Research. 2012;359:30–6.

[134] Xue JL, Zhao S, Liang RM, Yin X, Jiang SX, Su LH, et al. A biotechnological process efficiently co-produces two high value-added products, glucose and xylooligosaccharides, from sugarcane bagasse. Bioresource Technology. 2016;204:130–8.

[135] Zhu JJ, Zhu YY, Jiang FX, Xu Y, Ouyang J, Yu SY. An integrated process to produce ethanol, vanillin, and xylooligosaccharides from Camellia oleifera shell. Carbohydrate Research. 2013;382:52–7.

[136] Combo AMM, Aguedo M, Paquot M. Pectic oligosaccharides: production and potential applications. Biotechnologie Agronomie Societe Et Environnement. 2011;15(1):153–64.

[137] Strube ML, Ravn HC, Ingerslev HC, Meyer AS, Boye M. In situ prebiotics for weaning piglets: in vitro production and fermentation of potato galacto-rhamnogalacturonan. Applied and Environmental Microbiology. 2015;81(5):1668–78.

[138] Strube ML, Jensen TK, Meyer AS, Boye M. In situ prebiotics: enzymatic release of galacto-rhamnogalacturonan from potato pulp in vivo in the gastrointestinal tract of the weaning piglet. Amb Express. 2015;5.

[139] Olano-Martin E, Williams MR, Gibson GR, Rastall RA. Pectins and pectic-oligosaccharides inhibit Escherichia coli O157: H7 Shiga toxin as directed towards the human colonic cell line HT29. Fems Microbiology Letters. 2003;218(1):101–5.

[140] Combo AMM, Aguedo M, Quievy N, Danthine S, Goffin D, Jacquet N, et al. Characterization of sugar beet pectic-derived oligosaccharides obtained by enzymatic hydrolysis. International Journal of Biological Macromolecules. 2013;52:148–56.

[141] Khodaei N, Fernandez B, Fliss I, Karboune S. Digestibility and prebiotic properties of potato rhamnogalacturonan I polysaccharide and its galactose-rich oligosaccharides/oligomers. Carbohydrate Polymers. 2016;136:1074–84.

[142] Khodaei N, Karboune S. Enzymatic generation of galactose-rich oligosaccharides/oligomers from potato rhamnogalacturonan I pectic polysaccharides. Food Chemistry. 2016;197:406–14.

[143] Holck J, Hjerno K, Lorentzen A, Vigsnaes LK, Hemmingsen L, Licht TR, et al. Tailored enzymatic production of oligosaccharides from sugar beet pectin and evidence of differential effects of a single DP chain length difference on human faecal microbiota composition after in vitro fermentation. Process Biochemistry. 2011;46(5):1039–49.

[144] Combo AMM, Aguedo M, Goffin D, Wathelet B, Paquot M. Enzymatic production of pectic oligosaccharides from polygalacturonic acid with commercial pectinase preparations. Food and Bioproducts Processing. 2012;90(C3):588–96.

[145] Sabajanes MM, Yanez R, Alonso JL, Parajo JC. Pectic oligosaccharides production from orange peel waste by enzymatic hydrolysis. International Journal of Food Science and Technology. 2012;47(4):747–54.

[146] Babbar N, Baldassarre S, Maesen M, Prandi B, Dejonghe W, Sforza S, et al. Enzymatic production of pectic oligosaccharides from onion skins. Carbohydrate Polymers. 2016;146:245–52.

[147] de Assis CF, Araujo NK, Pagnoncelli MGB, Pedrini M, de Macedo GR, dos Santos ES. Chitooligosaccharides enzymatic production by Metarhizium anisopliae. Bioprocess and Biosystems Engineering. 2010;33(7):893–9.

[148] Jeon YJ, Kim SK. Production of chitooligosaccharides using an ultrafiltration membrane reactor and their antibacterial activity. Carbohydrate Polymers. 2000;41(2):133–41.

[149] Kim SK, Rajapakse N. Enzymatic production and biological activities of chitosan oligosaccharides (COS): a review. Carbohydrate Polymers. 2005;62(4):357–68.

[150] Lee HW, Park YS, Jung JS, Shin WS. Chitosan oligosaccharides, dp 2–8, have prebiotic effect on the Bifidobacterium bifidium and Lactobacillus sp. Anaerobe. 2002;8(6):319–24.

[151] Cheng CY, Li YK. An Aspergillus chitosanase with potential for large-scale preparation of chitosan oligosaccharides. Biotechnology and Applied Biochemistry. 2000;32:197–203.

[152] Izume M, Ohtakara A. Preparation of d-glucosamine oligosaccharides by the enzymatic-hydrolysis of chitosan. Agricultural and Biological Chemistry. 1987;51(4):1189–91.

[153] Kuo CH, Chen CC, Chiang BH. Process characteristics of hydrolysis of chitosan in a continuous enzymatic membrane reactor. Journal of Food Science. 2004;69(7):1–8.

[154] Kuroiwa T, Izuta H, Nabetani H, Nakajima M, Sato S, Mukataka S, et al. Selective and stable production of physiologically active chitosan oligosaccharides using an enzymatic membrane bioreactor. Process Biochemistry. 2009;44(3):283–7.

Chapter 9

Inhibitors and Activators of SOD, GSH-Px, and CAT

Xianyong Ma, Dun Deng and Weidong Chen

Additional information is available at the end of the chapter

http://dx.doi.org/10.5772/65936

Abstract

Reactive oxygen species (ROS) is harmful to our health, and SOD, CAT, and GPX are the major antioxidant enzymes that defend us from effects of ROS. In medicine, food, and dairy industries, antioxidant enzymes often surround complex environments. For better utilization of these enzymes, the inhibitors (including competitive inhibitors and noncompetitive inhibitors) and activators of SOD, CAT, and GPX are descripted in detail in this chapter. Also, the structure and catalytic mechanism of these antioxidants are summarized.

Keywords: reactive oxygen species, superoxide dismutase, catalase, glutathione peroxidase

1. Introduction

Reactive oxygen species (ROS) such as superoxide ($O_2^{-\cdot}$), hydrogen peroxide (H_2O_2), and hydroxyl radical (OH) are highly reactive oxidant species, which are by-products of intracellular metabolic processes, causing macromolecular damage. Reactive oxygen species are harmful to health and cause damage to crucial substances such as DNA, RNA, proteins, and lipids. Superoxide dismutase (SOD), catalase (CAT), and glutathione peroxidase (GPX) are three major antioxidant enzymes in animals. Superoxide dismutase (SOD), the first defense line of against oxygen-derived free radicals, catalyzes the dismutation of the superoxide anion ($O_2^{-\cdot}$) into hydrogen peroxide (H_2O_2), which isthen transformed into H_2O and O_2 by catalase. Glutathione peroxidase (GSH-PX or GPX) is a selenoprotein, which also protects the organism from oxidative damage; the main biochemical function of GPX is to reduce lipidic or nonlipidic hydroperoxides as well as H_2O_2 through oxidization of glutathione [1]. Although both enzymes can reduce H_2O_2, there are some differences between CAT and GPX. CAT, one of the enzyme with highest turnover rates,is the main enzyme involved in reduction of H_2O_2 via the Fenton reaction. This enzyme is almost exclusively expressed in peroxisomes [2]. GPX is specific for its hydrogen donor, but nonspecific for H_2O_2, and it degrades H_2O_2 using reduced glutathione in a powerful manner [3].

From the perspective of protection, improving antioxidase activity would be helpful for organisms to survive under various stresses, but from another respect, the antioxidase activity should be inhibited. Some anticancer agents, such as xenobiotics and radiation, act by producing ROS to kill tumor cells. Cells with high levels of antioxidant enzymes are resistant to these anticancer agents. Therefore, the levels of cellular antioxidant enzymes will influence the sensitivity of tumor cells to anticancer therapies [4]. Thus, more detailed information about an activator and inhibitor of SOD, CAT, and GPX should be known, for better utilization of these enzymes.

2. SODs

2.1. The structure and function of SOD

Superoxide dismutases (SODs, EC 1.15.1.1) are ubiquitous and one important class of antioxidant metalloenzymes against the harmful effects of superoxide free radicals. The main function of SODs was to decompose superoxide radicals into molecular oxygen and hydrogen peroxide inside cells, which reaction is as follows:

$$2O_2^{\cdot-} + 2H^+ \rightarrow O_2 + H_2O_2 \quad (1)$$

Based on the functional metal cofactors located at the active sites, four distinct classes of SODs have been found (**Figure 1**). SODs of Class I specifically require manganese or iron ion for catalytic activity (MnSOD and FeSOD) and enzymes that function with either of the two metal ions so-called cambialistic SODs. Class II is copper- and zinc-dependent enzymes (CuZn-SODs). Members of the two classes are found in both prokaryotes and eukaryotes. Nickel-containing SODs are mainly identified from marine actinomycetes and cyanobacteria [5, 6]. Enzymatic reactions of SODs depending on different metals indicate that SODs are developed by cells to offset the harmful effects of reactive oxygen species and match its surroundings.

Due to their antioxidative effects, SODs exhibited dramatic potential in medicine, cosmetic, food, agriculture, and chemistry industry. For example, considerable clinical experiments have shown that SODs could prevent oncogenesis and reduce the cytotoxic effects of anticancer drugs. Recently, SODs have found to prevent many diverse diseases such as cardiovascular diseases, diabetes, asthma, infertility, neurological disorders, and transplant rejection. SODs have also been successfully utilized as a major component in cosmetics for skin protection. In animal husbandry, SODs are considered to be one kind of strong antioxidative enzymes, which can reduce the oxidative stress of animal and prevent the oxidation of animal products, improve the quality of animal products such as meat, egg, or milk. In order to make better use

http://dx.doi.org/10.5772/65936

of SODs, their characteristics, the most suitable environment and influencing factors, etc., should be known especially the information of inhibitor and activator.

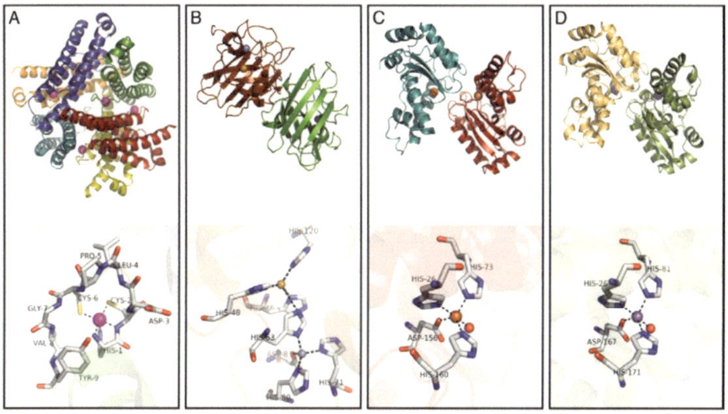

Figure 1. A comparison of the enzyme structures and active sites for the four SODs, (a) streptomyces coelicolor NiSOD (PDB, 1t6u), (b) human Cu/ZnSOD (PDB, 1pu0), (c) *E. coli* FeSOD (PDB, 1isa), and (d) MnSOD (PDB, 1vew). (adapted from reference [7]).

2.2. Inhibitors of SODs

2.2.1. Different inhibitors on sensitivity of SODs

Cu/Zn-SODs, Fe-SODs, and Mn-SODs are sensitive to different inhibitors, and we could distinguish these metal-specific SODs based on different inhibitory reactions (**Table 1**). The activity of Cu/Zn-SODs was inhibited by KCN, but Fe-SODs and Mn-SODs were not sensitive to KCN. Whereas, completely different phenomenon was observed when incubated in chloroform-ethanol (1:3, V/V) component solvent, as a result Fe-SODs and Mn-SODs activity were almost lost, while Cu/Zn-SOD was not sensitive to chloroform-ethanol. Fe-SODs and Mn-SODs are highly homologous and exhibit structural similarity, it is assumed that they originate from the same ancestry [8]. Fe-SODs and Mn-SODs can be distinguished by their different sensitivities to H_2O_2, because Mn-SOD activity was not sensitive to H_2O_2. In addition, NaN_3 also is used to detect the type of MnSOD, when the SOD was inhibited by neither KCN nor H_2O_2 [9].

SODs	Inhibitors			
	KCN	Chloroform-ethanol	H_2O_2	NaN_3
Cu/Zn-SOD	+	-	+	+
Fe-SOD	-	+	+	+
Mn-SOD	-	+	-	+

"+" indicate reaction and "-" indicate no reaction.

Table 1. Effect of inhibitors on different cofactors of SODs activity.

2.2.2. *Effects of metal ions on SOD activity*

Cu^{2+}, Zn^{2+}, Fe^{2+}, and Mn^{2+} are cofactors of superoxide dismutase, and they are vital to enzymatic activity. But heterogeneous expressed SODs were often existed in terms of apoenzyme or combined other metals present in a culture medium [10], thus might result in partial or total loss of activity. So that reconstituted metal ion is necessary to recovery activity. Apoenzyme was prepared by a metal removal procedure according reference [11] and simplified as follow:

1. Purified SOD was dialyzed against denature buffer: 20 mM 8-hydroxyquinoline, 2.5 mM guanidinium chloride, 5 mM Tris, 0.1 mM EDTA, pH 3.8, at 4°C for 18 h.

2. Purified SOD was dialyzed against 5 mM Tris buffer (pH 7.8), containing 1 mM metal ions for 18 h.

3. Excess metal ions were removed by dialysis against 5 mM Tris buffer (pH 7.8).

To be attention, only specific metal ions combined to SOD, the activity could be recovered. But for Fe-SOD and Mn-SOD, because they share highly homologous, Fe-substituted Mn-SODs also are active, but the activity is lower than Mn-reconstituted SOD.

SODs are metalloenzyme, and they will be inhibited by chelators [12, 13], such as EDTA and cuprizone (a copper-specific chelator). They should be avoided losing metal cofactors caused by chelators in the experiment. But a Mn-SOD from *Mycobacterium* sp. JC1 DSM 3803 was not sensitive to 10 mM EDTA [8], which may be due to the highly tightness of metal cofactors binding to enzymes.

Co^{2+}, Hg^{2+}, K^{+} and, Al^{3+} and other metal ions also show their inhibition effects on SOD activity. A Mn-SOD from deep-sea thermophile *Geobacillus* sp. EPT3 was activated only by Mn^{2+} among nine tested metal ions [12]. A Cu-/Zn-SOD from black soybean was also activated only by Cu^{2+} [13] and a similar result was shown in the report of Liu et al. [14]. But there are a few exceptions, for example, a manganese-containing superoxide dismutase was activated by Cu^{2+}, Zn^{2+}, and Al^{3+}.

2.2.3. *Singlet molecular oxygen inactivation of superoxide dismutase*

High reactive singlet molecular oxygen ($^1O_2$) is one kind of short-lived intermediate from oxidation reaction which oxidizes a variety of biological molecules easily, including lipids, nucleic acids, and proteins, and it also promotes deleterious processes such as lipid peroxidation, membrane damage, and cell death [15]. The biochemical production of singlet oxygen has been proposed to contribute to the destructive effects on a number of biological processes [16]. *In vivo*, singlet molecular oxygen is produced under normal and pathophysiological conditions. It is known to be particularly reaction with histidine, which often located at the active sites of SODs, so singlet molecular oxygen may result in prevention of the activities of those enzymes including CATs and GPXs. *In vitro*, singlet molecular oxygen could be produced by photoactived dyes, such as methylene blue or rose bengal [17], so inactivation of SODs, CATs, and GPXs should be avoid in application.

2.3. Different activator on activity of SODs

2.3.1. Effect of carbohydrates on activities of superoxide dismutase

Carbohydrates, such as maltose, sucrose, lactose, trehalose, glucose, D-fructose, D-trehalose, D-xylose, and so on could stabilize an enzyme structure. For example, trehalose plays a strong promotive effect on superoxide dismutase [18]. Trehalose is a kind of polyol compound with many hydroxyl groups, which has strong hydration ability and can change the free energy to the favorable direction in solution. The multihydroxyl structure of trehalose can connected with both the surface of the enzyme protein and the external water through hydrogen bonding, so that the structure of the enzyme is stable, and the enzyme activity was protected [19].

2.3.2. Polyethylene glycol modified SOD to improve its efficiency

Polyethylene glycols (PEGs) are considered to be a safety and nonimmunogenic materials. They also have multihydroxyl compounds and can be activated by many activators, such as cyanuric, dicycolhexylcarbodiimide, N-hydroxysucciniimide, and 1,1'-carbonyldiimidazole, then activated polyethylene glycol was conjugated with the ε-NH_2 group of SOD. PEG conjugated with SOD not only enhance the stability of the enzyme, but also avoid enzymatic immunogenicity. For example, Beckman et al. [20] reported that a superoxide dismutase conjugated with polyethylene glycol greatly increased endothelial the cell oxidant resistance and half-life of the enzyme.

2.3.3. Cyclodextrin modified SOD to enhance its stability

Cyclodextrins (CD), cyclic oligosaccharides containing six (α-CD), seven (β-CD) or eight (γ-CD) α-1-4-linked D-glucopyranose units have been used to stabilize enzymes in order to increase their activities and favor immobilization. On the one hand, superoxide dismutase modified by β-cyclodextrin could improve its performance. A superoxide dismutase was glycosylated by cyclodextrin-branched carboxymethylcellulose and its plasma half-life time was prolonged from 4.8 min to 7.2 h, its anti-inflammatory activity also increased by 2.2 times [21]. On the other hand, cyclodextrin and its derivative could synthesize SOD mimics. Puglisi et al. [22] reported a 6A,6B-Dideoxy-6A,6B-di[(N-salicylidene)amino]-β-cyclodextrin conjugated with a manganese(III) complex showed a SOD-like activity and a good solubility that favor its application.

3. CAT

3.1. The structure and function of CAT

CAT(EC 1.11.1.6) catalyzes the decomposition of hydrogen peroxide to water and oxygen, widely exists in animals, plants, and microorganisms. Its uniqueness lies in the enzymatic prosthetic group(ferriprotoporphyrin IX) that could catalyze the same reaction as the holoenzyme.

$$2H_2O_2 \rightarrow O_2 + 2H_2O \quad (2)$$

According to the significant catalytic activity, CATs can be divided into three distinct subgroups: typical catalases, a typical catalases and catalase-peroxidases [23]. Two subgroups, typical catalases and catalase-peroxidases, contain heme, but the third group has no heme, namely manganese catalases. Most of catalases belong to typical catalases, except catalases in the domain of Archaea. Although there are differences in the primary structure among these typical catalases, but the three-dimensional structure appears well conserved. Most of these hydroperoxidases are homotetramers with four prosthetic heme groups (**Figure 2(a)**, PDB, 1E93) [24].

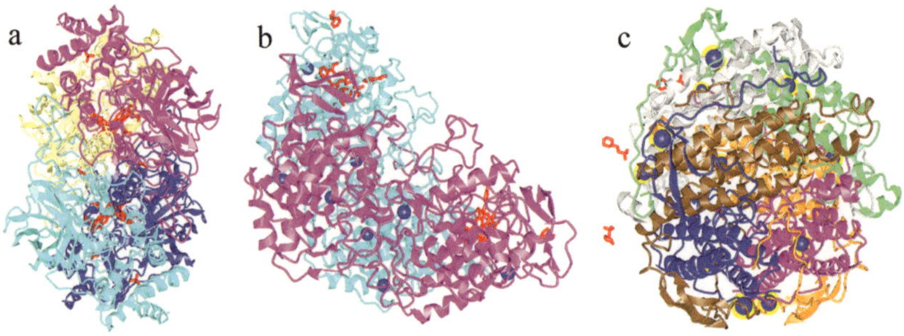

Figure 2. The three-dimensional structure of three types of catalases. (a) a typical catalase (depleted in iron) from *Proteus mirabilis*, (b) a catalase-peroxidase from *Synechococcus elongatus* PCC7942, (c) a manganese catalase from *Lactobacillus plantarum*. Color scheme: blue ball—Fe or Mn, red—heme group, others colors indicate different subunit.

Catalase-peroxidases may originate from ancestral, a relatively large momomeric unit comprising more than 700 amino acids, indicating that they probably from duplication of an ancestral gene [25]. And catalase-peroxidases show much higher sequence homology with heme peroxidases than with typical catalases. Recently, a three-dimensional structure of catalase-peroxidases has been obtained and is shown in **Figure 2(b)** (PDB, 3WXO) [26].

There is little structural information about manganese catalases up to now. Subunits with molecular weights around 30 kDa are recognized as tetramers or hexamers, and are remarkably stable at high temperatures [23, 27]. A crystal structure of a manganese catalase from *Lactobacillus plantarum* at 1.33a resolution is established (**Figure 2(c)**, PDB, 1O9I) [28].

3.2. Inhibitor of catalase

3.2.1. Noncompetitive inhibitor

3.2.1.1. Sodium azide, amine, and cyanide

Sodium azide (NaN_3), amine, and cyanide are nonspecific inhibitors of CAT. Catalase-peroxidases are very sensitive to NaN_3, a lower concentration of NaN_3 could lead to the enzyme

http://dx.doi.org/10.5772/65936

lose its activity by 50%. The inhibitory efficiency order was sodium azide>amine>cyanide>3-amino-1,2,4-triazole(**Table 2**) [29]. A similar result was found in the other typical monofunctional catalases [30, 31].

Inhibitors	Inhibitor concentration (mM) required for 50% inhibition	
	Catalase	Peroxidase
3-Amino-1,2,4-triazole	6	30
Hydroxylamine	0.02	1.5
Sodium azide	0.025	0.15
Potassium cyanide	1	1

Table 2. Effect of inhibitors on the catalase and peroxidase activity of the catalase-peroxidase of *Archaeoglobus fulgidus* [29].

3.2.1.2. 3-Amino-1,2,4-triazole

3-Amino-1,2,4-triazole (aminotriazole, ATZ) as a noncompetitive catalase-specific inhibitor is used to study on physiological changes in organisms [32]. Aminotriazole could combine catalase-H_2O_2 compound I, thus results in loss of enzymatic activity. In alcohol-induced liver injury, catalase plays a dual role. On the one hand, catalase could scavenge hydrogen peroxide originated from alcohol to water, but on the other hand, catalase decomposes alcohol that might be harmful to liver, some research studies [33] show that catalase is inhibited by ATZ, which attenuated alcohol-induced acute liver injury.

3.2.1.3. Salicylic acid

Salicylic acid acts as an electron donor for the peroxidative cycle of catalase, it is a noncompetitive inhibitor of catalases. It is interesting to note that different CAT salicylic acid exhibits different inhibitory property. CAT1 and CAT2 are two isoenzymes from maize (*Zea mays* L.). The Lineweaver-Burk plot of SA inhibition of CAT1 and CAT2 shows that CAT1 is noncompetitive manner, while CAT2 is inhibited in a competitive manner [34]. SA has a dual function on catalase, which means SA can both inhibit and activate its activity. Durner and Klessig [35] examined the effects of SA on the formation of the various redox states or reaction intermediates of catalase (**Figure 3**). The absorption spectrum of compounds I, II and III was different, thus various redox states or reaction intermediates of catalase can be distinguished spectroscopically by their absorption spectra in the Soret (near UV) region. Through the difference at the absorption spectra of intermediates, SA was confirmed acting as a one-electron donor that siphons compound I from the extremely fast catalytic cycle into the relatively slow peroxidative cycle (~1000 times slower) by promoting the formation of compound II [36, 37].

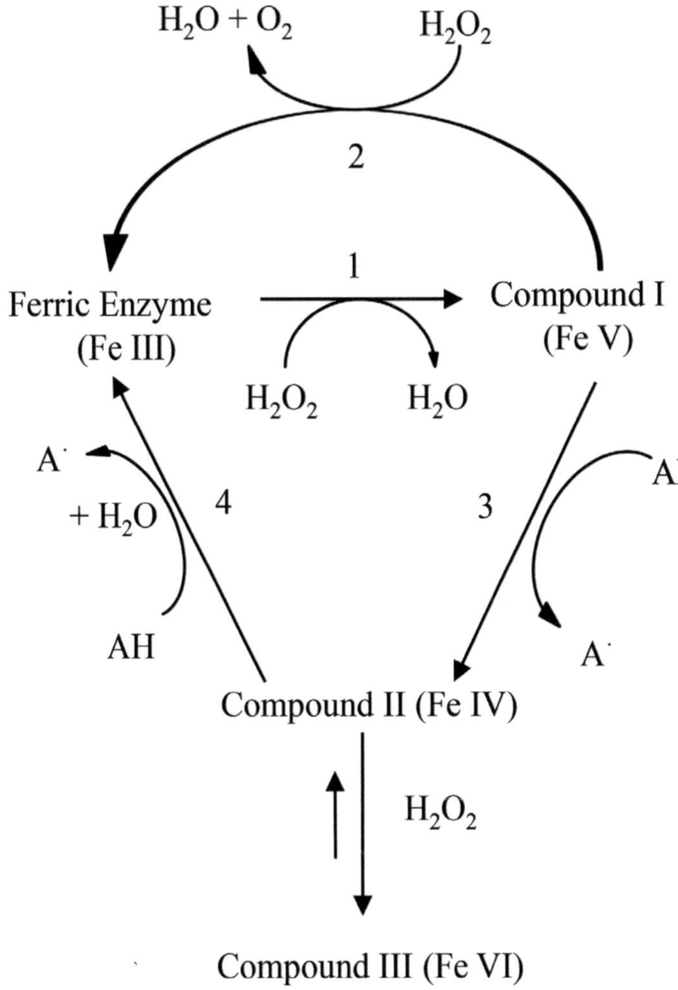

Figure 3. The reaction cycles of catalase [35].

3.2.2. Competitive inhibitor

3.2.2.1. Metal ions

Catalase mainly used in industrial sectors such as textiles, pulp, and paper, their work environment often with high concentration of metal ions. Previous studies have elaborated that catalase can be inhibited by certain metal ions (including Cu^{2+}, Zn^{2+}, and Ag^{+}), a process depends on the metal, concentration, the tissue, and species [38]. Lee et al. [39] compared several divalent metal ions on catalase-peroxidase (KatG) activity, only the manganese ion revealed some inhibitory effects on the recombinant KatG activity, and EDTA could relieve partly inhibited activity. This implies that manganese may competitively bind to near the heme group and be involved in the enzyme reaction.

http://dx.doi.org/10.5772/65936

3.2.2.2. ρ-Hydroxybenzoic acid is a competitive inhibitor of catalases from maize

Phenolic compounds, such as salicylic acid, aspirin, benzoic acid, o-coumaric acid, and ϱ-hydroxybenzoic acid play a role in the induction of abiotic stress resistance. But only ϱ-hydroxybenzoic acid showed the inhibitory effect on two catalases from maize in a competitive manner, the other compounds were in noncompetitive manner. Weak inhibition by ϱ-hydroxybenzoic acid was also found in both isozymes, only 15 and 9% activity was inhibited, respectively [34].

3.3. Activator of catalases

Metformin is a commonly used antidiabetic drug with AMP-activated protein kinase (AMPK)-dependent hypoglycemic activities. A recent study [40] shows that metformin can significantly enhance the activity of catalase. Although metformin bound to CAT byinteracting with hydrogen bonds…., metformin did not affect the expression level of catalase, just affecting its activities, such as Lys449, Val450, and Glu455 residues in murine CAT. The preliminary study indicated that metformin might be a new drug to alleviate oxidative injury and enhance the defense ability of antioxidants.

4. GPX

4.1. The structure and function of GPX

GPX is an important selenium-containing enzyme which protects cells from lipid peroxide damage and H_2O_2. GSH-Px widely existed in the body, there are eight family members: GPX1 is the most abundant selenoperoxidase and is ubiquitously expressed in almost all tissues; GPX2 expression is most prominent in the gastrointestinal tract cytoplasm; GPX3 is greatest expressed in the kidney, and also in various tissues, and is secreted into extracellular fluids as a glycoprotein; GPX4 is the only GPX enzyme that reduces phospholipid hydroperoxides, different with other members, it is not a tetramer, but a monomer; GPX6 was identified as a selenoprotein in the human genome by homology search. GPX1-4 and GPX6 are selenium-containing protein, but GPX5 does not contain selenocysteine or Se in active site [41, 42]. More recently, GPX7 (NPGPx) and GPX8 were discovered, but detailed information about these two kinds of enzymes is little up to now [43].

Molecular weight of GSH-PX in human red blood cell was 95,000 Da and that in bovine red blood cell was 83,000 Da, and they are all tetrameric selenoenzyme. A typical crystal structure of human glutathione peroxidase (2I3Y) is shown in **Figure 4**.

GSH-PX contains one selenocysteine per subunit and selenocysteine plays an important role in catalyst degradation of lipid peroxide, a widely accepted mechanism was proposed as follow:

$$E-CysSe^{-}+H^{+}+ROOH \rightarrow E-CysSeOH+ROH \quad (3)$$

$$E-CysSeOH+GSH \rightleftharpoons E-CysSe-SG+H_2O \quad (4)$$

$$E-CysSe-SG+GSH \rightleftharpoons E-CysSe^{-}+GSSH+H^{+} \quad (5)$$

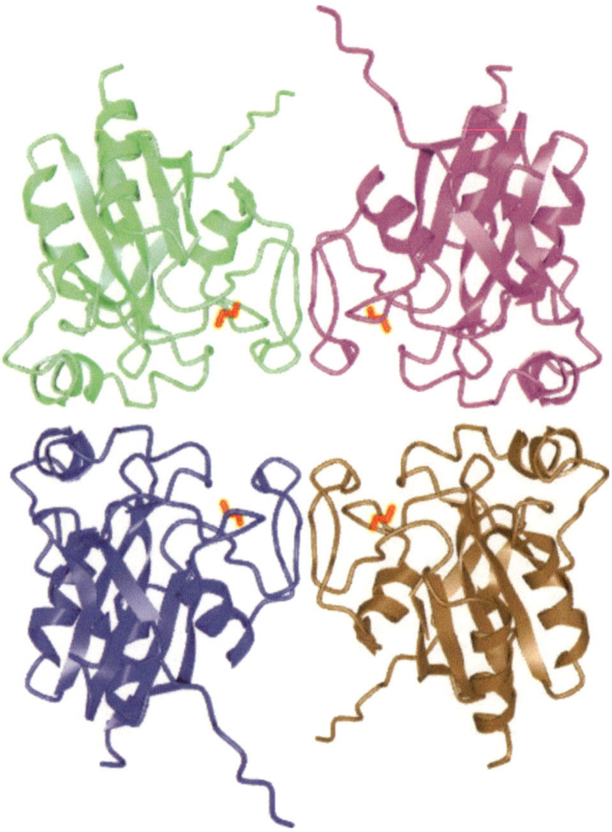

Figure 4. Three-dimensional structure of human glutathione peroxidase.

4.2. The effect of selenium on GPX activity

There are approximate 25 selenoproteins in humans, and selenium is an essential cofactor for these proteins, including the glutathione peroxidases. There is a complex relationship among utilization of selenium, GPX activity, and methylation. GPX synthesis utilizes selenium via selenocysteine and homocysteine is derived from *S*-adenosylhomocysteine, and the latter was formed as a result of methylation reactions including methylation of selenium [44]. From the view of selenium forms, the inorganic forms (such as sodium selenate and sodium selenite) of selenium were more effective on increasing the GPX activity compared with selenomethionine [44]. To be attention, different members of GPXs response differently to selenium deficiency, a phenomenon called the "hierarchy of selenoproteins." This means that upon selenium deprivation, some proteins decline fast, whereas others remain synthesized until selenium becomes severely deficient [45]. The expression of GPX1 mRNA, protein, and its activity in tissues is more sensitive than other selenoperoxidases or selenoproteins [46].

4.3. The effect of vitamin E on GPXs activity

DL-Alpha-tocopherol (vitamin E) also is an antioxidant, but it is different from selenium acting on GPX directly, vitamin E plays its antioxidant function through combining free radical, named "chain-breaking reaction." Vitamin E was considered to be the first line of defense against lipid peroxidation and free oxygen radicals that might suppress the enzymes, such as GPX. It seems that the sensitivity of GPXs was various. In vitamin E-deficient rat brain microsomes, phospholipid hydroperoxide glutathione peroxidase activity was significantly decreased but GPX activity was not affected. And in liver homogenate, phospholipid hydroperoxide glutathione peroxidase activity was approximately 20 times lower than that of GPX [47].

4.4. Inhibitors of GPXs

4.4.1. Competitive inhibitors

4.4.1.1. Misonidazole

GPX can combine to an electrophilic compound that might result in loss of its activity. More and more evidence shows that upregulation of the GPX system may serve to protect cancer cells from oxidative stress caused by anticancer drugs, thus block GPX that may help to treat cancer disease. A number of inhibitors of GPXs have been reported to use as therapeutics, such as thiol-containing inhibitors that bind covalently to a selenium atom in the active site [48], nonthiol inhibitors misonidazole [49]. However, thiols tend to combine ubiquitous multivalent metal ions and are easily oxidized, thus leading to nonspecific interactions with proteins. Recently, acylhydrazones have been reported as potential inhibitors of bovine glutathione peroxidase [50]. These inhibitors overcome the disadvantages of thiol-containing inhibitors, but the efficiency needs to be further improved.

4.4.1.2. Penicillamine and its analogues

D-Penicillamine is a drug to chelate metals in tissue and promotes its excretion in the urine. D-Penicillamine hydrochloride could competitively inhibit GSH-PX, that means the concentration of hydrogen peroxide and reduced glutathione were inversely proportion [51].

L-penicillamine hydantoin is an analogue of glutathione, but the acting configuration is different from D-penicillamine hydrochloride. After treated with L-penicillamine hydantoin, GPX activity was inhibited whatever the peroxide (H_2O_2, terl-butyl hydroperoxide orcumene hydroperoxide) used as substrate of the reaction. In the presence of 100 μM L-penicillamine hydantoin, the enzyme reactions catalyzed by glutathione peroxidase were inhibited, but neither glutathione transferases, nor glutathione reductase were affected by L-penicillamine hydantoin [52].

4.4.2. *Noncompetitive inhibitors*

4.4.2.1. *DL-Buthionine-[S, R]-sulfoximine*

DL-Buthionine-[*S*, *R*]-sulfoximine (BSO) can be used as an inhibitor to estimate the scavenging efficiency of H_2O_2 after GPX inhibition. The inhibitory effect did not act on GPX directly, but through suppress the synthesis of GSH by inhibiting γ-glutamylcysteine synthetase, that cause glutathione decreased sharply in many tissues, especially kidney, liver, and pancreas [53]. BSO has an obvious inhibitory effect, for example in human fibroblasts cells, after 500 μM BSO treated, the GSH levels decreased to154.0 ± 16.9 nmol/mg protein from 418.4 ± 13.1 nmol/mg protein [54].

4.4.2.2. *Gold(I) thioglucose*

Gold(I) thioglucose in the presence excess of glutathione (GSH) leads to strong and reversible inhibition of selenium-GPXs. Gold(I) could competitively combine in reduced form of selenocysteine in active sites, and gold(I) forms a dead-end complex with glutathione peroxidase resulting in suppression of GPXs. So glutathione peroxidase could be a target of gold drugs that used in the treatment of disease caused by excessive activity of GPXs, such as rheumatoid arthritis [55].

4.5. Activators of GPXs

To our knowledge, most literature studies on enhancing GPXs activity were about how to regulate expression of GPXs, study on the activator by acting the enzyme directly was few and most of them are GPX-mimetic compounds.

4.5.1. *Enhance activity of GPX mimetics*

For some GPX mimetics, its activity can be enhanced by electron-donating. naphthalene *peri*-diselenide mimeticswas increased by electron-donating methoxy substituents, while a further 100-fold increase was observed with the corresponding ditelluride. This was attributed to the ability of the methoxy group to stabilize the increasing positive charge at the selenium atom during the rate-determining step of the catalytic cycle, which involves the oxidation of Se(II) to Se(IV), thus improved their catalytic activity to levels comparable with their aliphatic counterparts [56]. Others report that 6-bromo-substituted diselenides also enhanced its activity by threefold [57]. Another strategy is to change the aqueous solubility of the mimetics. Diaryl selenides containing *o*-hydroxymethylene substituents function as peroxide-destroying mimetics of the antioxidant selenoenzyme glutathione peroxidase. Several selenide analogues

http://dx.doi.org/10.5772/65936

were attached to polyethylene glycol (PEG) oligomers greatly improved aqueous solubility and catalytic activity (10–100 folds) [58].

Fund support

This work was supported by grants from the Key project of Natural Science Foundation of Guangdong Province (2014A030311010), Guangdong Province Modern Agro-industry Technology Research System (2016LM1080)

Author details

Xianyong Ma[1,2,3*], Dun Deng[1,2,3] and Weidong Chen

*Address all correspondence to: 407986619@qq.com

1 Institute of Animal Science, Guangdong Academy of Agricultural Sciences, Guangzhou, China

2 Key Laboratory of Animal Nutrition and Feed Science in South China, Ministry of Agriculture, Guangzhou, China

3 Guangdong Key Laboratory of Animal Breeding and Nutrition, Guangzhou, China

References

[1] Kaynar H, Meral M, Turhan H, Keles M, Celik G, Akcay F. Glutathione peroxidase, glutathione-S-transferase, catalase, xanthine oxidase, Cu–Zn superoxide dismutase activities, total glutathione, nitric oxide, and malondialdehyde levels in erythrocytes of patients with small cell and non-small cell lung cancer. Cancer Letters. 2005;227: 133–139.

[2] Chance B, Sies H, Boveris A. Hydroperoxide metabolism in mammalian organs. Physiological Reviews. 1979;59: 527–605.

[3] Dorval J, Hontela A. Role of glutathione redox cycle and catalase in defense against oxidative stress induced by endosulfan in adrenocortical cells of rainbow trout (*Oncorhynchus mykiss*). Toxicology and Applied Pharmacology. 2003;192: 191–200.

[4] Zhong W, Yan T, Lim R, Oberley LW. Expression of superoxide dismutases, catalase, and glutathione peroxidase in glioma cells. Free Radical Biology and Medicine. 1999;27: 1334–1345.

[5] Eitinger T. In vivo production of active nickel superoxide dismutase from *Prochlorococcus marinus* MIT9313 is dependent on its cognate peptidase. Journal of Bacteriology. 2004;186: 7821–7825.

[6] Bryngelson PA, Arobo SE, Pinkham JL, Cabelli DE, Maroney MJ. Expression, reconstitution, and mutation of recombinant Streptomycescoelicolor NiSOD. Journal of the American Chemical Society. 2004;126: 460–461.

[7] Abreu IA, Cabelli DE. Superoxide dismutases—a review of the metal-associated mechanistic variations. Biochimica et Biophysica Acta (BBA) - Proteins and Proteomics. 2010;1804: 263–274.

[8] Nam JS, Yoon JH, Lee HI, Kim SW, Ro YT. Molecular cloning, purification, and characterization of a superoxide dismutase from a fast-growing Mycobacterium sp. Strain JC1 DSM 3803. The Journal of Microbiology. 2011;49: 399–406.

[9] Mayer BK, Falkinham JO. Superoxide dismutase activity of *Mycobacterium avium, M. intracellulare,* and *M. scrofulaceum*. Infection and Immunity. 1986;53: 631–635.

[10] Whittaker MM, Whittaker JW. Thermally triggered metal binding by recombinant *Thermus thermophilus* manganese superoxide dismutase, expressed as the apo-enzyme. Journal of Biological Chemistry. 1999;274: 34751–34757.

[11] Kirby T, Blum J, Kahane I, Fridovich I. Distinguishing between Mn-containing and Fe-containing superoxide dismutases in crude extracts of cells. Archives of Biochemistry and Biophysics. 1980;201: 551–555.

[12] Zhu YB, Wang GH, Ni H, Xiao AF, Cai HN. Cloning and characterization of a new manganese superoxide dismutase from deep-sea thermophile Geobacillus sp. EPT3. World Journal of Microbiology & Biotechnology. 2014;30: 1347–1357.

[13] Wang S, Shao B, Liu S, Ye X, Rao P. Purification and characterization of Cu, Zn-superoxide dismutase from black soybean. Food Research International. 2012;47: 374–379.

[14] Liu J, Wang J, Yin M, Zhu H, Lu J, Cui Z. Purification and characterization of superoxide dismutase from garlic. Food and Bioproducts Processing. 2011;89: 294–299.

[15] Halliwell B, Gutteridge JMC. Free Radicals in Biology and Medicine. Oxford University Press. 1999.

[16] Kanofsky JR. Singlet oxygen production by biological systems. Chemico-Biological Interactions. 1989;70: 1–28.

[17] Kim SY, Kwon OJ, Park J-W. Inactivation of catalase and superoxide dismutase by singlet oxygen derived from photoactivated dye. Biochimie. 2001;83: 437–444.

[18] Xueshan Y, Xiaopu Y, Wenli F. Effect of carbohydrates on activities of superoxide dismutase extracted from sheep blood. Food Science. 2011;32: 100–103.

[19] Sola-Penna M, Meyer-Fernandes JR. Stabilization against thermal inactivation promoted by sugars on enzyme structure and function: why is trehalose more effective than other sugars? Archives of Biochemistry and Biophysics. 1998;360: 10–14.

[20] Beckman JS, Minor RL, Jr., White CW, Repine JE, Rosen GM, Freeman BA. Superoxide dismutase and catalase conjugated to polyethylene glycol increases endothelial enzyme activity and oxidant resistance. Journal of Biological Chemistry. 1988;263: 6884–6892.

[21] Valdivia A, Pérez Y, Ramírez H, Cao R, Villalonga R. Improved pharmacological properties for superoxide dismutase modified with β-cyclodextrin–carboxymethylcellulose polymer. Biotechnology Letters. 2006;28: 1465–1470.

[22] Puglisi A, Tabbi` G, Vecchio G. Bioconjugates of cyclodextrins of manganese salen-type ligand with superoxide dismutase activity. Journal of Inorganic Biochemistry. 2004;98: 969–976.

[23] Zamocky M, Koller F. Understanding the structure and function of catalases: clues from molecular evolution and in vitro mutagenesis. Progress in Biophysics and Molecular biology. 1999;72: 19–66.

[24] Andreoletti P, Sainz G, Jaquinod M, Gagnon J, Jouve HM. High-resolution structure and biochemical properties of a recombinant Proteus mirabilis catalase depleted in iron. Proteins: Structure, Function, and Bioinformatics. 2003;50: 261–271.

[25] Welinder KG. Bacterial catalase-peroxidases are gene duplicated members of the plant peroxidase superfamily. Biochimica et Biophysica Acta. 1991;1080: 215–220.

[26] Kamachi S, Hirabayashi K, Tamoi M, Shigeoka S, Tada T, Wada K. The crystal structure of isoniazid-bound KatG catalase-peroxidase from *Synechococcus elongatus* PCC7942. FEBS Journal. 2015;282: 54–64.

[27] Allgood GS, Perry JJ. Characterization of a manganese-containing catalase from the obligate thermophile Thermoleophilum album. Journal of Bacteriology. 1986;168: 563–567.

[28] Whittaker MM, Barynin VV, Igarashi T, Whittaker JW. Outer sphere mutagenesis of *Lactobacillus plantarum* manganese catalase disrupts the cluster core. Mechanistic implications. European Journal of Biochemistry. 2003;270: 1102–1116.

[29] Kengen SW, Bikker FJ, Hagen WR, de Vos WM, van der Oost J. Characterization of a catalase-peroxidase from the hyperthermophilic archaeon *Archaeoglobus fulgidus*. Extremophiles. 2001;5: 323–332.

[30] Lee DH, Oh DC, Oh YS, Malinverni JC, Kukor JJ, Kahng HY. Cloning and characterization of monofunctional catalase from photosynthetic bacterium

Rhodospirillum rubrum S1. Journal of Microbiology and Biotechnology. 2007;17: 1460–1468.

[31] Shima S, Netrusov A, Sordel M, Wicke M, Hartmann CG, Thauer KR. Purification, characterization, and primary structure of a monofunctional catalase from *Methanosarcina barkeri*. Archives of Microbiology. 1999;171: 317–323.

[32] Ueda M, Kinoshita H, Yoshida T, Kamasawa N, Osumi M, Tanaka A. Effect of catalase-specific inhibitor 3-amino-1,2,4-triazole on yeast peroxisomal catalase in vivo. FEMS Microbiology Letters. 2003;219: 93–98.

[33] Ai Q, Ge P, Dai J, Liang TC, Yang Q, Lin L, Zhang L. The catalase inhibitor aminotriazole alleviates acute alcoholic liver injury. Sheng Li Xue Bao. 2015;67: 97–102.

[34] Horváth E, Szalai G, Pál M, Páldi E, Janda T. Differences between the catalase isozymes of maize (*Zea mays* L.) in respect of inhibition by various phenolic compounds. Acta Biologica Szegediensis. 2002;46: 33–34.

[35] Durner J, Klessig DF. Salicylic acid is a modulator of tobacco and mammalian catalases. Journal of Biological Chemistry. 1996;271: 28492–28501.

[36] Havir EA, McHale NA. Biochemical and developmental characterization of multiple forms of catalase in tobacco leaves. Plant Physiology. 1987;84: 450–455.

[37] Zamocky M, Herzog C, Nykyri LM, Koller F. Site-directed mutagenesis of the lower parts of the major substrate channel of yeast catalase A leads to highly increased peroxidatic activity. FEBS Letters. 1995;367: 241–245.

[38] Atli G, Alptekin O, Tukel S, Canli M. Response of catalase activity to Ag+, Cd2+, Cr6+, Cu2+ and Zn2+ in five tissues of freshwater fish *Oreochromis niloticus*. Comparative Biochemistry and Physiology: Toxicology & Pharmacology. 2006;143: 218–224.

[39] Lee HI, Yoon JH, Nam JS, Kim YM, Ro YT. Cloning, expression and characterization of the catalase-peroxidase (KatG) gene from a fast-growing Mycobacterium sp. strain JC1 DSM 3803. Journal of Biochemistry. 2010;147: 511–522.

[40] Dai J, Liu M, Ai Q, Lin L, Wu K, Deng X, et al. Involvement of catalase in the protective benefits of metformin in mice with oxidative liver injury. Chemico-Biological Interactions. 2014;216: 34–42.

[41] Lei XG, Cheng WH, McClung JP. Metabolic regulation and function of glutathione peroxidase-1. Annual Review of Nutrition. 2007;27: 41–61.

[42] Kryukov GV, Castellano S, Novoselov SV, Lobanov AV, Zehtab O, Guigo R, Gladyshev VN. Characterization of mammalian selenoproteomes. Science. 2003;300: 1439–1443.

[43] Toppo S, Vanin S, Bosello V, Tosatto SC. Evolutionary and structural insights into the multifaceted glutathione peroxidase (Gpx) superfamily. Antioxidants and Redox Signaling. 2008;10: 1501–1514.

[44] Bermingham EN, Hesketh JE, Sinclair BR, Koolaard JP, Roy NC. Selenium-enriched foods are more effective at increasing glutathione peroxidase (GPx) activity compared with selenomethionine: a meta-analysis. Nutrients. 2014;6: 4002–4031.

[45] Hofmann B, Hecht HJ, Flohe L. Peroxiredoxins. Biological Chemistry. 2002;383: 347–364.

[46] Bermano G, Nicol F, Dyer JA, Sunde RA, Beckett GJ, Arthur JR, Hesketh JE. Tissue-specific regulation of selenoenzyme gene expression during selenium deficiency in rats. Biochemical Journal. 1995;311 (Pt 2): 425–430.

[47] Bourre J-M, Dumont O, Clément M, Dinh L, Droy-Lefaix M-T, Christen Y. Vitamin E deficiency has different effects on brain and liver phospholipid hydroperoxide glutathione peroxidase activities in the rat. Neuroscience Letters. 2000;286: 87–90.

[48] Chaudiere J, Wilhelmsen EC, Tappel AL. Mechanism of selenium-glutathione peroxidase and its inhibition by mercaptocarboxylic acids and other mercaptans. Journal of Biological Chemistry. 1984;259: 1043–1050.

[49] Kumar KS, Weiss JF. Inhibition of glutathione peroxidase and glutathione transferase in mouse liver by misonidazole. Biochemical Pharmacology. 1986;35: 3143–3146.

[50] Wilde F, Lemmerhirt H, Emmrich T, Bednarski PJ, Link A. Microwave-assisted synthesis and evaluation of acylhydrazones as potential inhibitors of bovine glutathione peroxidase. Molecular Diversity. 2014;18: 307–322.

[51] Mercurio SD, Combs GF, Jr. Drug-induced changes in selenium-dependent glutathione peroxidase activity in the chick. Journal of Nutrition. 1985;115: 1459–1470.

[52] Mestdagh N, Poupaert J, Hénichart J-P, Vamecq J. Effect of l-penicillamine hydantoin, an analogue of glutathione, on rat liver glutathione peroxidase, reductase and transferase reactions. Biochemical Pharmacology. 1992;43: 1529–1537.

[53] Griffith OW, Meister A. Glutathione: interorgan translocation, turnover, and metabolism. Proceedings of the National Academy of Sciences of the United States of America. 1979;76: 5606–5610.

[54] Masaki H, Okano Y, Sakurai H. Differential role of catalase and glutathione peroxidase in cultured human fibroblasts under exposure of H_2O_2 or ultraviolet B light. Archives for Dermatological Research. Archiv für Dermatologische Forschung. 1998;290: 113–118.

[55] Chaudiere J, L. Tappel A. Interaction of gold(I) with the active site of selenium-glutathione peroxidase. Journal of Inorganic Biochemistry. 1984;20: 313–325.

[56] Press DJ, Back TG. Enhanced glutathione peroxidase activity of conformationally restricted naphthalene peri-dichalcogenides. Organic Letters. 2011;13: 4104–4107.

[57] Singh VP, Poon J-f, Butcher RJ, Lu X, Mestres G, Ott MK, Engman L. Effect of a bromo substituent on the glutathione peroxidase activity of a pyridoxine-like diselenide. The Journal of Organic Chemistry. 2015;80: 7385–7395.

[58] McNeil NMR, Press DJ, Mayder DM, Garnica P, Doyle LM, Back TG. Enhanced glutathione peroxidase activity of water-soluble and polyethylene glycol-supported selenides, related spirodioxyselenuranes, and pincer selenuranes. The Journal of Organic Chemistry. 2016;81: 7884–7897.

Chapter 10

Enzyme Dynamic in Plant Nutrition Uptake and Plant Nutrition

Metin Turan, Emrah Nikerel, Kerem Kaya,
Nurgul Kitir, Adem Gunes,
Negar Ebrahim Pour Mokhtari, Şefik Tüfenkçi,
M. Rüştü Karaman and K. Mesut ÇİMRİN

Additional information is available at the end of the chapter

http://dx.doi.org/10.5772/66938

Abstract

Soil contains enzymes, constantly interacting with soil constituents, e.g. minerals, rhizosphere and numerous nutrients. Enzymes, in turn, catalyse important biochemical reactions for rhizobacteria and plants, stabilize the soil by degrading wastes and mediate nutrient recycling.The available enzymes inside soil could originate from plants, animals or microbes. The enzymes that are produced from these organism could exhibit intracellular activities, at the cell membrane, interacting therefore with soil and its constituents, or extracellularly (so freely available). Therefore, vis-à-vis to plant nutrition, the (extra or sub) cellular localization has a key role. Typical major enzymes available in soil can be listed as dehydrogenases, hydrogenases, oxidases, catalases, peroxidases, phenol o-hydroxylase, dextransucrase, aminotransferase, rhodanese, carboxylesterase, lipase, phosphatase, nuclease, phytase, arylsulphatase, amylase, cellulase, inulase, xylanase, dextranase, levanase, poly-galacturonase, glucosidase, galactosidase, invertase, peptidase, asparaginase, glutaminase, amidase, urease, aspartate decarboxylase, glutamate decarboxylase and aromatic amino acid decarboxylase. An interesting strategy for improving the nutritional quality of the soil would be to inoculate microorganism to soil while giving attention to mineral or other compounds that affect enzyme activity in soil. Since, some elements or compounds could show both activation and inhibitory effect, such as Fe, Na, etc. metals, the regulation of their bioavailability is crucial.

Keywords: plant growth promoting rhizobacteria, amino acid, organic acid, nutrient element, hormone, plant physiology

1. Introduction

Soil contains, among many others, enzymes that are constantly interacting (regulating, being regulated by) with soil constituents, for example, minerals, rhizosphere and numerous nutrients. Enzymes, in turn, catalyse important biochemical reactions for rhizobacteria and plants, stabilize the soil by degrading wastes and mediate nutrient recycling [1].

The available enzymes inside soil could originate from plants, animals or microbes (bacteria or fungi). The enzymes that are produced from these organism could exhibit activities intracellular of the source organism, at its cell membrane, interacting therefore with soil and its constituents, or extracellularly (so freely available). Therefore, vis-à-vis to plant nutrition, or bioavailability of the macro- or micro-nutrients, the (extra or sub) cellular localization has a key role. Typical major enzymes available in soil can be listed as dehydrogenases, hydrogenases, oxidases, chief among those being glucose, aldehyde, urate, catechol, p-diphenol, ascorbate oxidases, catalases, peroxidases, phenol o-hydroxylase, dextransucrase, levan sucrase, aminotransferase, rhodanese, carboxylesterase, arylesterase, lipase, phosphatase, nuclease, nucleotidase, phytase, arylsulphatase, amylase, cellulase, laminarinase, inulase, xylanase, dextranase, levanase, poly-galacturonase, glucosidase, galactosidase, galactosidase, invertase, proteinase, peptidase, asparaginase, glutaminase, amidase, urease, inorganic pyrophosphatase, polymetaphosphatase, adenosine triphosphatase, aspartate decarboxylase, glutamate decarboxylase and aromatic amino acid decarboxylase [1].

An interesting strategy for improving the nutritional quality of the soil would be either inoculating microorganism to soil while giving attention to mineral or other compounds that affect enzyme activity in soil. Since, some elements or compounds could show both activation and inhibitory effect, such as Fe, Na, etc., metals, the regulation of their bioavailability is crucial.

Measurement of soil enzyme activity is important to determine soil characteristics, for further studies, such as, improving soil composition for plant growth using enzymes. A simple example can be given for proteases. Soil, when supplemented with proteases, would degrade proteins, thereby, increasing the amount of available nitrogen, which in turn is expected to improve plant nutrition. Similarly, soil supplemented with urease would increase bioavailable nitrogen level, and as such, this enzyme can be seen as a 'knob' for nitrogen regulation in soil and indirectly in plants. Finally, the use of enzymes, typically from microorganisms as plant growth promoting rhizobacteria (PGPR), is important not only from an economical perspective (improved crop yield), but also environmental point-of-view (reduced use of chemical fertilizers).

Enzymes are, at industrial scale, typically produced using either fungi or bacteria, either technology having advantages and disadvantages. While cultivation of bacteria is easier to handle (from both process and genetics perspective) and to scale up, fungi has typically larger portfolio of enzymes and the latter is more resilient to stress conditions, a characteristic of the production and application conditions.

Vis-à-vis plant nutrition, enzymes have crucial roles, tightly coupled to soil remediation as soil contains impurities in the form of heavy metals as well as polymers, for example, starch and cellulose residues, polyphosphate rocks, urea from N-cycle, oils and fats from either plants or animals that cannot be readily used by plants, in particular for nutrition. Enzymes are then responsible to break these residues into forms that renders them bioavailable to plants. The application depends on the soil type, the content of the above-listed polymers or substances. The conventional approach is to use directly plant growth promoting rhizobacteria (PGPR) to improve growth and yield. The mechanism of action of those is actually the use of key enzymes (not limited to the five enzymes listed here) for plant growth promoting effect, making nutrient-rich materials bioavailable, shortening composting time yielding highly rich soil, improving thereby plant nutrition and allowing soil remediation.

Taken together, the use of some key enzymes are promising for soil conditioning and plant nutrition. As a follow-up, to gather better soil environment for plants, information on both organisms and especially the enzymes that are produced is of great value. This chapter focuses on this idea and provides key properties for a handful of enzymes, relevant to plant nutrition. The focus is on amylase, cellulase, lipase, phosphatase, phytase and urease, some key properties thereof and list of applications relevant to plant nutrition.

2. Amylase

Amylases are enzymes hydrolyzing glycosidic bonds of polysaccharides. Usually these are classified into three sub-classes as α-amylase (E.C. 3.2.1.1), β-amylase (E.C. 3.2.1.2) and γ-amylase (E.C. 3.2.1.3). α-Amylase is responsible in endo-hydrolysis of (1-4)-α-D-glucosidic linkages, while β-amylase is responsible in the hydrolysis of (1-4)-α-D-glucosidic linkages in polysaccharides to remove maltose units from non-reducing ends. γ-Amylase, in contrast, is responsible in the hydrolysis of terminal (1-4)-α-D-glucose residues from non-reducing ends of the chains for releasing of β-D-glucose.

All the three versions of this enzyme are produced by bacteria and fungi. α-amylase have been reported by *Acinetobacter* spp., *Bacillus amyloliquefaciens, Bacillus licheniformis, Bacillus megaterium, Bacillus subtilis* and some thermophilic actinomycetes organisms as well, for example, *Thermomonospora curvata* and *Thermomonospora vulgaris* [2, 3], while β-amylase have been reported to be produced by *Bacillus cereus, Bacillus circulans, B. megaterium* and *Paenibacillus polymyxa* [4, 5]. Lastly,for γ-amylase, in addition to the *Bacillus* species, halophylic *Halolactibacillus* sp. and thermophilic organisms, for example, *Thermoactinomyces vulgaris* have been reported to produce this enzyme [6–8].

Amylases are reported to be active in a broad range of pH 1–13 [9, 10], yet β- and γ-amylases have narrower ranges. The optimum working pH range is reported to be from 2 [2] to 10.5 [11] for α-amylase, the other two being in a narrower range. As for the temperature, again α-amylases are active in a broad range or temperature from 20 [12] to 145°C [13]. Lastly, molecular weights range between 10 [14] and 240 kDa [15].

Despite the broad range of pH and temperature where the amylases are active, there is fairly long list of inhibitors for the microbiologically produced amylases: Ag^+, Ba^{2+}, Ca^{2+}, Co^{2+}, Cu^{2+}, Fe^{2+}, Hg^{2+}, Mg^{2+}, Mn^{2+}, Ni^{2+}, Sr^{2+}, Zn^{2+} [14]; Cd^{2+}, iodoacetate [16]; ethylenediaminetetraacetic acid (EDTA), K^+ [17]; Na^+, Triton X-100, Tween 20 [18]; phenylmethylsulfonyl fluoride (PMSF), 4-bromophenacyl bromide [19]; $Bi(NO_3)_3$, N-ethylmaleimide and sodium deoxycholate [20] are reported to be inhibitors. Interestingly, sodium dodecylsulfate (SDS), urea and 2-mercaptoethanol are reported to be both activating [21, 22] and inhibiting [17, 18] compounds.

The production of enzymes is typically performed in submerged fermentation, less often via solid state fermentation, typically under mesophilic conditions, moderate pH and temperature (30–50°C, mostly in 37°C; pH range of 3–9, mostly at 7) in chemically defined ($(NH_4)_2HPO_4$ as N-source, KH_2PO_4 as K-source) or complex media (yeast extract as N and K source), lactose, maltose glucose or starch as C-source, using chiefly *Bacillus* species [23–27]. Additionally, agro-wastes are also used as substrates and inducers as coconut oil cake, wheat/rice bran, spent brewing grain, cassava bagasse, jackfruit or tamarind seed powder, palm kernel, olive oil or mustard oil cake and rice husk [27].

3. Phosphatase

Phosphatases belong to the enzyme group responsible in the hydrolysis of ester-phosphate bonds which releases phosphates. These are sub-classified as phosphomonoesterases (EC 3.1.3.x), phosphodiesterases (EC 3.1.4.x), enzymes that hydrolyze phosphorus-containing anhydrides (EC 3.6.1.x), P-N bonds (EC 3.9.1.x) and various groups that act on this bonds. From an application point of view, these are grouped as alkaline, acid phosphatases and inorganic diphosphatases. The microbial producers of these enzymes are numerous, including *B. subtilis* [28], *Escherichia coli* [29] and *Pseudomonas aeruginosa* [30] for alkaline phosphatase; *Acidithiobacillus thiooxidans* [31], *E. coli* [32] and *Lactobacillus curvatus* [33] for acid phosphatase and *Geobacillus stearothermophilus* [34], *Rhodobacter capsulatus, Rhodopseudomonas palustris* [35] for Inorganic diphosphatase.

The large portfolio of phosphatases works in a broad range of pH and temperature. For the pH, the phosphatases are reported to work optimally between 2.5 [36] and 12.5 [37]. As for the temperature, active ranges are reported to be between 5 [38] and 95°C [39], while optimally, the enzyme works between 20 and 70°C [40, 41]. With different pockets or binding sites, there is also a range for the molecular weight, from 32.5 [42] to 128 [43] kDa.

Several agents are reported to inhibit the phosphatases. These are ascorbate, dithiothreitol, NaF, molybdate, $NaBH_4$, sodium lauryl sulfate, tartrate [31], 2-mercaptoethanol, $BaCl_2$, $CaCl_2$, hexametaphosphate, $HgCl_2$, $MnCl_2$, p-chloromercuribenzoate (PCMB), PMSF, tripolyphosphate and $ZnCl_2$ [33]. In contrast, some organic acids, for example, citrate, pyruvate, succinate [32], 1,10-phenanthroline, EDTA [33] have been found to stimulate enzyme activity.

4. Lipase

Lipases (EC 3.1.1.x) are enzymes degrading lipids. In literature, most of the studied and reported lipases are triacylglycerol lipases (EC 3.1.1.3), while additionally there are carboxylesterase (EC 3.1.1.1) which hydrolyze carboxylic ester bonds, arylesterase (EC 3.1.1.2) also acting on carboxylic esters but more specifically on phenolic esters, phospholipase A2 (EC 3.1.1.4) again hydrolyzing carboxylic esters specifically on phosphatidylcholine. It should be noted that distinguishing each of these enzymes is rather challenging as they have similar activities.

The producing organisms span the fungi and bacteria, in particular *B. subtilis* [44], *E. coli* [45] for EC 3.1.1.1 (carboxyl esterase), *Gluconobacter oxydans* [46] and *Lactobacillus casei* [47] for EC 3.1.1.2 (arylesterases) and *Acinetobacter calcoaceticus, B. subtilis, Chromobacterium viscosum, Micrococcus freudenreichii, Lactobacillus delbruckii, P. aeruginosa* and *Streptococcus lactis* [48] for EC 3.1.1.3 (triacylglycerol lipase)

Bacterial lipases has a pH working range between 4 [49] and 12 [50], while optimum pH is reported to vary between 6 [51] and 11 [52]. As for the temperature, there is a large range between 0 [53] and 100°C [54], while the optimum temperature for enzyme activity vary between 10 [55] and 90°C [50]. A span of molecular weights is reported for this enzyme (bacterial variants) from 11 [56] to 840 [57] kDa.

Metals ions such as Cu^{2+}, Fe^{2+}, Fe^{3+}, Hg^{2+}, Zn^{2+}, Ag^{+}, Co^{2+}, Ni^{2+}, Na^{+} and ascorbic acid are reported to have inhibitory effect on the carboxylesterase activity [58, 59] as well as sodium dodecylsulfate (SDS), diisopropylfluorophosphate, eserine, sodium fluoride [60] and phenylmethylsulfonyl fluoride (PMSF) [61]. Organic solvents such as acetone, EDTA, ethanol, isopropanol, PMSF and SDS [49] are reported to inhibit triacylglycerol lipases [62, 63]. Under lab conditions, Triton X-100, Tween-20, Tween-40, Tween-80 [64], 1,4-dioxane, acetone, dimethyl sulfoxide, ethanol and tetrahydrofuran [65] are reported activators to carboxyl esterases. Interestingly, acetone, Brij 52, cholic acid, deoxycholic acid, isopropanol, Dimethyl sulfoxide (DMSO), lithocholic acid, rhamnolipid and sodium deoxycholate are also reported as activators for triacylglycerol lipases [66].

For production of enzyme, apart from the above-listed organisms, see **Table 1**.

Microorganism	Media	Conditions	Production mode	References
Anaerovibrio zipolytica 5s	g/100 mL: 0.6 g Difco yeast extract; 0.75 g casein hydrolysate; 15 mL 0.3% (w/v) dipotassium hydrogen phosphate; 15 mL 0.3% (w/v) potassium dihydrogen phosphate; 0-1 mL 0.1% (w/v) resazurin and 10 mL 5% (w/v) glycerol, 0.5% (w/v) cysteine HCI and 6% (w/v) sodium bicarbonate	38°C	Batch via 300 mL vessels	[67]
Bacillus coagulans BTS-3	peptone (0.5%), yeast extract (0.5%), NaCl (0.05%), $CaCl_2$ (0.005%) and olive oil (1.0%, emulsified with gun)*, pH 8.5	48 h, 55°C 170 rpm,	Batch via 250 mL erlenmeyers (50 mL working volume)	[68]
Pseudomonas sp.	Ground soybean (2.0%), corn-steep liquor (2.0%), soluble starch (1.0%), K_2HPO_4 (0.5%) and $NaNO_3$ (0.5%) and the pH 9.0	30°C, 72 h, 150 rpm	Batch via 500 mL erlenmeyer (working volume of 50 mL)	[69]

* Besides olive oil, coconut oil, castor oil, groundnut oil, mustard oil, sunflower oil, Tween 20, Tween 80, cottonseed oil and soybean oil is studied as a carbon source. Beside peptone and yeast extract, gelatin and urea is also studied as organic nitrogen sources. Besides ammonium sulphate, ammonium nitrate, potassium nitrate and L-asparagine are also studied as inorganic nitrogen sources.

Table 1. Lipase production studies and the reported conditions.

5. Phytase

Phytases are enzymes that hydrolyze phytic acid which is an organic phosphorus source and makes inorganic usable phosphorus. Bacterially produced phytases are 3-phytase (EC 3.1.3.8), 4-phytase (EC 3.1.3.26) and protein-tyrosine-phosphatase (PTP, EC 3.1.3.48). Besides PTP, the other enzymes differentiate at which carbon they attack and take out the phosphorus in phytic acid. Several reports are available on the production of phytases. The organisms used are *Aerobacter aerogenes, B. amyloliquefaciens, B. subtilis, Enterobacter* sp., *E. coli, Klebsiella aerogenes, Lactobacillus amylovorus, Pseudomonas* sp., *Selenomonas ruminantium* [70] for three and four phytases and *B. subtilis, M. tuberculosis, S. aureus* [71–73]; typically grown under complex media (tryptone, yeast extract and NaCl and sugars, for example, lactose as inducer)

The activity of bacterially produced phytases change with pH, ranging from 2 [74] to 10 [75], while the optimum pH range is narrower (from 2.7 [76] to 8.5 [77]). As for the temperature, optimum working range is between 20 [78] and 80°C [79] due to the presence of some thermophilic organisms [70, 79]. The molecular weight range is found to be between 12.8 [80] and 700 [70] kDa, again depending on the producing host.

Similar to the other enzymes, several metal ions are reported to inhibit the phytase activity. These include Ba^{2+}, Cd^{2+}, Cu^{2+}, Li^{+}, Mg^{2+}, $Mn^{2+}$and Zn^{2+} [77, 81], while EDTA is considered as an activator compound [75]

6. Urease

An important enzyme for plant nutrition, in particular for N-cycle is Urease (EC 3.5.1.5), catalysing the conversion of urea to carbon dioxide and ammonia:

$$(NH_2)_2CO + H_2O = CO_2 + 2NH_3 \tag{1}$$

This enzyme is produced by bacteria, fungi as well as plants. Some bacterial producers are listed as *A. aerogenes, Arthrobacter oxydans, Bacillus pasteurii, Brevibacterium ammoniagenes, Brucella suis, E. coli, Helicobacter pylori, Proteus mirabilis, Providencia stuartii, S. ruminantium, Sporosarcina pasteurii, Staphylococcus saprophyticus* and *Ureaplasma urealyticum* [82–84], while the following organisms are reported to produce acid urease: *Arthrobacter mobilis, Lactobacillus fermentum* and *Streptococcus mitior* [82]. These are typically grown in batch mode, under complex (yeast extract, peptone and glucose) or chemically defined medium conditions, mezophilic temperatures, with urea as the inducer of the enzyme production [85, 86].

The pH range whereby the enzyme works optimally is 2–9 [87–90], while optimum temperature ranges from 20 to 70°C [91–94]. Molecular weights can vary from 11.1 [82] to 600 [90] kDa. Listed inhibitors are methylurea, thiourea, acetohydroxamic acid, phenylphosphorodiamidate, H_3PO_4, 2-mercaptoethanol, boric acid, lodoacetamide, lodoacetic acid, N-Ethylmaleimide, 5,5′-Dithiobis (2-nitrobenzoic acid) (DNTB) [95]; 12-hydroxytetradecanoc acid, 3-hydroxytetradecanoc acid, 6-hydroxytetradecanoc acid [96, 97] and several metal ions [98, 99]. Glycerol, n-octylglucoside, polyethylene glycol (PEG), sodium dodecyl sulfate (SDS),

Triton X-100 have some activatory effect in certain amounts [100]. It is worth noting that urease is nickel-containing metalloenzyme, as a result of which requires to a certain level nickel metal [101], as usual higher concentrations have inhibitory effect [99].

7. Cellulase

Cellulase (EC 3.2.1.4) is an important enzyme, naturally produced by bacteria, fungi and protozoa, in particular by necrophilic microorganisms, and is responsible to hydrolyze (1-4)-beta-D-glucosidic linkages in cellulose, which is by far the most abundant organic compound, totalling to almost 50% of the biomass synthesised by photosynthetic fixation of CO_2. Cellulases also degrade cellulose available in lichenin and cereal beta-D-glucans. As such, it is a key enzyme in degradation of the most abundant polymer on earth. The bacterial producers are listed as *Acetivibrio cellulolyticus, B. Subtilis, B. Amyloliquefaciens, Cellulomonas fimi, Pseudomonas fluorescens, Ruminococcus albus, Thermobifida fusca, Thermotoga maritima* [102–104].

Ag^{+}, Hg^{2+}, Mn^{2+}, iodoacetamide, N-bromosuccinimide [105]; Cu^{2+}, Pb^{2+}, Fe^{2+}, Sn^{2+}, ethylenediaminetetraacetic acid (EDTA) [106]; $NiCl_2$, $SrCl_2$ [7], sodium dodecyl sulphate (SDS) [107]; Cd^{2+}, Co^{2+}, Zn^{2+} [108] and 4-hydroxybenzoic acid, syringaldehyde, trans-cinnamic acid, vanillin [109] are shown to inhibit bacterial-originated cellulases. Arabitol, dithiothreitol, erythritol, glycerol, histamine [106]; N-ethylmaleimide [110]; CH_3COONa, NH_4Cl, NH_4NO_3 [111] and Ca^{2+} [107] are shown that activate enzyme. For production of the cellulase enzyme, reported conditions are listed in **Table 2**.

Bacteria	Media	Conditions	Production mode	References
Bacillus sp. *AC1*	Yeast 2.5 g/L, Tryptone 2.5 g/L, carboxymethyl cellulose (CMC) (low viscosity) 2.5 g/L, $(NH_4)_2SO_4$ 1 g/L, KH_2PO_4 0.5 g/L, K_2HPO_4 0.5 g/L and $MgSO_4$ 0.2 g/L	30°C, 2 d	Submerged fermentation	[112]
Bacillus sp. *NZ*	Carboxymethyl cellulose 5 g/L, peptone 5 g/L, yeast extract 5 g/L, KH_2PO_4 1 g/L, $MgSO_4.7H_2O$ 0.2 g/L, $NaCl_5$ g/L	45°C, 24–48 h, pH 9–10	Submerged fermentation via 250 mL erlenmeyers	[113]
Bacillus subtilis CBTK 106	10 g of banana fruit stalk with $Na_2HPO_4.2H_2O$ 1.1 g/L, $NaH_2PO_4.2H_2O$ 0.61 g/L, KCl 0.3 g/L, $MgSO_4.7H_2O$ 0.01 g/L	35°C, 72 h, pH 7.0, initial moisture content is 65%	Submerged fermentation via 250 mL erlenmeyers	[114]*

* Besides banana fruit stalk, wheat bran, rice bran and rice straw was tested as a substrate, but banana fruit stalk showed more cellulase activity. Also in this article with same media solid state and submerged fermentation was compared.

Table 2. Producing conditions of cellulase from bacteria.

8. The use of enzymes for plant nutrition

The use of enzymes for plant nutrition is typically mentioned within compost preparation context and optimization and/or speeding of this process. In an early work, Hankin et al. [115] studied several microorganisms from the extracellular enzyme production perspective from composting leaves and concluded that depending on the substrate available the microbial community produced tailor-made enzymes, and this production process was highly dependent on temperature [115]. The portfolio of enzymes produced covered all major enzymes. The temperature of the compost core increased significantly, when compared to the outer regions contacting with air. In general, Amylase is typically seen as one of the necessary enzymes to speed-up composting, yet of low importance. As for the use of the phosphatase enzyme for plant nutrition, there are several studies focusing on the soil phosphatase (both alkaline and acidic) activity, hinting the soil-bacteria collaboration. Tiquia et al. [116] reported the dynamics and enzyme activity during composting of poultry litter and yard trimmings, focusing on 19 different enzymes of different microbial groups in soil [116]. Similarly, Garcia et al. [117] after detailed biochemical analysis of biochemical parameters reported that highest phosphatase activity is found on sewage sludge [117].

The relation with enzymatic activity and compost state is so tight that enzymatic activity has been studied as indicator of composting process. Mondini et al. [118] reported the results of such study, whereby they concluded that drying the compost expectedly decreased the activity, but more importantly, measuring the activity of four enzymes (β-glucosidase, arylsulphatase, acid and alkaline phosphatase) in air-dried compost would be a fast and reliable method to follow composting process. Similar outcomes have been reported in Margesin et al. [119], focusing this time to lower temperatures. Herrmann and Shann [120] concluded that cellulase activity could be used as an indicator of stability, while lipase activity indicated compost maturity [120].

Lee et al. reported the positive effect of compost on lettuce growth as two to three times increase in fresh weight of the lettuce. They focused in particular to phosphatase and dehydrogenase activity [121]. Focusing more on the fats mixtures and the effect of lipase on co-composting sludges, Gea et al. [122] reported 85% reduction in fat content, with an initial fat content of 30%. The authors note that due to hydrophobic nature of the fats, the moisture content needs to be maintained above 40%. Krzywy-Gawrońska [123] focused on urease and dehydrogenase activity in compost-fertilized soil, in a 2-year field trial, and found increased level of organic carbon, nitrogen and phosphorus in fertilized soil, clearly pointing to highly nutritious soil.

9. Conclusions

Enzymes are key players in a plethora of biological processes, plant nutrition is no exception. Depending on the soil content and residues that it carries, different enzymes play key roles in rendering soil nutrient-rich; an immediate application is the composting process and PGPR-soil-plant interactions. This important area calls for further research not only on the plant side but also on the enzyme side and more importantly on applications to specific soil types. This knowledge will further facilitate decreased use of chemical fertilizers and will create avenues for organic farming practices.

Author details

Metin Turan[1]*, Emrah Nikerel[1], Kerem Kaya[1], Nurgul Kitir[1], Adem Gunes[2], Negar Ebrahim Pour Mokhtari[3], Şefik Tüfenkçi[4], M. Rüştü Karaman[5] and K. Mesut Çimrín[6]

*Address all correspondence to: m_turan25@hotmail.com

1 Genetics and Bioengineering Department, Engineering Faculty, Yeditepe University, Istanbul, Turkey

2 Soil and Plant Nutrition Science, Agricultural Faculty, Erciyes University, Kayseri, Turkey

3 Organic Farming Department, Islahiye Vocational School, Gaziantep University, Gaziantep, Turkey

4 Department of Agricultural Faculty, Bisystem Enginering, Yüzüncüyıl University, Van, Turkey

5 Vocational High School of Medical and Aramatic, Afyon Kocatepe University, Afyon

6 Department of Soil Science and Plant Nutrition, Faculty of Agriculture, Mustfa Kemal University, Hatay, Turkey

References

[1] Vaughan D, Malcolm RE. Soil Organic Matter and Biological Activity, Developments in Plant and Soil Sciences, Matinus Nijhoff/Dr. W. Junk, Springer, Netherlands, 1985, Vol. 16, ISBN 978-94-009-5105-1.

[2] Kindle KL. Characteristics and production of thermostable a-amylase, Appl.Biochem. Biotechnol. 1983; 8: 153–170.

[3] Allam AM, Hussein AM, Ragab AM. Amylase of the thermophilic actinomycete *Thermomonospora vulgaris*, Z. Allg. Mikrobiol. 1975; 15(6): 393–398.

[4] Niziolek S. Beta-amylase production by some *Bacillus cereus, Bacillus megaterium* and *Bacillus polymyxa* strains, Acta Microbiol. Pol. 1997; 46: 357–362.

[5] Arai M, Sumida M, Nakatani S, Murao S. A novel beta-amylase inhibitor, agric, Biol. Chem. 1983; 47: 183–185.

[6] Yu HY, Li X. Characterization of an organic solvent-tolerant thermostable glucoamylase from a halophilic isolate, *Halolactibacillus* sp. SK71 and its application in raw starch hydrolysis for bioethanol production, Biotechnol. Prog. 2014; 30: 1262–1268.

[7] Kumar P, Satyanarayana T. Microbial glucoamylases: characteristics and applications, Crit. Rev. Biotechnol. 2009; 29: 225–255.

[8] Ichikawa K, Tonozuka T, Mizuno M, Tanabe Y, Kamitori S, Nishikawa A, Sakano Y, Crystallization and preliminary X-ray analysis of *Thermoactinomyces vulgaris* R-47 maltooligosaccharide-metabolizing enzyme homologous to glucoamylase, Acta Crystallogr. Sect. F. 2005; 61: 302–304.

[9] Uchino F. A thermostable and unusually acidophilic amylase produced by a thermophilic acidophilic *Bacillus* sp., agric, Biol. Chem. 1982; 46: 7–13.

[10] Burhan A, Nisa U, Gokhan C, Omer C, Ashabil A, Osman G. Enzymatic properties of a novel thermostable, thermophilic, alkaline and chelator-resistant amylase from an alkaliphilic *Bacillus* sp. isolate ANT-6, Proc. Biochem. 2003; 38: 1397–1403.

[11] Ingle MB, Erikson RJ. Bacterial alpha-amylases, Adv. Appl. Microbiol. 1978; 24: 257–278.

[12] Pancha I, Jain D, Shrivastav A, Mishra SK, Shethia B, Mishra SVPM, Jha B. A thermoactive alpha-amylase from a *Bacillus* sp. isolated from CSMCRI salt farm, Int. J. Biol. Macromol. 2010; 47: 288–291.

[13] Konsoula Z, Liakopoulou-Kyriakides M, Perysinakis A, Chira P, Afendra A, Drainas C, Kyriakidis DA. Heterologous expression of a hyperthermophilic alpha-amylase in xanthan gum producing *Xanthomonas campestris* cells, Appl. Biochem. Biotechnol. 2008; 149: 99–108.

[14] Murakami S, Nishimoto H, Toyama Y, Shimamoto E, Takenaka S, Kaulpiboon J, Prousoontorn M, Limpaseni T, Pongsawasdi P, Aoki K. Purification and characterization of two alkaline, thermotolerant alpha-amylases from *Bacillus halodurans* 38C-2-1 and expression of the cloned gene in *Escherichia coli*, Biosci. Biotechnol. Biochem. 2007; 71: 2393–2401.

[15] Shen GJ, Saha BC, Lee YE, Bhatnagar L, Zeikus JG. Purification and characterization of a novel thermostable beta-amylase from *Clostridium thermosulphurogenes*, Biochem. J. 1988; 254: 835–840.

[16] Krishnan T, Chandra AK. Purification and characterization of alpha-amylase from *Bacillus licheniformis* CUMC305, Appl. Environ. Microbiol. 1983; 46: 430–437.

[17] Gangadharan D, Nampoothiri KM, Sivaramakrishnan S, Pandey A. Biochemical characterization of raw-starch-digesting alpha amylase purified from *Bacillus amyloliquefaciens*, Appl. Biochem. Biotechnol. 2009: 158: 653–662.

[18] Hmidet N, Maalej H, Haddar A, Nasri M. A Novel alpha-amylase from *Bacillus mojavensis* A21: Purification and biochemical characterization, Appl. Biochem. Biotechnol. 2010; 162: 1018–1030.

[19] Das K, Doley R, Mukherjee AK. Purification and biochemical characterization of a thermostable, alkaliphilic, extracellular alpha-amylase from *Bacillus subtilis* DM-03, a strain isolated from the traditional fermented food of India, Biotechnol. Appl. Biochem. 2004; 40: 291–298.

[20] Ray RR. Purification and characterization of extracellular beta-amylase of *Bacillus megaterium* B6, Acta Microbiol. Immunol. Hung. 2000; 47: 29–40.

[21] Bealin-Kelly FJ, Kelly CT, Fogarty WM. The alpha-amylase of the caldoactive bacterium *Bacillus caldovelox*, Biochem. Soc. Trans. 1990; 18: 310–311.

[22] Dheeran P, Kumar S, Jaiswal YK, Adhikari DK. Characterization of hyperthermostable alpha-amylase from *Geobacillus* sp. IIPTN, Appl. Microbiol. Biotechnol. 2010; 86: 1857–1866.

[23] Hewitt CJ, Solomons GL. The production of s-amylase (E.C.3.2.1.1.) by Bacillus amylofiquefaciens, in a complex and a totally defined synthetic culture medium, J. Ind. Microbiol. 1996; 17: 96–99.

[24] Meers JL. The regulation of alpha-amylase production in *Bacillus licheniformis*, Anton. Leeuw. 1972; 38: 585–590.

[25] Srivastava, RAK, Baruah, JN. Culture conditions for production of thermostable amylase by *Bacillus stearothermophilus*. Applied and Environmental Microbiology, 1986, 52(1), 179–184.

[26] Dey G, Mitra A, Banerjee R, Maiti BR. Enhanced production of amylase by optimization of nutritional constituents using response surface methodology, Biochem. Eng. J. 2001; 7: 227–231.

[27] Gangadharan D, Sivaramakrishnan S, Nampoothiri KM, Pandey A. Solid culturing of *Bacillus amyloliquefaciens* for alpha amylase production, Food Technol. Biotechnol. 2006; 44 (2): 269–274.

[28] Koksharov M, Lv C, Zhai X, Ugarova N, Huang E. *Bacillus subtilis* alkaline phosphatase IV acquires activity only late at the stationary phase when produced in *Escherichia coli*. Overexpression and characterization of the recombinant enzyme, Protein Expr. Purif. 2013; 90: 186–194.

[29] Yang K, Metcalf WW. A new activity for an old enzyme: *Escherichia coli* bacterial alkaline phosphatase is a phosphite-dependent hydrogenase, Proc. Natl. Acad. Sci. USA, 2004; 101: 7919–7924.

[30] Huang CT, Xu KD, McFeters GA, Stewart PS. Spatial patterns of alkaline phosphatase expression within bacterial colonies and biofilms in response to phosphate starvation, Appl. Environ. Microbiol. 1998; 64: 1526–1531.

[31] Tominaga N, Mori T. A sulfate-dependent acid phosphatase of *Thiobacillus thiooxidans*. Its partial purification and some properties, J. Biochem. 1974; 76: 397–408.

[32] Dassa E, Cahu M, Desjoyaux-Cherel B, Boquet PL. The acid phosphatase with optimum pH of 2.5 of *Escherichia coli*. Phys. Biochem. Study J. Biol. Chem. 1982; 257: 6669–6676.

[33] Magboul AAA, McSweeney PLH. Purification and properties of an acid phosphatase from *Lactobacillus curvatus* DPC2024, Int. Dairy J. 2000; 9: 849–855.

[34] Hachimori A, Takeda A, Kaibuchi M, Ohkawara N, Samejima T. Purification and characterization of inorganic pyrophosphatase from *Bacillus stearothermophilus*, J. Biochem. 1975; 77: 1177–1183.

[35] Klemme JH, Klemme B, Gest H. Catalytic properties and regulatory diversity of inorganic pyrophosphatases from photosynthetic bacteria, J. Bacteriol. 1971; 108: 1122–1128.

[36] Touati E, Danchin A. Cloning and characterization of the pH 2.5 acid phosphatase gene, appA cyclic AMP mediated negative regulation, Mol. Gen. Genet. 1987; 208: 499–505.

[37] Li J, Xu L, Yang F. Expression and characterization of recombinant thermostable alkaline phosphatase from a novel thermophilic bacterium *Thermus thermophilus* XM, Acta Biochim. Biophys. Sin. 2007; 39: 844–850.

[38] Dhaked RK, Alam SI, Dixit A, Singh L. Purification and characterization of thermolabile alkaline phosphatase from an antarctic psychrotolerant *Bacillus* sp. P9, Enzyme Microb. Technol. 2005; 36: 855–861.

[39] Dong G, Zeikus JG. Purification and characterization of alkaline phosphatase from *Thermotoga neapolitana*, Enzyme Microb. Technol. 1997; 21: 335–340.

[40] Wende A, Johansson P, Vollrath R, Dyall-Smith M, Oesterhelt D, Grininger M. Structural and biochemical characterization of a halophilic archaeal alkaline phosphatase, J. Mol. Biol. 2010; 400: 52–62.

[41] Pilar MC, Ortega N, Perez-Mateos M, Busto MD. Alkaline phosphatase-polyresorcinol complex: characterization and application to seed coating, J. Agric. Food Chem. 2009; 57: 1967–1974.

[42] Wang R, Ohtani K, Wang Y, Yuan Y, Hassan S, Shimizu T. Genetic and biochemical analysis of a class C non-specific acid phosphatase (NSAP) of *Clostridium perfringens*, Microbiology. 2010; 156: 167–173.

[43] Lahti R, Niemi T. Purification and some properties of inorganic pyrophosphatase from *Streptococcus faecalis*, J. Biochem. 1981; 90: 79–85.

[44] Streit TM, Borazjani A, Lentz SE, Wierdl M, Potter PM, Gwaltney SR, Ross MK. Evaluation of the side door in carboxylesterase-mediated catalysis and inhibition, Biol. Chem. 2008; 389: 149–162.

[45] Sanishvili R, Yakunin AF, Laskowski RA, Skarina T, Evdokimova E, Doherty-Kirby A, Lajoie GA, Thornton JM, Arrowsmith CH, Savchenko A, Joachimiak A, Edwards AM. Integrating structure, bioinformatics, and enzymology to discover function. BioH, a new carboxylesterase from *Escherichia coli*, J. Biol. Chem. 2003; 278: 26039–26045.

[46] Navarro-Gonzalez I, Sanchez-Ferrer A, Garcia-Carmona F. Molecular characterization of a novel arylesterase from the wine-associated acetic acid bacterium *Gluconobacter oxidans* 621H, J. Agric. Food Chem. 2012; 60: 10789–10795.

[47] Fenster KM, Parkin KL, Steele JL. Nucleotide sequencing, purification, and biochemical properties of an arylesterase from *Lactobacillus casei* LILA, J. Dairy Sci. 2003; 86: 2547–2557.

[48] Sharma R, Chisti Y, Banerjee UC. Production, purification, characterization, and applications of lipases, Biotechnol. Adv. 2001; 19: 627–662.

[49] Hasan F, Shah AA, Hameed A. Purification and characterization of a mesophilic lipase from *Bacillus subtilis* FH5 stable at high temperature and pH, Acta Biol. Hung., 2007; 58: 115-–132.

[50] Rathi P, Bradoo S, Saxena RK, Gupta R. A hyper-thermostable, alkaline lipase from *Pseudomonas* sp. with the property of thermal activation, Biotechnol. Lett. 2000; 22: 495–498.

[51] Ogino H, Hiroshima S, Hirose S, Yasuda M, Ishimi K, Ishikawa H. Cloning, expression and characterization of a lipase gene (lip3) from *Pseudomonas aeruginosa* LST-03, Mol. Genet. Genomics. 2004; 271: 189–196.

[52] Karadzic I, Masui A, Zivkovic LI, Fujiwara N. Purification and characterization of an alkaline lipase from *Pseudomonas aeruginosa* isolated from putrid mineral cutting oil as component of metalworking fluid, J. Biosci. Bioeng. 2006; 102: 82–89.

[53] Zhang J, Lin S, Zeng R. Cloning, expression, and characterization of a cold-adapted lipase gene from an antarctic deep-sea psychrotrophic bacterium, *Psychrobacter* sp. 7195, J. Microbiol. Biotechnol. 2007; 17: 604–610.

[54] Uttatree S, Winayanuwattikun P, Charoenpanich J. Isolation and characterization of a novel thermophilic-organic solvent stable lipase from acinetobacter baylyi, Appl. Biochem. Biotechnol. 2010; 162:1362–1376.

[55] Gatti-Lafranconi P, Caldarazzo SM, Villa A, Alberghina L, Lotti M. Unscrambling thermal stability and temperature adaptation in evolved variants of a cold-active lipase, FEBS Lett. 2008; 582: 2313–2318.

[56] Jovel SR, Kumagai T, Danshiitsoodol N, Matoba Y, Nishimura M, Sugiyama M. Purification and characterization of the second *Streptomyces phospholipase* A2 refolded from an inclusion body, Protein Expr. Purif. 2006; 50:82–88.

[57] Jaeger KE, Kharazmi A, Hoiby N. Extracellular lipase of *Pseudomonas aeruginosa*: biochemical characterization and effect on human neutrophil and monocyte function in vitro, Microb. Pathog. 1991; 10: 173–182.

[58] Goullet P, Picard B, Laget PF. Purification and properties of carboxylesterase B of *Escherichia coli*, Ann. Microbiol. 1984; 135: 375–387.

[59] Maqbool QU, Johri S, Rasool S, Riyaz-ul-Hassan S, Verma V, Nargotra A, Koul S, Qazi, GN. Molecular cloning of carboxylesterase gene and biochemical characterization of encoded protein from *Bacillus subtilis* (RRL BB1), J. Biotechnol. 2006; 125: 1–10.

[60] Riefler JF, Higerd TB. Characterization of intracellular esterase A from *Bacillus subtilis*, Biochim. Biophys. Acta. 1976; 429: 191–197.

[61] Pesaresi A, Devescovi G, Lamba D, Venturi V, Degrassi G. Isolation, characterization, and heterologous expression of a carboxylesterase of *Pseudomonas aeruginosa* PAO1, Curr. Microbiol. 2005; 50: 102–109.

[62] Ruiz, C., Blanco, A., Pastor, F.I., Diaz, P., 2002, Analysis of *Bacillus megaterium* lipolytic system and cloning of LipA, a novel subfamily I.4 bacterial lipase, FEMS Microbiol. Lett. 217: 263–267.

[63] Lee DW, Kim HW, Lee KW, Kim BC, Choe EA, Lee HS, Kim DS, Pyun YR. Purification and characterization of two distinct thermostable lipases from the gram-positive thermophilic bacterium *Bacillus thermoleovorans* ID-1, Enzyme Microb. Technol. 2001; 29: 363–371.

[64] Chen S, Qian L, Shi B. Purification and properties of enantioselective lipase from a newly isolated *Bacillus cereus* C71, Process Biochem. 2007; 42: 988–994.

[65] Lescic I, Vukelic B, Majeric-Elenkov M, Saenger W, Abramic M. Substrate specificity and effects of water-miscible solvents on the activity and stability of extracellular lipase from *Streptomyces rimosus*, Enzyme Microb. Technol. 2001; 29: 548–553.

[66] Sekhon A, Dahiya N, Tiwari RP, Hoondal GS. Properties of a thermostable extracellular lipase from *Bacillus megaterium* AKG-1, J. Basic Microbiol. 2005; 45: 147–154.

[67] Henderson C. A Study of the Lipase Produced by *Anaerovibrio lipolytica*, a rumen bacterium, J. Gen. Microbiol. 1971; 65: 81–89.

[68] Kumar S, Kikon K, Upadhyay A, Kanwar SS, Gupta R. Production, purification, and characterization of lipase from thermophilic and alkaliphilic *Bacillus coagulans* BTS-3, Protein Exp. Purif. 2005; 41: 38–44.

[69] Dong H, Gao S, Han S, Cao S. Purification and characterization of a *Pseudomonas* sp. lipase and its properties in non-aqueous media, Biotechnol. Appl. Biochem. 1999; 30: 251–256.

[70] Vohra A, Satyanarayana T. Phytases: Microbial sources, production, purification, and potential biotechnological applications, Crit. Rev. Biotechnol. 2003; 23(1):29–60.

[71] Mijakovic I, Musumeci L, Tautz L, Petranovic D, Edwards RA, Jensen PR, Mustelin T, Deutscher J, Bottini N. In vitro characterization of the *Bacillus subtilis* protein tyrosine phosphatase YwqE, J. Bacteriol. 2005; 187, 3384–3390.

[72] Noeren-Mueller A, Wilk W, Saxena K, Schwalbe H, Kaiser M, Waldmann H. Discovery of a new class of inhibitors of *Mycobacterium tuberculosis* protein tyrosine phosphatase B by biology-oriented synthesis, Angew. Chem. 2008; 47: 5973–5977.

[73] Vega C, Chou S, Engel K, Harrell ME, Rajagopal L, Grundner C. Structure and substrate recognition of the *Staphylococcus aureus* protein tyrosine phosphatase PtpA, J. Mol. Biol. 2011; 413: 24–31.

[74] Guerrero-Olazaran M, Rodriguez-Blanco L, Carreon-Trevino J, Gallegos-Lopez J, Viader-Salvado J. Expression of a *Bacillus phytase* C gene in *Pichia pastoris* and properties of the recombinant enzyme, Appl. Environ. Microbiol. 2010; 76: 5601–5608.

[75] Hong S, Chu I, Chung K. Purification and biochemical characterization of thermostable phytase from newly isolated *Bacillus subtilis* CF92, J. Appl. Biol. Chem. 2011; 54: 89–94.

[76] Dvorakova J. Phytase: source, preparation and exploitation, Folia Microbiol. 1998; 43: 323–338.

[77] Choi YM, Suh HJ, Kim JM. Purification and properties of extracellular phytase from *Bacillus* sp. KHU-10, J. Protein Chem. 2001; 20: 287–292.

[78] Li Z, Huang H, Yang P, Yuan T, Shi P, Zhao J, Meng K, Yao B. The tandemly repeated domains of a beta-propeller phytase act synergistically to increase catalytic efficiency, FEBS J. 2011; 278: 3032–3040.

[79] Shim JH, Oh BC. Characterization and application of calcium-dependent beta-propeller phytase from *Bacillus amyloliquefaciens* DS11, J. Agric. Food Chem. 2012; 60: 7532–7537.

[80] Tambe SM, Kaklij GS, Kelkar SM, Parekh LJ. Two distinct molecular forms of phytase from *Klebsiella aerogenes*: evidence for unusually small active enzyme peptide, J. Ferment. Bioeng. 1994; 77: 23–27.

[81] Wang Q, Fu S, Sun J, Weng X. Characterization of a thermostable alkaline phytase from *Bacillus licheniformis* ZJ-6 in *Pichia pastoris*, World J. Microbiol. Biotechnol. 2011; 27: 1247–1253.

[82] Krajewska B. Ureases I. Functional, catalytic and kinetic properties: a review, J. Mol. Catal. B: Enzym. 2009; 59:9–21.

[83] Nakano M, Iida T, Honda T. Urease activity of enterohaemorrhagic *Escherichia coli* depends on a specific one-base substitution in ureD, Microbiology. 2004; 150: 3483–3489.

[84] Dixon NE, Gazzola C, Blakeley RL, Zerner B. Metal ions in enzymes using ammonia or amides, Science. 1976; 191:1144–1150.

[85] Ramesh R, Aarthy M, Gowthaman MK, Gabrovska K, Godjevargova T, Kamini NR. Screening and production of a potent extracellular *Arthrobacter creatinolyticus* urease for determination of heavy metal ions, J. Basic Microbiol. 2013; 1–11.

[86] Zotta T, Ricciardi A, Rossano R, Parente E. Urease production by *Streptococcus thermophilus*, Food Microbiol. 2008; 25:113–119.

[87] Kakimoto S, Miyashita H, Sumino Y, Akiyama S. Properties of acid ureases from *Lactobacillus* and *Streptococcus* strains, Agric. Biol. Chem. 1990; 54: 381–386.

[88] del Mar Dobao M, Castillo F, Pineda M. Characterization of urease from the phototrophic bacterium *Rhodobacter capsulatus* E1F1, Curr. Microbiol. 1993; 27:119–123.

[89] Booth NJ, Beekman JB, Thune RL. *Edwardsiella ictaluri* encodes an acid-activated urease that is required for intracellular replication in channel catfish (*Ictalurus punctatus*) macrophages, Appl. Environ. Microbiol. 2009; 75:6712–6720.

[90] Evans DJ, Evans DG, Kirkpatrick SS, Graham DY. Characterization of the *Helicobacter pylori* urease and purification of its subunits, Microb. Pathog. 1991; 10:15–26.

[91] Contreras-Rodriguez A, Quiroz-Limon J, Martins AM, Peralta H, Avila-Calderon E, Sriranganathan N, Boyle SM, Lopez-Merino A. Enzymatic, immunological and phylogenetic characterization of *Brucella suis* urease, BMC Microbiol. 2008; 8: 121.

[92] Chen YYM, Clancy KA, Burne RA. *Streptococcus salivarius* urease: genetic and biochemical characterization and expression in a dental plaque streptococcus, Infect. Immun. 1996; 64: 585–592.

[93] Andrich L, Esti M, Moresi M. Urea degradation kinetics in model wine solutions by acid urease immobilised onto chitosan-derivative beads of different sizes, Enzyme Microb. Technol. 2010; 46: 397–405.

[94] Yang Y, Kang Z, Zhou J, Chen J, Du G. High-level expression and characterization of recombinant acid urease for enzymatic degradation of urea in rice wine, Appl. Microbiol. Biotechnol. 2014; 99: 301–308.

[95] Mobley HLT, Hausinger RP. Microbial Ureases: Significance, Regulation, and Molecular Characterization, Microbiol. Rev. 1989; 5(1): 85–108.

[96] Sokmen BB, Hasdemir B, Yusufoglu A, Yanardag R. Some monohydroxy tetradecanoic acid isomers as novel urease and elastase inhibitors and as new antioxidants, Appl. Biochem. Biotechnol. 2014; 172: 1358–1364.

[97] Khan WN, Lodhi MA, Ali I, Azhar-Ul-Haq I, Malik A, Bilal S, Gul R, Choudhary MI. New natural urease inhibitors from Ranunculus repens, J. Enzyme Inhib. Med. Chem. 2006; 21: 17–19.

[98] Xiao ZP, Ma TW, Fu WC, Peng XC, Zhang AH, Zhu HL. The synthesis, structure and activity evaluation of pyrogallol and catechol derivatives as *Helicobacter pylori* urease inhibitors, Eur. J. Med. Chem. 2010; 45: 5064–5070.

[99] Perez-Perez GI, Gower CB, Blaser MJ. Effects of cations on *Helicobacter pylori* urease activity, release, and stability, Infect. Immun. 1994; 62: 299–302.

[100] Gang JG, Yun SK, Hwang SY. *Helicobacter pylori* urease may exist in two forms: evidence from the kinetic studies, J. Microbiol. Biotechnol. 2009; 19:1565–1568.

[101] Dixon NE, Gazzola C, Blakeley RL, Zerner B. Jack Bean Urease (EC 3.5.1.5). A Metalloenzyme. A Simple Biological Role for Nickel?, J. American Chem. Soc. 1975; 97: 4131–4133.

[102] Robson LM, Chambliss GH. Cellulases of bacterial origin, Enzyme Microb. Technol. 1989; 11: 626–644.

[103] McCleary BV, Mangan D, Daly R, Fort S, Ivory R, McCormack N. Novel substrates for the measurement of endo-1,4-beta-glucanase (endo-cellulase), Carbohydr. Res. 2014; 385: 9–17.

[104] Yamane K, Suzuki H. Cellulases of *Pseudomonas fluorescens* var. cellulosa, Methods Enzymol. 1988; 160: 200–210.

[105] Mawadza C, Hatti-Kaul R, Zvauya R, Mattiasson B. Purification and characterization of cellulases produced by two *Bacillus* strains, J. Biotechnol. 2000; 83: 177–187.

[106] Au KS, Chan KY. Purification and properties of the endo-1,4-beta-glucanase from *Bacillus subtilis*, J. Gen. Microbiol. 1987; 133: 2155–2162.

[107] Saleem M, Akhtar MS, Yasmin R, Zahid M, Malik NN, Afzal M, Rajoka MI. Production, purification and characterization of beta-1,4-endoglucanase from a novel bacterial strain CTP-09 of a *Bacillus* sp, Protein Pept. Lett. 2008; 15: 402–410.

[108] Fukumori F, Kudo T, Horikoshi K. Purification and properties of a cellulase from alkalophilic *Bacillus* sp. No. 1139, J. Gen. Microbiol. 1985; 131: 3339–3345.

[109] Ximenes E, Kim Y, Mosier N, Dien B, Ladisch M. Inhibition of cellulases by phenols, Enzyme Microb. Technol. 2010; 46: 170–176.

[110] Ogura J, Toyoda A, Kurosawa T, Chong AL, Chohnan S, Masaki T. Purification, characterization, and gene analysis of cellulase (Cel8A) from *Lysobacter* sp. IB-9374, Biosci. Biotechnol. Biochem. 2006; 70: 2420–2428.

[111] Hirasawa K, Uchimura K, Kashiwa M, Grant WD, Ito S, Kobayashi T, Horikoshi K. Salt-activated endoglucanase of a strain of alkaliphilic *Bacillus agaradhaerens*, Anton. Leeuw. 2006; 89: 211–219.

[112] Li YH, Ding M, Wang J, Xu G, Zhao F. A novel thermoacidophilic endoglucanase, Ba-EGA, from a new cellulose-degrading bacterium, *Bacillus* sp.AC-1, Appl. Microbiol. Biotechnol. 2006; 70: 430–436.

[113] Nizamudeen, S., & Bajaj, B. K. (2009). A novel thermo-alkalitolerant endoglucanase production using cost-effective agricultural residues as substrates by a newly isolated Bacillus sp. NZ. Food technology and Biotechnology, 47(4), 435–440.

[114] Krishna C. Production of bacterial cellulases by solid state bioprocessing of banana wastes, Bioresour. Technol. 1999; 69: 231–239.

[115] Hankin L, Poincelot RP, Anagnostakis SL. Microorganisms from composting leaves: ability to produce extracellular degradative enzymes, Microbial. Ecol. 1975; 2(4): 296–308.

[116] Tiquia SM, Wan HC, Tam NF. Microbial population dynamics and enzyme activities during composting, Compost Sci. Util. 2002; 10(2): 150–161.

[117] Garcia C, Hernandez T, Costa C, Ceccanti B, Masciandaro G, Ciardi C. A study of biochemical parameters of composted and fresh municipal wastes, Bioresour. Technol. 1993; 44(1): 17–23.

[118] Mondini C, Fornasier F, Sinicco T. Enzymatic activity as a parameter for the characterization of the composting process, Soil Bio. Biochem. 2004; 36(10): 1587–1594.

[119] Margesin R, Cimadom J, Schinner F. Biological activity during composting of sewage sludge at low temperatures, Int. Biodeterior. Biodegradation. 2006; 57(2): 88–92.

[120] Herrmann RF, Shann JR. Enzyme activities as indicators of municipal solid waste compost maturity, Compost Sc. Util. 1993; 1(4): 54–63.

[121] Lee YJ, Kim BK, Lee BH, Jo KI, Lee NK, Chung CH, Lee YC, Lee JW. Purification and characterization of cellulase produced by *Bacillus amyloliquefaciens* DL-3 utilizing rice hull, Biores. Technol. 2008; 99: 378–386.

[122] Gea T, Ferrer P, Alvaro G, Valero F, Artola A, Sánchez A. Co-composting of sewage sludge: fats mixtures and characteristics of the lipases involved. Biochem. Eng. J. 2007; 33(3): 275–283.

[123] Krzywy-Gawrońska E. Enzymatic activity of urease and, degydrogenase in soil fertilized with GWDA compost with or without a PRP SOL addition, Pol. J. Environ. Studs. 2012; 21(4): 949–955.

Chapter 11

Enzyme Inhibitors and Activators

Olga D. Lopina

Additional information is available at the end of the chapter

http://dx.doi.org/10.5772/67248

Abstract

Enzymes are very effective biological catalysts that accelerate almost all metabolic reactions in living organisms. Enzyme inhibitors and activators that modulate the velocity of enzymatic reactions play an important role in the regulation of metabolism. Enzyme inhibitors are also useful tool for study of enzymatic reaction as well as for design of new medicine drugs. In this chapter, we focused on the properties of enzyme inhibitors and activators. Here we present canonical inhibitor classification based on their kinetic behavior and mechanism of action. We also considered enzyme inhibitors that were used for design of various types of pharmacological drugs and natural inhibitors as a plausible source for design of future drugs. Mechanisms of action of enzyme activators and some features of allosteric modulators are considered.

Keywords: enzyme conformational mobility, classification of enzyme inhibitors, enzyme activators and inhibitors, mechanism of action

1. Introduction

Enzymes (E) is a group of biologically active polymers (mainly proteins) that catalyze almost all metabolic reactions in all living organisms. Enzymes are able to accelerate chemical reaction dividing it into separate steps. Because each step of enzymatic reaction has a value of activation energy significantly lower than the value of activation energy for the same chemical reaction, enzymes can increase a rate of reaction 10^6–10^{18} folds. According to contemporary hypothesis, high conformational mobility of the enzymes allows them to adopt their active sites to substrate(s) and intermediates of the reaction in the best way [1, 2]. Multiple conformers of enzymes with close values of free energy preexist in the solution simultaneously. Along the reaction way, a conformer is picked out, the structure of which can stabilize definite intermediate that makes a reaction more thermodynamically profitable [3].

This hypothesis is supported by unsuccessful attempts to create catalytically effective low molecular enzymes having needed active site (molecular mass should be higher than 10,000 Da) or enzyme with correct active site but with restricted conformational mobility (antibodies with needed active site, so-called *abzymes* [4]). In the context of described hypothesis, enzyme inhibitors being bound to the enzyme "freeze" it in definite conformation that makes impossible selection of conformers participating in numerous steps of enzymatic reaction. By this way, inhibitors stop enzymatic reaction.

On the other hand, the binding of enzyme activators may lead to the creation of more profitable conformers that can be more effective in carrying out definite steps of the reaction. Therefore, they will accelerate enzymatic reaction. Taking into account this information about enzymes in this chapter, we consider contemporary knowledge about enzyme inhibitors and activators.

2. Enzyme inhibitors

2.1. Definition, classification, and main properties

Enzymes are different chemical compounds that are combined into a group because of their only feature—they can suppress enzyme activity. The suppression of the activity is the result of the binding of inhibitor to the enzyme molecule that arrests catalytic reaction. Because enzymes catalyze most part of chemical reactions in living organisms, the enzyme inhibitors play an important role in the development of different sciences (biochemistry, physiology, pharmacy, agriculture, ecology) as well as the technologies (production of pharmaceutical drugs, insecticides, pesticides, chemical weapons, etc.).

Many pharmacological drugs are enzyme inhibitors. The group of well-known pharmaceutical agents with name nonsteroidal antiinflammatory drugs (NSAIDs) includes inhibitors of enzyme cyclooxygenase that catalyzes a first step of synthesis of biologically active compounds prostaglandins that are responsible for the development of pain, inflammation, fever, contraction of smooth muscle, formation of blood clots, and others [5].

All inhibitors may be combined in different groups in accordance with their chemical structure: *ions of metals* (Hg^{+}, Fe^{2+}, Cu^{+}, Pb^{2+}), *organic compounds* (e.g., N-ethylmaleimide, diisopropyl phosphofluoridate, oligomycin), and *large bioorganic molecules*, (peptides, proteins, etc). However, this classification does not reflect mechanism of their interaction with enzyme.

In accordance with the mode of action, enzyme inhibitors may be divided into two different groups (*reversible* and *irreversible* inhibitors). Reversible inhibitors, in turn, may be combined in four groups in accordance with kinetic behavior (competitive, uncompetitive, noncompetitive, and mixed inhibitors) [6].

The mechanism of action of enzyme inhibitors includes a step of enzyme-inhibitor complex formation (EI complex) that has no (or low) enzyme activity. An *irreversible* inhibitor dissociates from this complex very slow because it is tightly bound to the enzyme. Mainly this mode of inhibition is connected with the formation of covalent bond or hydrophobic interaction between enzyme and inhibitor. Irreversible inhibitors usually react with the enzyme and change it chemically. These inhibitors often contain reactive functional groups that modify amino acid residues of enzyme that are essential for its activity. They also can provide inhibition affecting the enzyme conformation. An example of irreversible inhibitor is N-ethylmaleimide that covalently interacts with SH-group of cysteine residues of enzyme molecules, like peptidase (insulin-degrading enzyme) [7], 3-phosphoglyceraldehyde dehydrogenase [8], or hydrophobic compound from group of cardiotonic steroids that at the last bind to Na,K-ATPase using hydrophobic interactions [9]. Another well-known irreversible inhibitor is diisopropyl phosphofluoridate that modifies OH-group of serine residue in active site of such enzymes as chymotrypsin and other serine proteases [10, 11] or acetylcholine esterase in cholinergic synapsis of the nervous system being a potent neurotoxin [12]. Inhibition of this enzyme causes an increase in the acetylcholine neurotransmitter concentration that results in muscular paralysis and death. Inhibitor of cyclooxygenase aspirin (acetyl salicylic acid) covalently modifies OH-group of serine residue located in a close proximity to the active site of cyclooxygenase [13].

Irreversible inhibition is different from irreversible enzyme inactivation. Irreversible inhibitors are generally specific for one class of enzymes and do not inactivate all proteins. In contrast to denature agents such as urea, detergents do not destroy protein structure but specifically alter the active site of the target enzyme.

Consequently because of tight binding, it is difficult to remove an irreversible inhibitor from the EI complex after its formation [14]. So, we can refer some chemical compound to irreversible enzyme inhibitor, if after the formation of EI complex, the dilution of it with significant amount of water (100–200 excess) does not restore enzyme activity.

Irreversible inhibitors display time-dependent loss of enzyme activity. Interaction of irreversible inhibitor with enzyme is a bimolecular reaction:

$$E + I \xrightarrow{k_I} EI, \tag{1}$$

where E is enzyme, I is inhibitor, EI is complex of enzyme-inhibitor, and *ki* is a constant of the velocity of this reaction.

However, usually the action of irreversible inhibitors is characterized by the constant of observed pseudo-first order reaction under conditions when concentration of inhibitor is significantly higher than concentration of the enzyme. The value of pseudo-first order rate of inhibition may be measured by plotting of the ln of enzyme activity (in % relatively enzyme activity in the absence of inhibitor) vs. time. Tangent of slope angle of straight line obtained by this way will be equal to value of constant of pseudo-first order inhibition. The value of rate constant of bimolecular reaction for irreversible inhibition may be then calculated by dividing the obtained value of constant of pseudo-first order reaction per inhibitor concentration.

Reversible inhibitor binds to the enzyme reversibly [6, 14]. It means that there is equilibrium between the formation and dissociation of EI complex:

$$E + I \underset{k_2}{\overset{k_1}{\leftrightarrow}} EI \tag{2}$$

where k_1 is a constant of the velocity of direct reaction and k_2 is a constant of the velocity of reverse reaction. The effect of reversible inhibitors is characterized by the constant of dissociation of EI complex that is equal to [E] [I]/[EI] or k_1/k_2.

Usually *reversible inhibitor* binds to the enzymes using non-covalent interactions such as hydrogen or ionic bonds. Different types of reversible inhibition are produced depending on whether these inhibitors bind to the enzyme, the enzyme-substrate complex, or both.

One type of reversible inhibition is called *competitive* inhibition. In this case, there are two types of complexes: enzyme inhibitor (EI) and enzyme substrate (ES); complex EI has no enzyme activity. The substrate and inhibitor cannot bind to the enzyme at the same time. This inhibition may be reversed by the increase of substrate concentration. However, the value of maximal velocity (Vmax) remains constant. The value of apparent Km will increase; however, the value of maximal velocity (Vmax) remains constant (**Figure 1**). It can be competitive inhibition not only in relation to substrate but also to cofactors, as well as to activators.

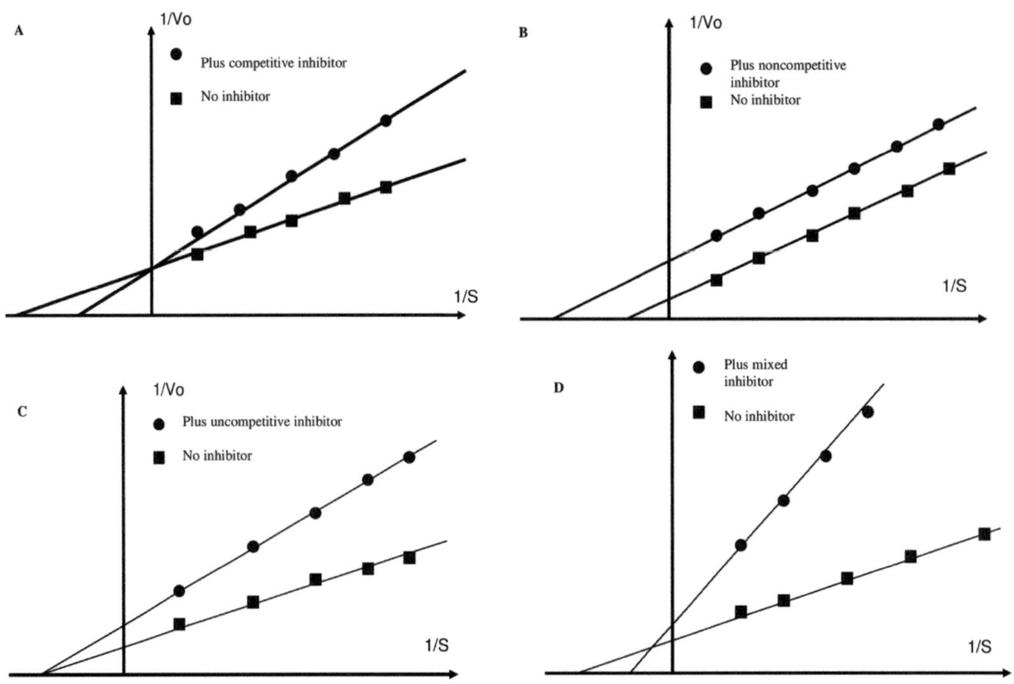

Figure 1. Kinetic test for reversible inhibitor classification. Double reciprocal plot ($1/V_o$) vs. (1/s) for competitive (A), uncompetitive (B), noncompetitive (C), and mixed (D) enzyme inhibition [14].

Another type of reversible inhibition is *uncompetitive* inhibition. In this case, the inhibitor binds only to the substrate-enzyme complex; it does not interfere with the binding of substrate with active site but prevents the dissociation of complex enzyme substrate: it resulted in the dependence of the inhibition only upon inhibitor concentration and its *Ki* value. This type of inhibition results in Vmax decrease and Km decrease (**Figure 1, B**).

http://dx.doi.org/10.5772/67248

The third type of inhibition is *noncompetitive*. This type of inhibition results in the inability of complex enzyme (E) inhbitor (I) substrate (EIS) to dissociate giving a product of reaction. In this case, inhibitor binds to E or to ES complex. The binding of the inhibitor to the enzyme reduces its activity but does not affect the binding of substrate. As a result, the extent of the inhibition depends only upon the concentration of the inhibitor. In this case, Vmax will decrease, but Km will remain the same (**Figure 1**, **C**).

In some cases, we can see *mixed* inhibition, when the inhibitor can bind to the enzyme at the same time as to enzyme-substrate complex. However, the binding of the inhibitor effects on the binding of the substrate and vice versa. This type of inhibition can be reduced, but not overcome by the increase of substrate concentrations. Although it is possible for mixed-type inhibitors to bind in the active site, this inhibition generally results from an allosteric effect of inhibitor (see below). An inhibitor of this kind will decrease Vmax, but it will increase Km (**Figure 1**, **C**).

Special case of enzyme inhibition is inhibition by the excess of substrate or by the product. This inhibition may follow the competitive, uncompetitive, or mixed patterns. Inhibition of enzyme by its substrate occurs when a dead-end enzyme-substrate complex forms. Often in the case of substrate inhibition, a molecule of substrate binds to active site in two points (e.g., by the "head" and by the "tail" of molecule). At high concentrations, two substrate molecules bind in active site the following manner: one substrate molecule binds using the "head" and another molecule using the "tail." This binding is nonproductive and substrate cannot be converted to the product (**Figure 2**). An example of such inhibition is inhibition of acetyl cholinesterase by the excess of acetylcholine [15].

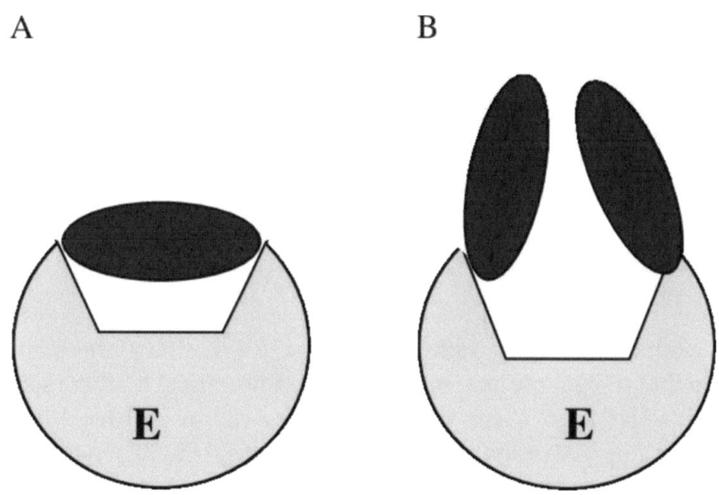

Figure 2. Enzyme inhibition by substrate. Productive binding of one substrate molecule with two points of enzyme active site (A) and unproductive binding of two substrate molecules with the same site (B).

Competitive inhibitors mainly interact with enzyme active site preventing binding of real substrate. Classical example of competitive inhibition is inhibition of fumarate hydratase by maleate that is a substrate analog (**Figure 3**). Enzyme is highly stereospecific; it catalyzes the hydration of the trans-double bound of fumarate but not maleate (cis-isomer of fumarate). Maleate binds to active site with high affinity preventing the binding of fumarate. Despite the binding maleate to active site, it cannot be converted into the product of reaction. However, maleate occupies active site making it inaccessible for real substrate and providing by this way the inhibition [16].

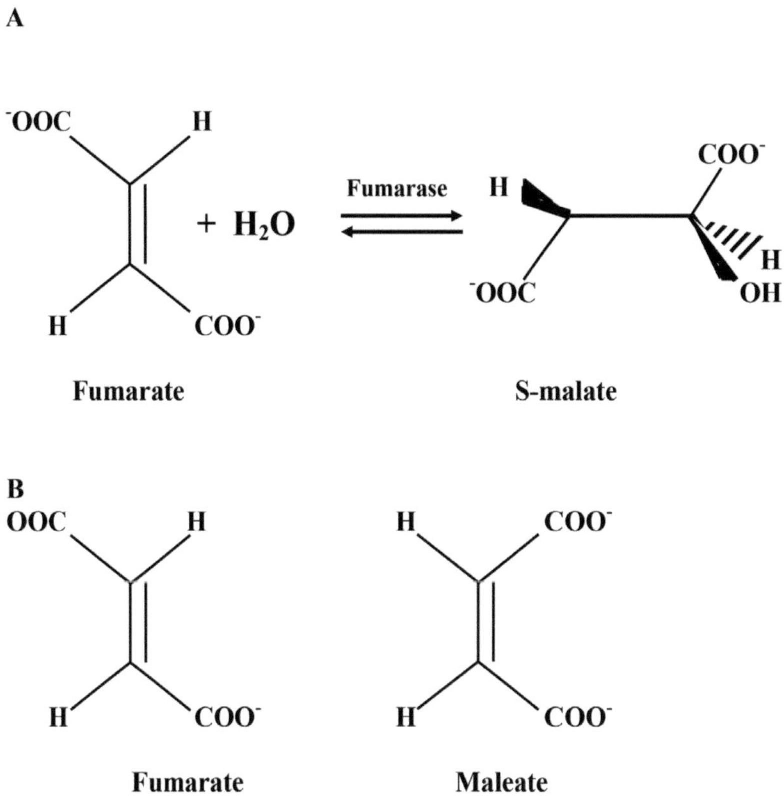

Figure 3. Example of enzyme competitive inhibitors. A reaction catalyzing by fumarate hydratase (A) and comparison of structure of fumarate (substrate of reaction) and maleate (enzyme competitive inhibitor) (B) [16].

Some reversible inhibitors bind so tightly to the enzyme that they are essentially irreversible. It is known that proteolytic enzymes of the gastrointestinal tract are secreted from the pancreas in an inactive form. Their activation is achieved by restricted trypsin digestion of proenzymes. To stop activation of proteolytic enzymes, the pancreas produces trypsin inhibitor. It is a small protein molecule (it consists of 58 amino acid residues) [17]. This inhibitor binds directly to trypsin active site with Kd value that is equal to 0.1 pM. The binding is almost irreversible; complex EI does not dissociate even in solution of 6 M urea. The inhibitor is a very effective analog of trypsin substrates; amino acid residue Lys-15 of inhibitor molecule interacts with aspartic residue located in a pocket of enzyme surface destined for substrate binding, thereby preventing its binding and conversion into the product (**Figure 4**).

http://dx.doi.org/10.5772/67248

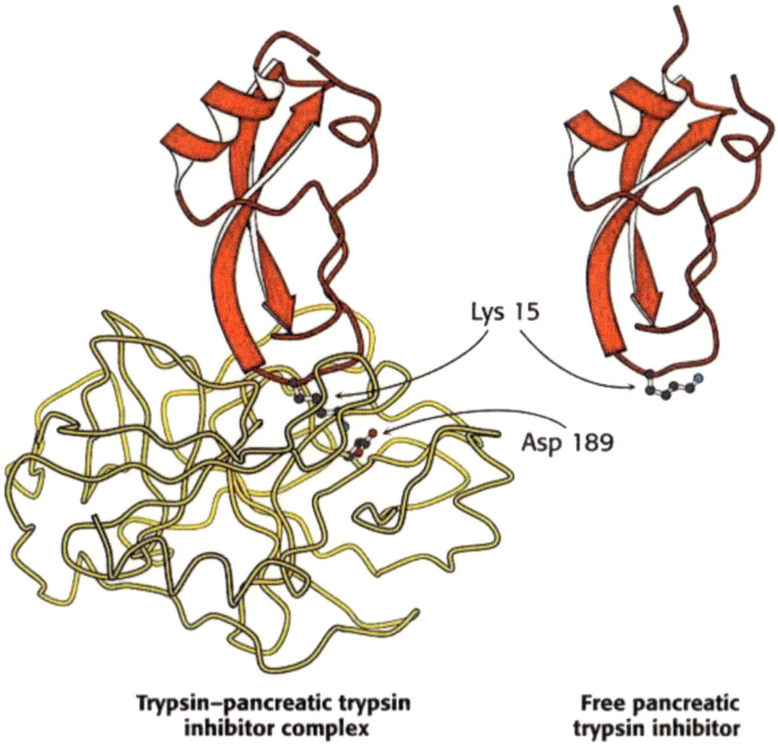

Figure 4. Structure of complex pancreatic trypsin inhibitor—trypsin and free trypsin inhibitor [17].

2.2. Irreversible inhibitors as a tool for study of enzymes: enzyme active sites labeling by irreversible inhibitors

To obtain information concerning the mechanism of enzyme reaction, we should determine functional groups that are required for enzyme activity and located in enzyme active site. First approach is to reveal a 3D structure of enzyme with bound substrate using X-ray crystallography. Alternative and/or additional approach is to use group-specific reagent that simultaneously is irreversible inhibitor of the enzyme. It can covalently bind to reactive groups of enzyme active site that allow to elucidate functional amino acid residues of the site. Modified amino acid residues may be found later after achievement of complete enzyme inhibition, enzyme proteolysis, and identification of labeled peptide(s).

Irreversible inhibitors that can be used with this aim may be divided into two groups: (1) group-specific reagents for reactive chemical groups and (2) substrate analogs with included functional groups that are able to interact with reactive amino acid residues. These compounds can covalently modify amino acids essential for activity of enzyme active site and in such a manner can label them.

One from the most known group-specific reagent that was used to label functional amino acid residue of enzyme active site of protease chymotrypsin was diisopropyl phosphofluoridate [18]. It modified only 1 from 28 serine residues of the enzyme. It means that this serine residue is very reactive. Location of Ser-195 in active site of chymotrypsin was confirmed in investigation carried out later, and the origin of its high reactivity was revealed. Diisopropyl phosphofluoridate was also successfully used for identification of a reactive serine residue in the active site of acetylcholinesterase [12].

To reveal reactive SH-group in active site of various enzymes, different SH-reagents were used, among them ^{14}C-labeled N-ethylmaleimide, iodoacetate, and iodoacetamide. Using these reagents, cysteines were revealed in the active sites of some dehydrogenase, cysteine protease, and other enzymes.

The second approach is the application of reactive substrate analogs. These compounds are structurally similar to the substrate but include chemically reactive groups, which can covalently bind to some amino acid residues. Substrate analogs are more specific than group-specific reagents. Tosyl-L-phenylalanine chloromethyl ketone, a substrate analog for chymotrypsin that is able to bind covalently with histidine residue and irreversibly inhibit enzyme, makes possible identification of Hys-57 in chymotrypsin active site [19].

2.3. Natural enzyme inhibitors

Many cellular enzyme inhibitors are proteins or peptides that specifically bind to and inhibit target enzymes. Numerous metabolic pathways are controlled by these specific compounds that are synthesized in organisms. Very interesting example of these inhibitors is protein serpins. It is a large family of proteins with similar structures. Most of them are inhibitors of chymotrypsin-like serine protease [20, 21].

Serine proteases (e.g., mentioned above chymotrypsin) possess a reactive serine residue in active site and have similar mechanisms of catalysis. Cleavage of peptide bond by these proteases is a two-step process. Reactive serine residue of the protease active site that looses H^+ and becomes nucleophilic one in the beginning of catalytic act attacks substrate peptide bond. This results in the release of new N-terminal part of protein substrate (first product) and in the formation of a covalent ester bond between the enzyme and the second part of substrate (see Ref. [16]). The second step of catalysis of usual substrates leads to the hydrolysis of ester bond and to the release of the second product (C-terminal part of protein substrate). If serpin is cleaved by a serine protease, it undergoes conformational transition before the hydrolysis of ester bond between enzyme and the second part of substrate (serpin). The change of serpin conformation leads to the "freezing" of intermediate (complex of enzyme with covalently attached second part of serpin is retained for several days) [21]. Therefore, serpins are irreversible inhibitors with unusual mechanism of action. They have named "suicide inhibitors," because each serpin molecule can inactivate a single molecule of protease and kills itself during the process of protease inhibition.

Considering enzyme inhibitors we should keep in mind that many living organisms are in the state of "chemical war." Fungi are fighting with bacteria for food using antibiotics. Most immobile organisms like plants and some sea invertebrates use different poisons to defense themselves from being eaten; some vertebrates (like snakes) and invertebrates (e.g., bee and wasps) use poisons not only for defense but also to get food. If we will analyze the composition of these poisons, we can find in their content a lot of various enzyme inhibitors. They were selected during the evolution to stop many metabolic processes in organisms of victims that lead to their death.

Poisons of plants and invertebrates were used as medicine drugs during thousands of years. But only in the twentieth century, it became clear that the poisons contain various enzyme inhibitors as well as the blockers of some other biological molecules (channels, receptors, etc.) For example, bee venom includes melittin, peptide containing 28 amino acids. This peptide can interact with many enzymes suppressing their activities; in particular, it binds with protein calmodulin [22] that are activator of many enzymes. Special studies have shown that melittin structure imitates structure of some proteins (to be exact, some part of protein molecules) that can interact with target enzyme to provide their biological function [23].

Another example of natural inhibitors is cardiotonic steroids that were found initially in plants (digoxin, digitonin, ouabain) and in the mucus of toads (marinobufagenin, bufotoxin, etc.). These compounds are irreversible inhibitors of Na,K-ATPase that is enzyme transporting Na^+ and K^+ through the plasma membrane of animals against the electrochemical gradients. In the end of the twentieth century, it was shown that cardiotonic steroids are presented in low concentrations in the blood of mammals including human beings. The increase of concentration of these compounds in the blood may be involved in the development of several cardiovascular and renal diseases including volume-expanded hypertension, chronic renal failure, and congestive heart failure [24].

Natural poisons are a powerful instrument for investigation of enzyme function, and analysis of their action is necessary for these studies. It might be also a model for design of new inhibitors and activators that will imitate natural compounds with such properties.

2.4. Enzyme inhibitors as pharmaceutical agents

We have mentioned above nonsteroidal anti-inflammatory drugs that are the inhibitors of cyclooxygenase. This group of compounds (the most prescribed drugs in the world, the oldest among them is aspirin) was successfully used for more than one century around the whole world for treatment of patients with fever, cardiovascular diseases, joint pain, etc. [5]. Among these drugs are both irreversible and reversible inhibitors that slow down production of prostaglandins that control many aspects of inflammation, smooth muscle contraction, and blood clotting. But there are many other groups of drugs that are by nature of inhibitors of some enzymes; the following groups of enzyme inhibitors are developed now by pharmaceutical companies and have very important therapeutic significances [24].

Inhibitors of angiotensin-converting enzyme (ACE). ACE catalyzes a conversion of inactive decapeptide angiotensin I into angiotensin II by the removal of a dipeptide from the C-terminus of angiotensin I. Angiotensin II is a powerful vasoconstrictor. Inhibition of ACE results in the decrease of angiotensin I concentration and in the relaxation of smooth muscles of vessels. Inhibitors of ACE are widely used as drugs for treatment of arterial hypertension [25].

Proton pump inhibitors (PPIs). Proton pump is an enzyme that is located in the plasma membrane of the parietal cells of stomach mucosa. It is a P-type ATPase that provides proton secretion from parietal cells in gastric cavity against the electrochemical gradient using energy of adenosine triphosphate (ATP) cleavage. PPIs are groups of substituted benzopyridines that in acid medium of stomach are converted into active sulfonamides interacting with cysteine residues of pump [26]. Therefore, PPIs are acid-activated prodrugs that are converted into drugs inside the organisms. PPIs are introduced in therapeutic practice in 80th years of the twentieth century. Since this time, the drugs are successfully used for treatment of gastritis, gastric and duodenal ulcer, and gastroesophageal reflux disease.

Statins represent a group of compounds that are analogs of mevalonic acid. They are inhibitors of 3-hydroxy-3-methylglutaryl-CoA reductase, an enzyme participating in cholesterol synthesis. Statins are used as drugs preventing or slowing the development of atherosclerosis [27]. Because of the existence of some adverse effects, statins may be recommended for patients that cannot achieve a decrease of cholesterol level in the blood through diet and changes in lifestyle.

Antibiotic penicillin covalently modifies the enzyme transpeptidase, thereby preventing the synthesis of bacterial cell walls and thus killing the bacteria [28].

Methotrexate is a structural analog of tetrahydrofolate, a coenzyme for the enzyme dihydrofolate reductase, which catalyzes necessarily step in the biosynthesis of purines and pyrimidines. Methotrexate binds to this enzyme approximately 1000-fold more tightly than the substrate and inhibits nucleotide base synthesis. It is used for cancer therapy [29].

New promising direction of anticancer therapy that is connected with suppression of protein kinases controlling the cellular response to DNA damage is now on the step of development. Selective inhibitors of these enzymes are now being tested in clinical trials in cancer patients [30].

Breakthrough in treatment of patients with acquired immune deficiency syndrome (AIDS) that is provoked by human immunodeficiency virus (HIV) was achieved recently using two different types of enzyme inhibitors. Nucleoside reverse transcriptase inhibitors and protease inhibitors are now recommended for treatment of patients with this decease. These inhibitors affect also some other viral infections and demonstrated anticancer activity. Presented here list of enzyme inhibitors that are used in therapy of numerous deceases that is far from being complete. But even mentioned above, points demonstrate how useful and important are therapeutic application of theoretical knowledge obtained as result of study of enzyme inhibitors [31].

Sciences around the world are involved in a search of new inhibitors of known enzymes that have therapeutic significance. An example of this complex research is a work devoted to design, synthesis, and study of new inhibitors of carbonic anhydrase, an enzyme that is involved in the development of such symptoms and deceases as edema, glaucoma, obesity, cancer, epilepsy, and osteoporosis (see Ref. [32]).

3. Enzyme activators

3.1. Definition and mechanisms of action

Enzyme *activators* are chemical compounds that increase a velocity of enzymatic reaction. Their actions are opposite to the effect of enzyme inhibitors. Among activators we can find ions, small organic molecules, as well as peptides, proteins, and lipids.

http://dx.doi.org/10.5772/67248

There are many enzymes that are specifically and directly activated by small inorganic molecules, mainly by cations such as Ca^{2+} which is a the second messenger (among enzymes activated by Ca^{2+}, we can find different regulatory enzymes, in particular phospholipases II, protein kinases C, adenylyl cyclases, etc.). These enzymes usually have special site for Ca^{2+} binding; the binding of Ca^{2+} with it results in the change of enzyme conformation that increase enzyme activity [33].

Cations can bind not only with enzyme but also with the substrate increasing its affinity to the enzyme that activate enzyme. For example, magnesium ions interact with ATP or with other nucleotides that are negatively charged molecules, decreasing their charge that provides effective binding of nucleotides in substrate binding site of various enzymes and increasing their activity.

In some cases, activation of enzymes is due to the elimination of enzyme inhibitors. In total this effect looks as enzyme activation. Some cations including heavy metal cations inhibit definite enzymes. Small organic compounds like ethylene glycol-bis(β-aminoethyl ether)-N,N,N′,N′-tetraacetic acid (EGTA) and ethylenediaminetetraacetic acid (EDTA) that are known as chelating agents bind these inhibitory cations and by this way can eliminate their inhibitory effect.

Special group of activators can produce activation of target enzymes only after the formation of complex with another molecule. This complex, in turn, binds to enzyme and increases the velocity of enzymatic reaction. The most well-known example of such type of activators is Ca-binding protein calmodulin (calcium-modulated protein) that is expressed in all eukaryotic cells. Calmodulin is a small protein containing 148 amino acids (16.7 kDa). Its molecule consists of two symmetrical globular domains each with two Ca-binding motifs (EF-hand) located on N- and C-domains that are jointed by flexible linker. Flexibility of calmodulin molecule and the presence of nonpolar grooves in the middle part of the protein allow it to bind a large variety of proteins [33]. The binding of Ca^{2+} to calmodulin changes its conformation. These, in turn, make complex calmodulin-Ca^{2+} suitable for interaction with target enzymes (calmodulin-dependent protein kinases and phosphatases,Ca-ATPase of plasma membrane, etc.), by this manner increasing their activity. Therefore calmodulin is considered as a participant of calcium signal transduction pathway that provides enforcing and prolongation of the effect of Ca^{2+} as a second messenger [34].

3.2. Allosteric enzyme modulators

Inhibitors and activators (modulators) that bind to enzymes not in the active site but in special center located far enough from it have name allosteric modulators. Their binding to allosteric sites induces the change of enzyme conformation that affects both the structure of active site and enzyme conformational mobility leading to the decrease or to the increase of enzyme activity. Just as enzyme active site is specific in relation to substrate, the allosteric site is specific to its modulator [16].

Many metabolic pathways are regulated through the action of allosteric modulators. Enzymes in metabolic pathways work sequentially, and in such pathways, a product of one reaction becomes a substrate for the next one. The rate of whole pathway is limited by the rate of the lowest reaction. Allosteric regulators often are a final product of whole metabolic pathway that activates enzymes catalyzing a limiting step of the whole pathway. Enzymes in a metabolic pathway can be inhibited or activated by downstream products. This regulation represents negative and positive feedbacks that slow metabolic pathway when the final product is produced in large amounts or accelerate it when a final product is presented in low concentration. Therefore, allosteric modulators are important participants of such negative and positive feedbacks in metabolic pathways or between them making metabolism self-controlled.

For example, ATP and citrate are inhibitors of phosphofructokinase that is a key enzyme of glycolytic pathway. One product of glycolysis is ATP. Another product is pyruvate that after the conversion into acetyl-CoA is condensed with citrate opening cycle of citrate acids (Krebs cycle). Reactions of this cycle produce reduced nicotinamide adenine dinucleotide reduced (NADH) and flavinadeninidinucleotide reduced (FADH2), oxidation of which is coupled with massive production of ATP in mitochondria. Availability of ATP or citrate inhibits glycolysis preventing glucose oxidation (negative feedback). Inhibition of phosphofructokinase by ATP or by citrate occurs by allosteric manner [35]. Described negative feedback control maintains a steady concentration of ATP in the cell. It should be noted also that metabolic pathways are regulated not only through inhibition but also through activation of the key enzymes. Mentioned above phosphofructokinase is activated by adenosine diphosphate (ADP), adenosine monophosphate (AMP), and fructose-2,6-bisphospate that represents positive feedback control.

Enzymes that are regulated by allosteric modulators are usually presented by several interacting subunits (they are called oligomers). A very interesting example of regulation of the activity of oligomeric enzymes is c-AMP-dependent protein kinase that is an important regulatory enzyme participating in the phosphorylation of serine and threonine residues of target proteins changing by this way their activity. This enzyme consists of four subunits; two of them are catalytic and two are regulatory. Cyclic AMP (c-AMP) is allosteric activator of this enzyme. Catalytic subunit being bound to the regulatory one is inactive. Binding of two c-AMP molecules to allosteric sites of each regulatory subunit induces their conformation transition that results in dissociation of the tetrameric complex and in activation of catalytic subunits [36]. Decrease of c-AMP concentration leads to its dissociation from the allosteric site and to association of regulatory and catalytic subunits with subsequent inactivation of catalytic subunits. By this way, c-AMP activity depends upon the c-AMP concentration in the cell.

4. Conclusions

Enzyme inhibitors and activators are a number of various chemical compounds that can slow down (or even stop) and activate enzymes, natural protein catalysts. They include inorganic compounds (often anions), different organic compounds (mainly containing reactive groups that can modify amino acids of protein), natural proteins, lipids, and carbohydrates. Mechanism of inhibitor and activator action on the enzyme activity includes a step of their binding to the enzyme, after which a step of the change of enzyme conformation often follows.

Inhibitors are a good tool for study of enzyme reaction mechanisms. Many natural inhibitors especially obtained from plants and invertebrates often imitate natural proteins or some of

http://dx.doi.org/10.5772/67248

their motifs that participate in the protein-protein interactions in the cell that are important for metabolic regulation. Among enzyme activators and inhibitors, one can highlight a group of allosteric modulators that participate in feedback regulation of metabolic pathways. And finally, we should note a practical significance of enzyme inhibitors that are a base for the design of different classes of pharmaceutical drugs, pesticides, and insecticides.

Author details

Olga D. Lopina

Address all correspondence to: od_lopina@mail.ru

Department of Biochemistry, School of Biology, M.V. Lomonosov Moscow State University, Moscow, Russia

References

[1] McAuley M, Timson DJ. , 2017; 181(1):83-90.

[2] Yang LW, Bahar I. Coupling between catalytic site and collective dynamics: a requirement for mechanochemical activity of enzymes. Structure. 2005; **13**:893–904. DOI: 10.1016/j.str. 2005.03.015.

[3] Herschlag D, Natarajan A. Fundamental challenges in mechanistic enzymology: progress toward understanding the rate enhancement of enzyme. Biochemistry. 2013; **52**:2050–67. DOI: 10.1021/bi4000113.

[4] Nevinsky GA, Buneva VN. Natural catalytic antibodies in norm, autoimmune, viral, and bacterial diseases. Sci World J. 2010; **10**:1203–33. DOI: 10.1100/tsw.2010.98.

[5] Simon LS. Nonsteroidal antiinflammatory drugs and their effects. J Clin Rheumatol. 1996; **2**(3):135–40. PMID: 19078048.

[6] Cleland WW. The kinetics of enzyme-catalyzed reactions with two or more substrates or products. II. Inhibition: nomenclature and theory. Biochim Biophys Acta. 1963; **67**:173–87. DOI: 10.1016/0926-6569(63)90226-8.

[7] Malito E, Ralat LA, Manolopoulou M, Tsay JL, Wadlington NL, Tang W-J. Molecular bases for the recognition of short peptide substrates and cysteine-directed modifications of human insulin-degrading enzyme. Biochemistry. 2008; **47**(48):12822–34. DOI: 10.1021/bi801192h.

[8] Malhotra OP, Srivastava DK, Kayastha AM, Srinivasan. Inactivation of glyceraldehyde-3-phosphate dehydrogenase with SH-reagents and its relationship to the protein quaternary structure. Indian J Biochem Biophys. 1993; **30**(5):264–9. PMID: 8144169.

[9] Ogawa H, Shinoda T, Cornelius F, Toyoshima C. Crystal structure of the sodium potassium pump (Na^+,K^+-ATPase) with bound potassium and ouabain. Proc Natl Acad Sci. 2009; **106**(3):13742–47. DOI: 10.1073/pnas.0907054106.

[10] Oosterbaan RA, Kunst P, Cohen JA. The nature of the reaction between diisopropylfluorophosphate and chymotrypsin. Biochim Biophys Acta. 1955; **16**(2):299–300.

[11] Schmidt AE, Sun MF, Ogawa T, Bajaj SP, Gailani D. Functional role of residue 193 (chymotrypsin numbering) in serine proteases: influence of side chain length and beta-branching on the catalytic activity of blood coagulation factor XIa. Biochemistry. 2008; **47**(5):1326–35. DOI: 10.1021/bi701594j.

[12] Leuzinger W. The number of catalytic sites in acetylcholinesterase. Biochem J. 1971; **123**(2):139–41.

[13] Lecomte M, Laneuville O, Ji C, DeWitt DL, Smith WL. Acetylation of human prostaglandin endoperoxide synthase-2 (cyclooxygenase-2) by aspirin. J Biol Chem. 1994; **269**(18):13207–15. PMID: 8175750.

[14] Berg JM, Tymoczko JL, Stryer L. Biochemistry. 7th ed. W.H. Freeman and Company, New-York, USA. 2012: 229.

[15] Colović MB, Krstić DZ, Lazarević-Pašti TD, Bondžić AM, Vasić VM. Acetylcholinesterase inhibitors: pharmacology and toxicology. Curr Neuropharmacol. 2013; **11**(3):315–35. DOI: 10.2174/1570159X11311030006.

[16] Nelson DL, Cox MM. Lehninger Principles of Biochemistry. 4th ed. 2005: 1216. DOI: 10.1002/bmb.2005.494033010419.

[17] Krowarsch D, Cierpicki T, Jelen F, Otlewski J. Canonical protein inhibitors of serine proteases. Cell Mol Life Sci. 2003; **60**(11):2427–44. DOI: 10.1007/s00018-003-3120-x.

[18] Ogata S, Misumi Y, Tsuji E, Takami N, Oda K, Ikehara Y. Identification of the active site residues in dipeptidyl peptidase IV by affinity labeling and site-directed mutagenesis. Biochemistry. 1992; **31**(9):2582–7. PMID: 1347701.

[19] Tsilikounas E, Rao T, Gutheil WG, Bachovchin WW. ^{15}N and ^{1}H NMR spectroscopy of the catalytic histidine in chloromethyl ketone-inhibited complexes of serine proteases. Biochemistry. 1996; **35**(7):2437–44. DOI: 10.1021/bi9513968.

[20] Janciauskiene S. Conformational properties of serine proteinase inhibitors (serpins) confer multiple pathophysiological roles. Biochim Biophys Acta. 2001; **1535**(3):221–35.

[21] Pike RN, Bottomley SP, Irving JA, Bird PI. Serpins: finely balanced conformational traps. IUBMB Life. 2002; **54**(1):1–7. DOI: 10.1080/15216540213825.

[22] Kataoka M, Head JF, Seaton BA, Melittin binding causes a large calcium-dependent conformational change in calmodulin. Proc Natl Acad Sci USA. 1989; **86**:6944–8.

[23] Raynor RL, Zheng B, Kuo JF. Membrane interactions of amphiphilic polypeptides mastoparan, melittin, polymyxin B, and cardiotoxin. Differential inhibition of protein kinase C, Ca^{2+}/calmodulin-dependent protein kinase II and synaptosomal membrane Na,K-ATPase, and Na^+ pump and differentiation of HL60 cells. J Biol Chem. 1991; **266**(5):2753–8. PMID: 1847132.

[24] Buckalew VM. Endogenous digitalis-like factors: an overview of the history. Front Endocrinol (Lausanne). 2015; **6**:49. DOI: 10.3389/fendo.2015.00049.

[25] Hsiao FC, Tung YC, Chou SH, Wu LS, Lin CP, Wang CL, Lin YS, Chang CJ, Chu PH. Fixed-dose combinations of renin-angiotensin system inhibitors and calcium channel blockers in the treatment of hypertension: a comparison of angiotensin receptor blockers and angiotensin-converting enzyme inhibitors. Medicine (Baltimore). 2015; **94**(51):e2355. DOI: 10.1097/MD.0000000000002355.

[26] Jain KS, Shah AK, Bariwal J, Shelke SM, Kale AP, Jagtap JR, Bhosale AV. Recent advances in proton pump inhibitors and management of acid-peptic disorders. Bioorg Med Chem. 2007; **15**(3):1181–205.

[27] Satoh M, Takahashi Y, Tabuchi T, Minami Y, Tamada M, Takahashi K, Itoh T, Morino Y, Nakamura M. Cellular and molecular mechanisms of statins: an update on pleiotropic effects. Clin Sci (Lond). 2015; **129**(2):93–105. DOI: 10.1042/CS20150027.

[28] Katz AH, Caufield CE. Structure-based design approaches to cell wall biosynthesis inhibitors. Curr Pharm Des. 2003; **9**(11):857–66. DOI: 10.2174/1381612033455305.

[29] McGuire JJ. Anticancer antifolates: current status and future directions. Curr Pharm Des. 2003; **9**(31):2593–613. DOI: 10.2174/1381612033453712.

[30] Garrett MD, Collins I. Anticancer therapy with checkpoint inhibitors: what, where and when? Trends Pharmacol Sci. 2011; **32**(5):308–16. DOI: 10.1016/j.tips.2011.02.014.

[31] Rutherford GW, Horvath H. Dolutegravir plus two nucleoside reverse transcriptase inhibitors versus efavirenz plus two nucleoside reverse transcriptase inhibitors as initial antiretroviral therapy for people with HIV: a systematic review. PLoS One. 2016; **11**(10):e0162775. DOI: 10.1371/journal.pone.0162775.

[32] Fidan I, Salmas RE, Arslan M, Sentürk M, Durdag S, Ekinci D, Sentürk E, Cosgun S, Supuran CT. Carbonic anhydrase inhibitors: design, synthesis, kinetic, docking and molecular dynamics analysis of novel glycine and phenylalanine sulfonamide derivatives. Bioorg Med Chem. 2015; **23**(23):7353–8. DOI: 10.1016/j.bmc.2015.10.009.

[33] Villarroel A, Taglialatela M, Bernardo-Seisdedos G, Alaimo A, Agirre J, Alberdi A, Gomis-Perez C, Soldovieri MV, Ambrosino P, Malo C, Areso P. The ever changing moods of calmodulin: how structural plasticity entails transductional adaptability. J Mol Biol. 2014; **426**(15):2717–35. DOI: 10.1016/j.jmb.2014.05.016.

[34] Brini M, Carafoli E. Calcium signalling: a historical account, recent developments and future perspectives. Cell Mol Life Sci. 2000; **57**(3):354–70. DOI: 10.1007/PL00000698.

[35] Sola-Penna M, Da Silva D, Coelho WS, Marinho-Carvalho MM, Zancan P. Regulation of mammalian muscle type 6-phosphofructo-1-kinase and its implication for the control of the metabolism. IUBMB Life. 2010; **62**(11):791–6. DOI: 10.1002/iub.393.

[36] Turnham RE, Scott JD. Protein kinase A catalytic subunit isoform PRKACA; history, function and physiology. Gene. 2016; **577**(2):101–8. DOI: 10.1016/j.gene.2015.11.052.